# The City as a Global Political Actor

T0187872

This book engages with the thorny question of global urban political agency. It critically assesses the now popular statement that in the context of paralysed and failing nation state governments, cities can and will provide leadership in addressing global challenges.

Cities can act politically on the global scale, but the analysis of global urban political agency needs to be firmly embedded in the field of urban studies. Collectively, the chapters in this volume contextualize urban agency in time and space and pluralize it by looking at how urban agency is nurtured through coalitions between a wide range of public and private actors. The authors develop and critically assess the conceptual underpinnings of the notion of global urban political agency from a variety of theoretical and disciplinary perspectives. The second part contains several (theoretically informed) empirical analyses of global urban political agency in cities around the globe.

This book geographically expands analysis by looking beyond global cities in diverse contexts. It is highly recommended reading for scholars in the fields of international relations and urban studies who are looking for an interdisciplinary and empirically grounded understanding of global urban political agency, in a diversity of contexts and a plurality of forms.

**Stijn Oosterlynck** is Associate Professor in Urban Sociology at the University of Antwerp, Sociology department.

**Luce Beeckmans** is a full-time post-doctoral researcher at the Department of Architecture & Urban Planning of Ghent University, Belgium.

**David Bassens** is Associate Professor of Economic Geography at the Vrije Universiteit Brussels, where he is Associate Director of Cosmopolis: Centre for Urban Research.

**Ben Derudder** is Professor of Human Geography at Ghent University's Department of Geography, and an Associate Director of the Globalization and World Cities (GaWC) research network.

**Barbara Segaert** is a scientific coordinator at the University Centre Saint-Ignatius Antwerp (UCSIA), where she develops academic programmes on various topics of contemporary relevance to society.

**Luc Braeckmans** is Director of Academic Affairs of the University Centre Saint-Ignatius Antwerp (UCSIA). He is a professor in the Department of Philosophy at the University of Antwerp, where he is also chair of the Centre for Andragogy.

# Routledge Studies in Urbanism and the City

This series offers a forum for original and innovative research that engages with key debates and concepts in the field. Titles within the series range from empirical investigations to theoretical engagements, offering international perspectives and multidisciplinary dialogues across the social sciences and humanities, from urban studies, planning, geography, geohumanities, sociology, politics, the arts, cultural studies, philosophy and literature.

**Rebel Streets and the Informal Economy**
Street Trade and the Law
*Edited by Alison Brown*

**Mega-events and Urban Image Construction**
Beijing and Rio de Janeiro
*Anne-Marie Broudehoux*

**Urban Geopolitics**
Rethinking Planning in Contested Cities
*Edited by Jonathan Rokem and Camillo Boano*

**Contested Markets, Contested Cities**
Gentrification and Urban Justice in Retail Spaces
*Edited by Sara González*

**The City as a Global Political Actor**
*Edited by Stijn Oosterlynck, Luce Beeckmans, David Bassens,*
*Ben Derudder, Barbara Segaert and Luc Braeckmans*

For more information about this series, please visit: www.routledge.com/series/RSUC

# The City as a Global Political Actor

Edited by Stijn Oosterlynck,
Luce Beeckmans, David Bassens,
Ben Derudder, Barbara Segaert and
Luc Braeckmans

Routledge
Taylor & Francis Group

LONDON AND NEW YORK

First published 2019
by Routledge
2 Park Square, Milton Park, Abingdon, Oxon OX14 4RN

and by Routledge
52 Vanderbilt Avenue, New York, NY 10017, USA

First issued in paperback 2020

*Routledge is an imprint of the Taylor & Francis Group, an informa business*

*British Library Cataloguing-in-Publication Data*
A catalogue record for this book is available from the British Library

*Library of Congress Cataloging-in-Publication Data*
Names: Oosterlynck, Stijn, editor. | Beeckmans, Luce, editor. | Bassens, David, editor. | Derudder, Ben, editor. | Segaert, Barbara, editor. | Braeckmans, Luc, 1953–editor.
Title: The city as a global political actor / [edited by Stijn Oosterlynck, Luce Beeckmans, David Bassens, Ben Derudder, Barbara Segaert, and Luc Braeckmans].
Description: Abingdon, Oxon; New York, NY: Routledge, 2018. | Series: Routledge studies in urbanism and the city | Includes bibliographical references and index.
Identifiers: LCCN 2018025188 | ISBN 9781138573574 (hbk: alk. paper) | ISBN 9780203701508 (ebk) | ISBN 9781351330725 (mobi/kindle) | ISBN 9781351330732 (epub) | ISBN 9781351330749 (pdf)
Subjects: LCSH: Metropolitan government. | Cities and towns–Political aspects. | International relations. | Globalization–Political aspects.
Classification: LCC JS50 .C57 2018 | DDC 320.8/5–dc23
LC record available at https://lccn.loc.gov/2018025188

ISBN 13: 978-0-367-58430-6 (pbk)
ISBN 13: 978-1-138-57357-4 (hbk)

Typeset in Times New Roman
by codeMantra

# Contents

# Illustrations

## Figures

## Tables

# Contributors

**Michele Acuto** is Chair of Global Urban Politics in the Faculty of Architecture, Building and Planning at the University of Melbourne, where he is also Director of AURIN (the Australian Urban Research Infrastructure Network), and the Connected Cities Lab. He is also a Senior Fellow of the Chicago Council on Global Affairs and the Bosch Foundation Global Governance Futures Program. Michele was previously Director of the City Leadership Lab and Professor of Diplomacy and Urban Theory at University College London, and Barter Fellow of the Oxford Programme for the Future of Cities at the University of Oxford. He also taught at the University of Canberra, University of Southern California, Australian National University and National University of Singapore. Outside academia, Michele worked for the Institute of European Affairs in Dublin, the International Campaign to Ban Landmines (ICBL), the Kimberley Process for conflict diamonds, the European Commission's response to pandemic threats. He has also worked for several years on city leadership and city networks with, amongst others, Arup, World Health Organization, World Bank Group, the C40 Climate Leadership Group, and UN-Habitat. Michele is the author of several articles, publications and policy documents on the link between urban governance and international politics, having recently served as co-chair of the Nature Sustainability Expert Panel on 'science and the future of cities'.

**Sabine Barthold** is a PhD Candidate in Urban Studies at the Center for Metropolitan Studies and Fellow of the German National Academic Foundation at TU Berlin. She studied at the New School for Social Research and obtained a degree in Sociology and Political Science from Dresden University. Her research interests range from post-socialist urban transformations to urban climate governance and eco-modernization in global cities and draws from theories of Urban Sociology as well as Critical Geography and Urban Political Ecology. Her current research is focussed on inter-urban knowledge and policy transfer within the C40 Cities Climate Leadership Group and the ways that city networks engage as global actors in international climate governance.

**Eric Corijn** is cultural philosopher and social scientist, Professor of Urban Studies at the Vrije Universiteit Brussel, founder of Cosmopolis, Centre for Urban Research, vice-chair of the Brussels Studies Institute and Director of the Brussels Academy.

**Carola Fricke** is a researcher and lecturer for Human Geography at the University of Freiburg. She received her PhD from the Berlin University of Technology. Her PhD project explored the evolution of metropolitan policies in various contexts, including the European Union, France and Germany. She moreover researches urban and regional policies in Europe from a comparative perspective, as well as cross-border governance in the field of spatial planning. She is part of the International Metropolitan Research Consortium (IMRC), a research collective focussed on comparing processes of metropolitanization within Europe and globally. Methodologically, her work on policy analysis incorporates an interpretive-constructivist perspective, a spatial dimension and a qualitative empirical approach.

**Jorn Koelemaij** is a PhD candidate in Social and Economic Geography at Ghent University. Having a background in Human Geography and Urban Studies (University of Amsterdam), his research project focusses on three main issues: (1) global city aspirations and policies in the non-Western world; (2) the geography of transnational capital investment in real estate and its impact on (mobile) urban development strategies; and (3) the applicability and comparability of the concept of neoliberal urbanism in different historical-institutional contexts.

**Ludovic Lepeltier-Kutasi** is a PhD Student in Social Geography (UMR CITERES-EMAM, University of Tours), fellow at the CEFRES French Research Center in Humanities and Social Sciences of Prague. Currently he is finishing his PhD thesis about the role of former state housing in contemporary urban dynamics in Budapest, focussing on the case of Magdolna, in the 8th district. His approach is inductive, characterized by the consideration of fine scales of observation and ethnographic methods.

**Enzo Mingione** is Professor of Sociology at the University of Milano-Bicocca. He has been the President of the Research Committee on Urban and Regional Development, one of the founders of the *International Journal of Urban and Regional Research* and a trustee of the *Foundation of Urban and Regional Studies*. His main fields of interest are urban sociology, poverty, social politics, social exclusion, labour market, informal sector and economic sociology. Among his books are *Social Conflict and the City*, Blackwell, Oxford (1981); *Beyond Employment*, together with Nanneke Redclift (eds), Blackwell, Oxford (1985); *Fragmented Societies*, Blackwell, Oxford (1991); (Ed) *Urban poverty and the Underclass*, Blackwell, Oxford (1996).

**Laura Nkula-Wenz** is a researcher and lecturer in urban geography. She is based in Cape Town, where she currently works with the African Centre for Cities on delivering the MA in Critical Urbanisms. She obtained her doctorate in Human Geography from the University of Münster as a fellow of the German National Academic Foundation. Her empirical research has focussed on Cape Town, addressing questions of urban governance and the construction of local political agency through the lens of how the popular creative city paradigm has been renegotiated in a city of the so-called 'Global South'. As a postdoctoral fellow of the work package '*Circulation of Models and Heterogeneity of Development*' within the Laboratory of Excellence 'Territorial and Spatial Dynamics' (LabEx DynamiTe), her most recent work focussed on questions of urban experimentation and policy learning with regards to urban safety and security models.

**Gergely Olt** works at the Hungarian Academy of Sciences at the Centre for Social Sciences, Institute for Sociology as a fellow of the 'young researcher' programme. Currently he is finishing his PhD thesis at the ELTE Doctoral School of Sociology about the transformation of a derelict inner city neighbourhood into a 24/7 party district since 2006, especially about the conflicts of this process and the possibilities of interest articulation of different social groups and actors. His research interest turned recently to comparative urban studies to understand the case of Budapest better in light of other cities.

**Elisabeth Peyroux** is a senior researcher at the National Centre for Scientific Research (CNRS), based at the UMR Prodig, Paris. She is an urban geographer with an interest in urban governance, urban restructuring and socio-spatial transformation in South Africa. Her research focusses on urban policy mobility in the fields of governance, planning and urban regeneration, as well as on cities' international relations strategies. She is currently working on 'smart urbanism' in Southern cities, looking at the role of geo-technologies in urban governance and planning. She coordinates a Working Group on '*Circulation of Models and Heterogeneity of Development*' within the Cluster of Excellence 'Territorial and Spatial Dynamics' (Labex DynamiTe).

**Gilles Pinson** is a professor of political science in Sciences Po Bordeaux (France) and a researcher in the Centre Emile Durkheim. His research deals with urban policies and politics, urban and metropolitan governance and the transformations of the relations between territorial states and cities. He is currently working on the politicization of inter-municipal bureaucrats in French metropolis and the construction of the local markets of the 'smart city'. Among his main publications: *Debating the Neoliberal City* (edited with Christelle Morel Journel), Routledge, 2017; *The Neoliberal City: Theory, Evidence, Debates* (coordinated with

Christelle Morel Journel, *Territory Politics Governance*, 4(2), 2016); 'Networked Cities and Steering States. Urban Policy Circulations and the Reshaping of State-Cities Relationships' (with Vincent Béal and Renaud Epstein, *Environment and Planning C: Politics and Space*, January 2018); 'From the Governance of sustainability to the management of climate change: reshaping urban policies and central–local relations in France' (with Vincent Béal, *Journal of Environmental Policy & Planning*, 7(3), 2015, 402–419); 'When mayors go global: international strategies, urban governance and leadership' (with Vincent Béal, *International Journal of Urban and Regional Research*, 38(1), 2014, 302–317); *Gouverner la Ville par Projet. Urbanisme et Gouvernance des Villes Européennes* (Presses de Sciences Po, 2009).

**Jennifer Robinson** is Professor of Human Geography at University College London. Formerly Professor of Urban Geography at the Open University, she is author of a much-praised recent book, *Ordinary Cities*, which offers a critique of the divisions in urban studies between Western and 'Third world' cities and argues for a truly cosmopolitan approach to understanding cities. Her empirical research has focussed on South Africa, including studies of segregation and state power as well as the politics of urban development. More recently she has been researching the politics of urban development in London. She is currently working on methodologies for comparative international urban research, with recent papers in the *International Journal of Urban and Regional Research* (2011, Cities in a World of Cities) and *Progress in Human Geography* (2016, Thinking Cities through Elsewhere). She is currently leading a comparative research project on large-scale urban developments in London, Johannesburg and Shanghai, with Prof Phil Harrison (University of the Witwatersrand) and Prof Fulong Wu (UCL).

**Kevin Ward** is Professor of Human Geography, School of Environment, Education and Development and Director of the Manchester Urban Institute (www.mui.manchester.ac.uk) at the University of Manchester. His research focusses on the geographies of urban policy making, state reorganization, and the politics of urban and regional development. Author and editor of numerous journal articles and books, including *Researching the City* (Sage, 2014), *Urban Theory: New Critical Perspectives* (Routledge, 2017, with Mark Jayne) and *Cities under Austerity: Restructuring the US Metropolis* (SUNY Press, 2018, with Mark Davidson), he is also currently Editor in Chief of the journal *Urban Geography*.

# Editors

**David Bassens** (PhD Geography, Msc Geography, MA Social and Cultural Anthropology) is Associate Professor of Economic Geography at the Vrije Universiteit Brussels where he is Associate Director of Cosmopolis: Centre for Urban Research. His research interests lie in two broad areas: (1) geographies of globalized urbanization: world-city formation and world-city-networks, urban political economies, post-colonial urban critiques, mobility of urban policies; and (2) geographies of finance and financialization: financial geography, economic geography, regulation theory, varieties of capitalism, and the sociology of finance. His PhD research dealt with geographies of Islamic Finance. David is also Associate Director of Financialization at the Globalization and World Cities Research Network (GaWC) and acting Treasurer and Executive Committee member of the Global Network on Financial Geography (FINGEO).

**Luce Beeckmans** is a post-doctoral research fellow funded by the Flanders Research Foundation (FWO) and affiliated to Ghent University (Department of Architecture and Urban Planning, head office), University of Leuven (Interculturalism, Migration and Minorities Research Centre) and Antwerp University (Urban Studies Centre) in Belgium. She has a keen interest in the circulation of spatial knowledge between Europe and Africa, both within the framework of colonization and development cooperation and as a result of transnational migration. Luce Beeckmans graduated as a civil engineer-architect in 2005 at Ghent University (Belgium), Department of Architecture & Urban Planning. She then worked as an architect and urban planner at the office of Stéphane Beel Architects. From 2007 she was a research assistant at the Groningen Research Institute for the Study of Culture (ICOG), Faculty of Arts, Groningen University (the Netherlands), where she obtained a PhD in the Arts: History of Architecture and Urbanism. In her doctoral dissertation 'Making the African City' she studied the urban (planning) history in Africa from a comparative and interdisciplinary perspective. Her dissertation has been awarded the Jan van Gelder Prize for best art-historian publication of the year (2013). Luce Beeckmans has published articles in a broad range of journals (*Progress in Planning*, *Urban History*, *The Journal of Architecture*,...) and presented her research on several international conferences (SAH, EAUH, EAHN, ASA, ...). She moreover contributed to a number of exhibitions (such as Afropolis and 'Wonen in Diversiteit'). Since 2013 she has been a full-time post-doctoral researcher at the Department of Architecture & Urban Planning of Ghent University, where she coordinated the research group Labo S (Labo Stedenbouw) for two years. In 2016 she was granted a FWO post-doctoral fellowship in which she further develops her research track on spatial manifestations of African diaspora in European mid-sized cities. Luce Beeckmans is a member of the executive board of the Ghent Africa Platform (GAP) and one of the promotors of the Centre for the Social Study of Migration and Refugees (CESSMIR).

**Luc Braeckmans** (PhD in philosophy) is director of academic affairs of the University Centre Saint Ignatius Antwerp (UCSIA). He is a professor at the Department of Philosophy at the University of Antwerp, where he is also chair of the Centre for Andragogy. The focus of his research is in the field of philosophy of religion, philosophy of education and metaphysics.

**Ben Derudder** is Professor of Urban Geography at Ghent University's Department of Geography, and an Associate Director of the Globalization and World Cities (GaWC) research network. His research focusses on the conceptualization and empirical analysis of transnational urban networks in general, and its transportation and production components in particular. He also has a keen interest in the potential of social network analysis methods in geographical research. His work on transnational urban networks has been published in leading academic journals, and he has co-edited a number of books on this topic, including the *International Handbook of Globalization and World Cities* (Edward Elgar, 2011, with P.J. Taylor, F. Witlox & M. Hoyler). A second edition of the *World City Network: a Global Urban Analysis* (Routledge, 2016, together with P. Taylor) has been published with Routledge in 2016.

**Stijn Oosterlynck** is Associate Professor in Urban Sociology at the University of Antwerp, Sociology department and is chair of the Antwerp Urban Studies Institute. He teaches courses on urban studies, poverty and social inequality. His research is concerned with local social innovation and welfare state restructuring, the political sociology of urban development, urban renewal and community building, new forms of solidarity in diversity and civil society innovation. He currently coordinates a large-scale research project on civil society innovation around diversity and politicization (CSI Flanders) and an international research network on solidarity in diversity. He supervises PhD research projects on local social innovation and welfare state restructuring, urban governance and dynamics of in/exclusion, urban migration integration policies, migration and spatial planning, spatial planning, new forms of solidarity in diversity, social urban renewal and collective housing.

**Barbara Segaert** holds a Master's degree in Oriental Studies, Islamic Studies and Arab Philology (KU Leuven), Belgium and a Master's in the Social Sciences (Open University), UK. Since 2002 she has been scientific coordinator at the University Centre Saint-Ignatius Antwerp, where she develops academic programmes on various topics of contemporary relevance to society.

# 1 An urban studies take on global urban political agency

*David Bassens, Luce Beeckmans, Ben Derudder and Stijn Oosterlynck*[1]

## Introduction

With the overall aim to enrich the academic debate ignited by the book entitled *If Mayors Ruled the World* by political scientist Benjamin Barber (Barber, 2013), this introduction intends to engage with the thorny question of global urban political agency that is at the heart of the different contributions to this edited volume. Barber's account presents mayors as the saviours of global democracy in the context of what he sees as largely dysfunctional and paralysed nation state governments. Although this point of view is both provocative and meaningful, we view it to be simplistic and also potentially constraining the debate. Against this backdrop, as a team of urban anthropologists, urban planners, urban geographers, and urban sociologists, we review what the burgeoning field of urban studies has to offer to the international relations literature that in the past few years has discovered—and highlighted—the increasingly high profile roles of cities and city governments in international affairs.

As explained by Acuto (2013, see also this volume), political science and especially its international relations variant has generally been inattentive to the role of cities in global politics for the most part of the 1990s and 2000s, or has rarely attributed agency to urban actors. Only recently has the debate in political science circles started to delve into questions around forms of urban agency, but mostly framed as a matter of how mayors may marshal urban leadership in moving beyond the stasis in existing global governance structures. We draw on the rich literature on urban politics in urban studies to more carefully weigh and specify the agency in and of cities at the global level, which the 'pro-urban literature' (as epitomized for example by Barber's contribution and Glaeser's *The Triumph of the City* (Glaeser, 2011)) suggests is crucial in constructing progressive outcomes for key global challenges. Our perspective is that political agency, arguably the *raison d'être* of political science, largely supersedes political science's rather narrow focus on formal political agents and institutions proper. There is, then, a clear-cut need to develop a more pluralist understanding of urban political agency, and our starting point is to look for 'analytical support in the vast urban studies scholarship' (Acuto, 2013, p. 483), a challenge that provides the rationale for bringing together a range of contributions from political science and urban studies alike.

Our key argument is that grounding the analysis of global urban political agency more firmly in urban studies may help achieve a more nuanced and contextualized answer to the question of how cities (can) act politically on the global scale. Urban studies is a highly 'contextual' discipline: it takes the spatial and temporal specificity of the urban processes under analysis very seriously, in part in opposition to the methodological nationalism of large parts of the social sciences. At the same time, it maintains a sensitivity for how these particularities are related to a wider set of cases that would allow for generalization, process-tracing, and concept-building. Moreover, with its eye for geo-historical specificities, urban studies tend to critically reflect on the 'sites' where scholarly knowledge is generated, suggesting that theories on global urban agency cannot be based solely on a limited number of paradigmatic cases in the Global North.

In this introduction we aim to set the scene for the contributions in this volume by outlining what urban studies has on offer to develop a conceptually, methodologically, and empirically richer and sharper understanding of urban agency on the global scale. More specifically, the chapter underlines three insights from urban studies that feature prominently in the subsequent chapters in this edited volume:

(i) **Contextualizing urban agency** in time and space. As far as time is concerned, this implies being sensitive to the path-dependency of socially produced urban agency. This means theorizing urban agency through space, thereby reflecting on the preferential attachment of political processes to the urban as a site of innovation and mobilization, but also analysing how agency operates across multiple spatial scales and in and through networks and territories;

(ii) **Pluralizing urban agency** by seeing it as socially produced, i.e. as nurtured through and emerging from the forging of coalitions of public and private actors, which each bring in relevant resources to develop the collective capacity to act. Urban agency, then, is not just a matter of mayors and city governments;

(iii) **Expanding geographically** our understanding of how urban agency can make a difference on the global scale, which requires us to look beyond global cities and draw on a diversity of contexts.

While by no means exhaustive, the above three elements are, in our reading, the key insights that emerge from the different chapters offered in the volume. Each of the chapters provides a conceptual or empirical elaboration of one or more of these insights. The volume in general, and the introduction in particular, should thus be read as a preparation for those scholars in political science and urban studies who are interested in peeking over the disciplinary fence, and who wish to continue the debate in a much more intellectually diverse (and challenging) setting. The next section critically reviews the dominant political science perspectives on urban political agency. This stylized review is then used in the subsequent section to advocate the three potential insights emerging from urban studies, after which we conclude with a brief overview of how the different chapters in this volume underpin the relevance of these insights.

## Where and what is urban political agency? A view from political science and policymaking

Benjamin Barber's *If Mayors Ruled the World* was timely in that it codified real-world changes in governance structures, while at the same time fuelling the 'urban turn' in the political sciences community. Interestingly, Barber's key message also bridged the fuzzy divide between policymaking and political science, as it became clear that the very discourses of urban renaissance were also picked up somewhat uncritically by mayors around the world. New York's former mayor Michael Bloomberg, for example, amply referenced by Barber himself, explains why cities are key to fighting climate change in a piece in Foreign Affairs (Bloomberg, 2015). His line of reasoning is quite straightforward and echoes Barber's argument when claiming that cities don't 'do' ideology, but focus on pragmatic solutions to whatever issues are emerging. The influence of cities is thought to emerge particularly in 'countries that suffer from bureaucratic paralysis and political gridlock' (Bloomberg, 2015, p. 117). The urban picture being painted here is rosy across the board: cities are considered to be financially healthy and resilient, and likely to cooperate with one another.

Nonetheless, although Bloomberg's reading of the central role cities should play in the global political constellation is well aligned with Barber's thinking, there is a subtle, yet important, shift in that he clothes Barber's liberal-progressive ideas in neoliberal urban growth terminology: cities have a 'competitive advantage', a framing that allows Bloomberg to cast them first and foremost as business environments. For example, when considering the need to tackle environmental issues, he asserts that 'dirty air is a major liability for a city's business environment' (Bloomberg, 2015, p. 120). As a result, what the city as a political agent can or should be is stretched to include mostly private economic perspectives and agents.

This strongly economy-driven perspective on the need and desirability of 'urban agency' is also apparent in work by scholars who are themselves enrolled in policymaking circuits. In an online piece in *Quartz*, Noah Toly, both an academic and a fellow at the Chicago Council on Global Affairs, presents an economistic narrative that chimes well with a neoliberal perspective (Toly, 2016). He observes that the rise of cities as 'global actors' is in essence an outcome of their economic ascendancy, as 42 of the largest 100 economic entities are major cities. Toly goes on to claim that cities' dominance parallels the impact of the multinational corporations [...] so when Mexico City and Guangzhou have economic outputs comparable to the world's seven largest corporations—Wal-Mart, Royal Dutch Shell, China Petroleum & Chemical, Exxon Mobil, British Petroleum, PetroChina, and Volkswagen—it's worth examining how they too might exercise similar influence.

Even though Toly is sensitive to issues of uneven development within and between cities and regions, little critical reflection is offered to think through how such a reductionist reading of cities may in fact co-produce the very patterns of inequality that are observed (cf. Smith, 2001).

Overall, we have two chief concerns with this economistic interpretation of urban political agency. Our first concern is that there is an urgent need to critically assess the 'urban renaissance' idea as it circulates in policy circles and the public debate from an academic research perspective. There is considerable bias and selectivity at work in the way the role of cities in today's globalizing world is presented in the policy-oriented literature, a bias which shows up for example in the one-sided focus on market-oriented economic processes or in the lack of engagement with evolving scholarly ideas and literature on the changing relationship between states and cities (e.g. Brenner, 1998; Taylor, 2013). A social scientific engagement with the idea of global urban political agency demands a more nuanced consideration of the spatio-temporal context from which global political urban agency emerges and the plurality of actors involved in constructing this agency. A second concern is that the current focus on large metropolitan areas and global cities may produce a divide between the largest urban economies generating policy and agency and 'other' cities that are seen as neither having nor needing such agency (cf. Roy, 2009). The end result may well be a geographically circumscribed understanding of where urban political agency is found and needed, and relatedly, how we understand its scope and nature.

When trying to locate the *scholarly* debate on global urban political agency, one can quite logically start by reviewing contributions from political science and international relations (IR), both of which have increasingly centred on cities and their capacities as global actors. A first perspective is offered by the literature on city diplomacy. Although emblematic for a rising appreciation of the urban scale, as exemplified by Tavares' (2016) overview of city diplomacy practices, that scale is often merely a new empirical layer added to the study of IR. Agency is seen as vested in governance capacity coded in law, hence making it mostly a business of mayors and governors who act at (re-)emerging city-regional scales. Although foregrounding cities, urban agency is primarily defined in relation to the central state, where devolution to the city-regional scale offers a space of subnational governance albeit with a globally-oriented twist. Such a subnational view also underpins other contributions to the field. As Amen et al. (2011) state in the introduction of *Cities and Global Governance: New Sites of International Relations*, 'evidence about the global role of cities', which they see as running counter to a Westphalian world order, warrants 'theorizing the role sub-national formal and informal political actors play in global governance' (Amen et al., 2011, p. 4). As the title suggests, however, the focus of these scholars is mostly about cities as being *sites* of governance, i.e. 'wherein the multiple processes of the international system develop and manifest themselves' (Amen et al., 2011, p. 5), even though the chapters in the volume paint a more nuanced picture. Overall, Amen et al.'s prime objective, it appears, is to unpack the city as a container of now-global activities, focusing on matters of governance rather than (urban) politics, struggle, transition, and opposition. Taken together, the above strand of literature, while adding a new layer to the field of IR, does little to problematize, contextualize, and pluralize the issue of urban agency.

A second strand of literature, which mostly centres on the nexus between cities and climate change, offers a more contextualized and pluralist take on urban agency as it explicitly acknowledges the struggles and conflicts embedded in the current urban renaissance. As Bulkeley (2013, p. 9) explains:

> Urban authorities have a significant but varied role in relation to urban planning, building codes, the provision of transportation and the supply of energy, water, and waste services [...]. Given these powers and their democratic mandate as the local level of government, municipalities can therefore be seen as in a position to address the challenges of mitigating and adapting to climate change.

Urban agency, in this view, is located in the authority of a democratically-elected local government, but also includes private and civil society actors and their (social) innovations with whom municipalities should seek to develop strategic partnerships. Kern and Bulkeley (2009), in turn, focus on how and to what end Transnational Municipal Networks (TMNs) govern in the context of multilevel developments in Europe. Their research focus carries some similarities with the rescaling debate in urban studies (e.g. Brenner, 1998) as they pay attention to the uneven enrolment of European cities in these networks. Kern and Bulkeley (2009) primarily discuss the emergence of TMNs as a spatial adaptation to changing European decision-making structures in which city networks can perform an auxiliary function. These functions revolve around communication, best practice sharing, benchmarking, and the like, but also include the influencing and lobbying of national and European institutions. The implication is that while TMNs emerge in part as a top-down consequence of Europeanization, their actions also feed back into decision-making at higher scales.

Bouteligier (2013) adds another layer of complexity as she broadens the scope of TMNs by providing case studies of for-profit firms (e.g. environmental consultancy firms and environmental NGOs next to the networks of municipalities per se. The attention thus shifts from the issue of multilevel governance to networked forms of governance, which she sees as typical for the current phase of globalized urbanization, in which 'cities can be the places of experiments, demonstration, leadership and innovation' (Bouteligier, 2013, p. 12). Cities have the capacity to act globally because of the networked properties of the actors they host, which echoes assertions about centralized control capacities in the context of globalization raised in the literature on global and world cities (e.g. Sassen, 1991; Castells, 1996; Taylor and Derudder, 2016). This requires a

> governance perspective, which acknowledges that multiple actors (public, civil society, and market actors) at multiple levels (from the local to the global) are now involved in governing, often through hybrid constellations that exist next to each other without hierarchical order.
>
> (Bouteligier, 2013, p. 13)

Taken together, such a specification of urban agency in relation to climate change policy does already contextualize and pluralize it compared to more straightforward notions of city diplomacy. However, what both contributions on city diplomacy and cities in climate change appear to lack, is a clear contextualization of urban agency *in relation to* the restructuring of globalizing national states – relations that can be considered to take place on a theoretical continuum between full antagonism and full cooperation.

A third strand of literature in IR and political science is far more attentive to the nature of the relationship between nation states and cities. In the book *Global Cities and Global Order* (2016), Curtis argues that the rise of the global city can only be understood in the context of wider processes of change in 'the international political order' and 'international society'. He analyses how national states themselves have in the past few decades supported the emancipation of cities from the control of national states. Clark and Moonen's (2016) book on city leadership titled *World Cities and Nation States* starts from the observed paradox that in the age of cities, all political structures are still embedded in inherited nation state systems. In their book, they consequently focus on how world cities and nation states are reorganizing their relationships and in particular on how nation states are adjusting to the age of cities. Clark and Moonen argue that a spirit of partnership between cities and nation states is essential. They discuss a number of factors that shape this changing relationship (e.g. the specific nation state system and the degree of political polarization) and highlight the path dependencies that are at work in how cities and nation states reconfigure their relationship. Overall, Clark and Moonen stress the ongoing relevance of the nation states. However, an important shortcoming of Clark and Moonen's work is that they tend to restrict 'global urbanization' to 'world cities', i.e. cities well connected to the global capitalist economy such as London, Tokyo, Paris, and Sao Paulo.

But in our view, the most integrated and complete take on the global urban political agency question to date is offered by Ljungkvist in her book titled *The Global City 2.0: From Strategic Site to Global Actor* (Ljungkvist, 2016). Drawing on the complementary strengths of IR and urban studies, Ljungkvist argues that the global city literature in urban studies (Sassen, 1991) fails to see the global city as a 'strategic actor in world politics' because it is economically determinist and approaches global cities mainly as 'strategic sites in the global economy'. In order to understand how global cities are 'actorized', Ljungkvist takes an 'inside out' perspective and explores 'how the Global City's role in the globalized world is constructed and narrated *locally*' (Ljungkvist, 2016, p. 2). Global cities claim political authority in foreign and security affairs (and not just a role in the globalized economy as in the Global City literature) on the basis of a locally developed collective identity. According to Ljungkvist, 'Global City reflexivity' is crucial here: cities start referring to themselves as global cities and interacting with the world through policies and practices developed on behalf of their urban societies. They do so in a wider structural context that may enable but also constrain their actions and in that sense global cities are always socially situated collective actors. Ljungkvist usefully adopts role theory to connect the reflexive agency of global cities with the structural nature of the 'global action context' and distinguishes three ideal-typical global city roles: the neoliberal role, the governance role, and the security/risk role.

In our view, even though her work is very much the odd one out, Ljungkvist's contribution opens up an exciting path for theorizing and analysing urban agency at the crossroads of urban studies and political science and IR. While the latter contributions are generally empirically very rich, we would argue that three key questions, informed by debates in urban studies, typically remain below the radar. First, while there is ample focus on shifts in multilevel governance, there is typically little reflection on the more structural conditions (i.e. macro-political/economic changes) amidst which these changes take place. Little is said about how new forms of governance reflect the search for stability and regulation in a context of globalization. Key issues of state rescaling (Brenner, 1998) are often taken as a mere backdrop against which the resurgence of cities takes place. If we agree that globalization engenders reterritorialization, the rise of cities is much more than a simple shift in governance structures, yet is reflexive of a wider struggle to reposition urban, regional, and national economies in the global economy. Second, such a repositioning takes place against place-based opportunities and constraints that make an urban renaissance a political affair that goes beyond the narrow field of urban policy, urban governance or even urban management. What a city is and for whom it acts is oftentimes not immediately clear, and a critical analysis should go beyond the study of the narrow confines of official institutions and their (elected) representatives. Rather, the political is also embodied in the daily lives of citizens and their grassroots organization that may be and in many cases definitively is detached from the 'official' circuits as they reject, criticize, and oppose them (see also Boudreau, 2017). Third and finally, the debate on urban agency in political science and IR literature is generally not reflexive about the concepts it uses to capture, analyse, and explain the processes it observes. Yet, most of our ideas about governance are tainted with (implicitly) normative stances on what good and just governance is. Quite clearly, there is a firm ideologically informed belief in representative democracy and its rootedness in (urban) institutions and their linkages with civil society. Still, these very concepts are geo-historically produced and are hence context dependent. A simple projection of democratic principles typically does little to explain how urban governance de facto comes about, especially when leaving the context of cities in countries with democratic elections. The next section offers a more detailed overview of how the broad field of urban studies may inform research on urban agency at a global scale.

## Insights from urban studies

### Contextualizing urban agency

In their efforts to generalize their observations on urban agency, political scientists and IR scholars (*pace* Ljungkvist, 2016) typically offer little reflection on why it is that some (and not other) actors in some cities (and not others) can act on a global scale and what the different modalities and conditions of that agency are across time and space. Urban studies, with its capacity to bring in both the particular context as an explanation and a sense of comparison and generalizability (Peck, 2015), can help remedy the universalizing tendencies of our political science interlocutors. A useful concept in this regard is the notion of positionality, which captures 'the shifting, asymmetric, and path-dependent ways in which the futures of places depend on their interdependencies with other places' (Sheppard, 2002, p. 308). What undergirds this concept of 'geographic situatedness' is a view that space is not a mere container, but a product of social relations that has dynamic and malleable properties. Spatial scales such as the urban are not pre-given, but are the outcome of power struggles in a given geo-historical setting. Typically, what is perceived as structural conditions are the products of hegemonic socio-spatial relations that will act as a context in which less powerful individuals and collectives act.

The rise of the global scale is very much the hegemonic scale on which the current power relations between (actors in) cities play out, even though that scale is in itself the product of very particular socio-spatial relations (finance, migration, trade and investment, air travel, the spreading of diseases, climate change, etc.). In Brenner's (1998) view globalization does not negate the production of scale, but is crucially about the reshuffling of the existing hierarchy between different spaces, most notably the eroding of the dominance of nation state spaces vis-à-vis supra-national spaces such as the European Union or the North American Free Trade Area. According to Swyngedouw amongst others, globalization also entails the increasing

importance of urban and regional spaces in regulating socio-spatial processes under capitalism, a phenomenon which he labelled 'glocalization'
(Swyngedouw, 1997). A first corollary is that the typical objects of study in
political science and IR – elected governments, administrations, mayors,
diplomats, consultancy circuits – are but one element in a wider mode of
regulation of capitalist social relations and their spatially uneven articulation across space.

Moreover, these new governance articulations are prone to inertia and
resilience, are sites of political struggle and vested interests, and hence interact with what comes before/after, i.e. they are path-dependent (Oosterlynck,
2012). The current urban renaissance, in other words, cannot be understood
without an appreciation of the wider restructuring of scales of regulation
or, more narrowly defined, governance over time. This certainly rings true
when studying the internal governance structures of the 'post-contexts' –
post-socialist countries, post-colonial countries, post-apartheid countries –
but also has a meaning for the alleged heartlands of the global economy
such as Europe when teasing out the relations between urban scales and the
European level.

A second consequence is that the capability to act politically is to an important degree structured by the positionality of the place from which one
acts (Sheppard, 2002). This positionality hinges on multiple dimensions.
First, it will depend on the power distribution within territorial jurisdic-
tions, including the power of cities relative to regional and national hinterlands and nested governance structures. Second, political agency will be
moderated by the above-mentioned interscalar relations, as some actors
may be more successful in scale jumping between the urban and the (supra-)
national scales (Cox, 1998). Third, urban density and concentration underlines power as a product of the agglomeration of human action and the ability to mobilize and project power out of this concentrated activity. Fourth
and finally, the power to act from a place is moderated by the insertion of
individuals and organizations in (deterritorialized) socio-spatial relations
that connect cities. While key voices in urban studies project a condition
of planetary urbanization (Brenner and Schmid, 2015) where 'the urban'
largely overflows cities sensu stricto, that condition does not undo the fact
that the capacity to act politically at a global scale is articulated unevenly in
space. Doing politics or being political is namely not just a social act; it is in
essence a socio-spatial act (Nicholls, 2008).

A third implication of thinking through positionality is that we need to
be critical about whose interests are served when 'cities' – as a collective –
act politically on a global scale. We cannot assume that the capabilities and
resources to jump scales to the global level are evenly distributed amongst
all citizens and urban actors; particular elite interests may in fact dominate 'the' urban agendas, as the Bloomberg example so clearly shows. In
that respect the discursive urban turn in political sciences – emblemized
by the intimate connections between Bloomberg and Barber – should raise

concerns about the clear and present danger that the urban renaissance is the keystone of a neoliberal urban project. It is, in other words, very important not to reify cities as unitary actors, yet to utilize plural definitions of urban political agency and the oppositions and coalitions they bring along – an issue we turn to in the next section.

### Pluralizing urban agency

Although equating urban agency with the mayor and/or city government is still prominent in the popularizing literature, notably in Benjamin Barber's work, many scholars have criticized this tendency and argued for a more pluralist understanding of this urban agency. There is less consensus, however, on how to construct a pluralist understanding of urban agency. In order to arrive at such a pluralistic understanding, we need to move beyond narrow political science understandings of urban politics in terms of electoral processes and the formal political decision-making arena and adopt a properly sociological definition in which formal decision-making is embedded in broader, societal dynamics. Savage, Warde and Ward (2003, p. 153) propose such a definition when defining urban politics as 'what local state agencies do and [...] the external, mobilized social groups, which try to influence their policies'. Especially the reference to 'external, mobilized social groups', opens up a discussion on how power is produced not only in elections, political parties, and city councils, but also in coalitions that are forged between different public and private for-profit and not-for-profit actors in the wider urban society.

This insight is not new in urban studies and can be traced back to the community power debate in the 1950s and 1960s (Hunter, 1953; Dahl, 1961). Community power analysts rejected the Chicago School orthodoxy in which the growth and socio-spatial organization of cities is governed by an 'unconscious biotic struggle for existence'. They put political decision-making central in their analysis of the dynamics of cities and the making of urban space. The community power debate was marked by two opposing positions. On one side of the debate are 'elitists', who argue that cities are run by a small elite that tends to be heavily dominated by local businessmen. On the other side of the debate are 'pluralists', who claim that power in cities is more diffuse and that each policy sector tends to be dominated by one specific group. The implication of this is that business interests are represented by only one elite amongst many others. The importance of this debate for thinking through urban agency is that community power scholars drew attention to how the development of cities is shaped by collective decision-making processes and that the power of certain organized interest groups, notably the local business world, in shaping collective decision-making should certainly not be underestimated. Although the post-war community power scholars were instrumental in shifting the focus to politics and decision-making as an integral part of urban dynamics and from the very beginning approached decision-making from a pluralist perspective, the focus on urban politics is decisively local and power is seen as an attribute of people rather than as a generative force that is produced through forging alliances and building coalitions.

The debate on urban political agency was revived in the early 1980s, notably by the publication of Peterson's book, *City Limits*. In his book, Peterson shifts attention away from the local embedding of urban politics. He argues that cities have to compete with each other for companies and people. Urban development choices and decisions therefore do not reflect the needs and preferences of the urban electorate but of supra-local economic elites. Important here is how Peterson draws attention to the wider (national) context for urban politics, but he does so in a way which leaves little or no scope for urban agency (except for 'allocational policies'), even to the point of claiming that urban politics is marginal. Peterson's book was only the first in a range of highly influential contributions that approach urban politics from a political economy perspective. Urban growth coalition and urban regime theory revived the community power debate and its two camps, with the former gravitating towards the elitist position and the latter towards the pluralist stance. These two approaches follow Peterson in taking into account the larger political-economic context in which city governments operate, but they oppose the idea that city governments merely follow an externally imposed logic. They focus instead on internal struggles and how these mediate external political-economic logics. They also move beyond community power scholarship by firmly rejecting the latter's methodological individualism and seeing power as a feature of social relations.

While the urban growth coalition theory developed by Logan and Molotch (1987) focuses on how place-bound interests bring various localized public and private actors together in a pro-growth coalition, urban regime theory is explicitly based on a rethinking of the nature of power in urban politics. Power, according to urban regime proponent Stone, is not so much about the capacity of one actor to make other actors comply to its wishes – it is rather created by bringing a set of actors with unequal resources together and making them work towards shared goals (Stone, 1993). To this rather US-based tradition in urban political economy, one can add the literature on urban entrepreneurialism (Harvey, 1989) and urban governance in the context of state spatial rescaling (Brenner, 2004). However, both these approaches tend to focus strongly on how the political strategies emerging from the contradictions of the capitalist accumulation process shape urban governance rather than on the complex processes through which urban agency is created. Finally, and somewhat separate from the above traditions, there is the neo-Weberian approach to the city (Bagnasco and Le Galès, 2000), which is mostly Europe-based, and sees the city as a 'collective actor', but one 'always in the making' (see Pinson, this volume). For neo-Weberians, this process of making the city a collective actor is shaped by a wide range of social groups and not just (capitalist) economic interests.

Overall then, urban studies approaches to urban politics may help us to see how agency in cities is produced not only in the electoral process, but far beyond it in the forging of coalitions and alliances between a wide range of public and private actors, each of which have their social basis in urban society. The main distinction here is between those approaches that stress how economic interests always tend to dominate and other approaches that see power as essentially diffused throughout urban society and argue that the city is governed by a multiplicity of 'local elites'. For our purposes, however, both sets of approaches can be called pluralist in the sense that in any case urban political agency is built on coalitions of actors – coalitions that are not just there, but require work to be maintained. However instructive they may be, these urban studies approaches to urban politics do not suffice to conceptualize *global* urban political agency. This is the case for two reasons. Firstly, as Ljungkvist states, even if recent approaches to urban politics are duly taking macro-level pressures into account (although mainly of an economic kind), the focus is not on the 'global and transnational dimensions of such local agency', but on how 'local political agency matters for local and urban outcomes' (Ljungkvist, 2016, p. 29). Secondly, their theoretical and empirical insights are mainly derived from US and European contexts. Theorizing global urban political agency requires us to take into account a geographically more encompassing set of experiences with urban agency, a topic to which we now turn.

### Expanding urban agency geographically

Because of the sizable attention paid throughout the 1990s to research into the specific urban experiences in/of putative 'global cities' such as New York, London, and Tokyo (cf. Sassen, 1991), the 'global urban studies' research agenda initially tended to implicitly set up these global urban experiences as a scientific (and often also policy) norm. Following this norm, being part of a global urban network of finance and informational capital has long been considered a key factor in the categorization of cities as 'global', thus potentially suggesting that urban developments and experiences of cities that are less connected in such networks are less relevant (Robinson, 2002; Roy, 2009). However, from the early 2000s onwards, inspired by postcolonial thinking that seeks to underline the parochial character of universal knowledge claims (Spivak, 1988), urban research has tried to undo the imprint of such a narrow perspective on what constitutes the field of 'global urban studies' (Sheppard, Leitner and Maringanti, 2013; Harrison and Hoyler, 2018). Since the world is urbanizing at a fast pace, a key question according to various, hybrid, contrasting, and alternative models has been how to include a broader variation of urban experiences into the 'global urban studies' research agenda.

Most of the postcolonial writings have come to the fore in the wake of the work of Robinson (2002, 2006), who introduced the concept of 'ordinary cities' as a framework for a new 'urban theory that draws inspiration from the complexity and diversity of city life, and from urban experiences and urban scholarship across a wide range of different kinds of cities' (Robinson, 2006, p. 1; see also Robinson's chapter in this book). By doing so, she raised three inter-related concerns that can be paraphrased as follows (see van Meeteren, Derudder and Bassens, 2016): (i) the normalization of the conceptual models that make sense of specific cities leading to the subjugation of alternative accounts of what constitutes globalized urbanization; (ii) the normalization of a focus on specific sets of processes and actors leading to an inadequate understanding of urban actors and processes in what are very different cities; and (iii) the tendency for research to be translated into context-less conceptual models, facilitating these to be recast into an aspirational standard for urban theory-making and urban planning around the world. Robinson's (2002, 2006) writings set in motion a broader research agenda criticizing the (restricted) geography of urban theory-making, above all sharply noting the 'enduring divide between "First World" or "Global North" cities (read: global cities) that are seen as models, generating theory and policy, and "Third World" or "Global South" cities (read: mega-cities) that are seen as problems, requiring diagnosis and reform' (Roy, 2009, p. 820).

The subsequent call for devising theoretical frameworks that allow re-thinking the 'epistemologies and methodologies of urban studies' (Roy, 2011, p. 224) has been repeatedly acknowledged over the past few years (e.g. Sheppard, Leitner and Maringanti, 2013; Ren and Keil, 2018). However, and perhaps more importantly, this critique has been used as the starting point for a quest for such alternative accounts. In addition to Roy's (2011) own attempts to recast the epistemological categories commonly used to make sense of the 'informal' forms of living of the urban poor, there has been a series of efforts to analyse how cities in the Global South are being 'worlded' in very different ways (cf. Roy and Ong, 2011). An example is the work of Vasudevan (2015), who argues that it is possible to build a conceptual model of the city (including Western cities) from the perspective of squatters and slum-dwellers in cities in the 'Global South' (see also Pieterse, 2008).

Although this geographical expansion and diversification in the knowledge base has undoubtedly enriched the field and resulted in a geographical broadening of the field, it can be argued that this has led sometimes to the reaffirmation of old binaries. While Roy (2009), for example, aims at dismantling these dualisms when pleading for a 'strategic essentialism', i.e. a unified approach to representing subjugated urban processes and actors based on shared experiences in spite of strong differences, the way she implicitly equates First World/Global North with 'global cities' and Third World/Global South with 'megacities' (p. 820) reaffirms these binaries. Also, while it is indeed a valid claim that 'the distinctive experiences of the cities of the Global South can generate productive and provocative theoretical frameworks for all cities' (Roy, 2009, p. 820), for instance by showing that power and resistance are not always simply opposed to each other but are often simultaneously at work (Roy, 2009, p. 827), this line of reasoning in many ways reaffirms the subaltern position given to cities in the

Global South, as if research into cities of the Global South always needs to be validated by what it yields for our understanding of cities in the Global North. Finally, by so sharply opposing 'global cities' to 'mega-cities', and thus focusing on city size, the range of city-types covered by urban studies today is rather limited. Geographically expanding the knowledge base can and should also entail inter alia the experiences of post-socialist cities (see chapters in this book), mid-sized cities (Dehaene, Havik and Notteboom, 2012), 'not-yet-cities' (De Boeck, Cassiman and Van Wolputte, 2010), and many other examples that do not fit in either of these categories. Precisely because of their less exceptional state, and with many people living in these mundane urban typologies, they probably deserve much more attention. We thus agree with Bunnell and Maringanti (2010) and Robinson (chapter in this book), who call attention to the need for alternative practices or ways of doing global urban research that move well beyond strategic essentialism, in particular when essentialism simply becomes an equivalent for (geographical) binarism, thus valorizing the diverse, situated practices and engagements of a much broader range of actors in a much broader range of cities. Moreover, by moving away from the prevailing core-periphery model in urban studies, also more light will be shed on South-South experiences in the making, modelling, and inter-referencing of the urban (Roy and Ong, 2011).

Overall, then, geographical expansion of the knowledge base in 'global urban studies' calls for a continuing provincializing of that knowledge base, not only by being informed by one other urban experience, mostly from the Global South, but from all possible urban experiences. This has two complementary implications. First, in the strict sense, it implies acknowledging that concepts and processes may take on a different meaning or be outright irrelevant in other geographical contexts (e.g. 'civil society', 'state actors'; see Bayat, 2000; Appadurai, 2002; Chatterjee, 2006). But second, and more broadly, it also emphasizes that conceptual exchanges and analogies (cf. Nijman, 2007) should be multi-directional and extend well beyond well-worn Global North/South divides.

## Structure of the book

Collectively, the chapters in this volume variously speak to our overall objective of contextualizing, pluralizing and geographically expanding our analysis of urban political agency. The contributions in the first part of this volume are more theoretical in nature. The authors in this part develop and/or critically assess the conceptual underpinnings of the notion of global urban political agency from a variety of theoretical perspectives. The second part contains a number of (theoretically informed) empirical analyses of global urban political agency in cities around the globe. We will now briefly introduce each contribution, also paying attention to how they work towards one or more of the three aforementioned goals of contextualizing, pluralizing, and expanding global urban political agency.

While most contributions to this volume are quite critical of Barber's book, in his chapter Corijn presents a sympathetic account arguing for the need to further specify Barber's general ideas, even though he is keenly aware of the need to contextualize and pluralize global urban political agency. In his essay, Corijn argues that there is above all a need for more integration and a more inclusive approach to the mushrooming urban policy networks such as the C40 network discussed by Barthold in this volume. Observing that these networks represent but a section of the global urbanity and tend to adopt a specific framing/focus, they fall short of emancipating the urban in terms of setting the global political agenda. This leads to a call for Barber's 'global parliament of mayors' to purposefully work on four registers at the same time, i.e. global urban agenda setting, intercity regulation, economic cooperation, and urban governance to produce added value over the many existing bodies.

Mingione, in turn, contextualizes the emergence of 'the urban' on the global (political) agenda in the ongoing transformations of welfare capitalism. This entails emphasizing and theorizing the darker side of cities, the side often downplayed or ignored in the pro-urban literature. He notes that the interconnection between social fragmentation and downscaling not only makes the city the main arena for new forms of political organization and experiences of social innovation, but also the many conflicts, inequalities, exclusion, and discrimination that come with this. Drawing on Polanyi's theory of the 'double movement' and the notion of 'democratic-welfare-capitalism', he posits that the increasing importance of cities does not necessarily mean that it is at this scale that an alternative social and political order is emerging, but rather that the main tensions of the present transitions are mostly happening on the urban scale. Based on this, he affirms the increasing importance of the urban scale and discusses the difficulties of translating the 'nation-state-based' parameters and narratives to explain inequalities and resistance at the urban level.

Pinson's chapter contains perhaps the clearest depiction of the need to contextualize urban political agency, in particular in relation to the fate of territorial states. His starting point is that accounts of the city as a political actor too often assume a zero-sum relationship between the rising political prominence of cities and the assumed decline of territorial states. Drawing on research on urban governance and state-city relations in France, Italy, and the UK, he develops the compelling (and for many probably counterintuitive) argument suggesting that those cities that emerge as political actors are often located in states that have themselves managed to reposition themselves in a context of globalization and affirmation of subnational and supranational political levels. Echoing insights from the rescaling literature arguing that rescaling entails more than simply redistributing, he thus shows that there is no simple trade-off between cities and states (see also Brenner, 1998), but rather a rich and complex array of options that shapes how cities (can) affirm themselves as political actors.

Ward critically reflects on what appears to be growing orthodoxy suggesting that the city is now the primary site for policy experiments, ranging from 'leftist' interventions such as the introduction of living wages to 'right-wing' experiments with downsizing/rightsizing of government. It is this meta-narrative of the 'urban age' and the 'urban triumph' that undergirds the emergence of what might be termed the rise of cities as 'formal' political actors. Ward's key intervention is that he seeks to pluralize our understanding of how cities become actors by emphasizing the less visible but at least as crucial 'informal' political work that takes place to bring alive these formal structures through more ad hoc, ephemeral, fleeting, incremental and informal ways in which cities perform governing globally. This entails recognizing that urban policymaking and the politics should not be taken as given; the insight that urban politics is constantly re-shaped based on systems of comparing, borrowing, exchanging, imitating, learning, reinterpreting, and translating; the recognition that a wide range of actors have been drawn into the making of urban policy; and being sensitive to the dangers of overemphasizing the agency-like qualities of cities as political actors in the context of the wider transformations in economic, political, and social systems.

Extending her earlier work on the need to pluralize and geographically extend our understanding of what cities are/do, in her chapter Robinson seeks to stretch analyses of the city as a political actor by drawing on insights from the range of political dynamics at work across different urban contexts as urban actors and institutions seek to launch themselves on the global stage or to localize 'global' processes. She advances postcolonial ideas that call attention to be equally inspired by urban processes important to urban areas located in different regions, shaped by diverse histories of globalization, highly disparate levels of poverty and wealth, and different regulatory and institutional forms. Such a more diverse comparative imagination stretching across a range of different cities can help examine and stretch analyses of the multiple and differentiated nature of 'the global' in relation to which 'the urban' might have political agency.

Acuto's chapter stands out in the sense that it provides a detailed account of power and politics with the overall objective of opening up a more thorough engagement between political science and IR theory and urban studies. One conspicuous argument emerging from this is that setting up purely 'parallel' urban political circuits will probably result in fragmented structures, as emphasized by Corijn. He thus calls for a thorough rethinking of the role and status of cities within global governance in general, and the United Nations (UN) in particular. While recognizing that the UN is of course flawed in many ways, it still remains pivotal in convening, spurring, and sanctioning global action, and this increasingly includes urban-centred action (e.g. Revi, 2017). Building a better understanding of this context and of the complex governance arrangements it offers may perhaps be the best available tool for shaping the politics of global urban governance. In line with Pinson, therefore, Acuto sees no simple trade-off between cities and states, but ample opportunities for the emergence of cities as global political actors to draw on the more classical governance and political structures offered by the state machinery.

Opening the second, more empirically focused part of the book, Barthold critically assesses the global political agency of cities in multilevel climate governance through the lens of the C40. C40 is an organization of mayors of global cities that facilitates knowledge, technology, and best practice sharing between cities and promotes and defends the interests of global cities in global climate negotiations and strategies. Barthold pluralizes urban agency by showing how urban climate governance (and C40 in particular) challenges the classical international political system of nested (state) hierarchies and how the C40 network brings together a variety of global urban economic, cultural, and political elites around the globally circulating concept of urban 'sustainability'. She also contextualizes this global strategy of 'greening' of urban politics by arguing that already powerful global cities in the C40 network use and develop their symbolic power to set technical and political norms and thus create a global hierarchy of cities built on criteria derived from a market-based version of urban sustainability. The 'we are all in this together' rhetoric of C40 obscures the structural relations of power that emerged from a history of colonialism and endure in the unequal development between North and South today. Attending to the need of expanding urban agency, Barthold concludes that for many less affluent cities global environmental programs and climate initiatives have become a 'worlding' strategy and an indicator of how '(eco-)modern' they are.

Fricke, in her chapter on 'metropolitan regions' in the European Union's supra-national policy arena, starts from a more classical political science approach to supra-national urban political agency. She explores the increasing visibility of metropolitan regions both as policy issues and as institutional scales in the EU. Fricke identifies three different understandings of 'the metropolitan', namely an internal functional, an external economic, and a governance conceptualization, and shows that metropolitan regions as urban political actors do not have a formalized role in the EU. However, through networking and lobbying, metropolitan regions have succeeded in establishing themselves as an issue and a scale for policy implementation in the context of EU policies. Here, she adds an interesting urban studies perspective by explaining the tentative incorporation of metropolitan regions into EU policy by referring to processes of rescaling and reframing. Fricke's analytical focus on processes of scale construction and shifting policy frames implies a pluralization and contextualization (albeit in a more restrictive formal-institutional sense than discussed in the 'Insights from urban studies' section of this introductory chapter) of urban political agency.

Peyroux expands our understanding of global urban political agency in geographical terms through a case study of the Johannesburg IR Strategy. With this strategy, Johannesburg aims to strengthen its international status and influence through inter alia city-to-city cooperation, participation in transnational urban networks and direct engagement with international organizations. Her chapter enriches scholarly debates on global urban political agency by bringing attention to the geopolitical dimension of international city strategies from a Global South perspective. In this context, Peyroux stresses the growing economic and diplomatic power of Global South countries and cities, particularly within the BRICS, and the changing origin and nature of urban expertise associated with this shift of power. In line with a pluralized understanding of urban agency, Peyroux argues for a networked reading of political agency that is attentive to how Johannesburg is building the capacity to influence the production of urban norms and values based on South Africa's urban experiences.

Like Peyroux, Wenz mobilizes a South African case study to disrupt the hegemonic status of a few Euro-American global cities as privileged sites of urban theory-making and geographically expand our understanding of global urban political agency. Looking at high profile international accolades and the hosting of their associated events such as the (first African) World Design Capital (in 2014) in Cape Town, Wenz proposes the notion of 'worlding' as a conceptual tool to trace the intricate politics through which 'Southern' cities interact with and become part of 'global forms in circulation'. In line with this volume's ambition to pluralize urban agency, Wenz focuses on unraveling both the 'extrospective' and 'introspective' politics that fed into the process of worlding Cape Town. The World Design Capital allowed Cape Town's local elites to strategically couple the extrospective politics of urban boosterism to an introspective politics of persuading the local population through a carefully crafted self-image. Finally, by paying due attention to local contention and the city's historically ambiguous relationship with its continental location, Wenz is also keen to contextualize Cape Town's global urban agency.

In the final two chapters of the book, the notion of global urban political agency is further expanded with a focus on the politics of urban development in Eastern Europe. In their chapter, Olt and Lepeltier-Kutasi are concerned with urban democracy. They critically assess Barber's claim that municipal political leaders are best placed to democratically represent the interests of urbanites. They argue that global business interests and the national socio-political context are just as important as democratically elected urban political leaders in democratic struggles in cities, hence contextualizing urban political agency. In Budapest, which is the empirical focus of their contribution, the heritage of the state socialist dictatorship and the process of post-socialist transformation deeply shape urban democracy and neo-patrimonial power and property relations frequently undermine the legal authority of the state.

Koelemaij, finally, geographically expands and pluralizes urban agency by scrutinizing the mode of implementation of a typical neoliberal infrastructural urban megaproject, but in the less researched region of Eastern Europe. By focusing on the non-transparent layers of governance underlying the urban development project, he highlights yet another plurality of urban actors involved in improving a city's global profile. Koelemaij starts his analysis from the well-established theories of neoliberal urbanism and state rescaling. Although he observes some of the neoliberal logics identified in these theories, he contextualizes global political urban agency in the case of Belgrade by drawing attention to the specificities of the post-socialist context of Serbia and its authoritative neo-statism. According to Koelemaij, this specific context explains why the theoretically predicted downscaling of state-functions is not happening here and the national government – or more specifically the prime minister and his closest political confidants – plays a dominant role here, in alliance with transnational capital and using non-transparent and informal strategies to push this project through.

## Note

1 The authors' names appear in alphabetical order. The order does not reflect a hierarchy in contribution.

## References

Acuto, M., 2013. City leadership in global governance. *Global Governance: A Review of Multilateralism and International Organizations*, July–September 2013, 19(3), pp. 481–498.

Amen, M., Toly, N.J., McCarney, P.L. and Segbers, K., 2011. *Cities and global governance. New sites for international relations*. London and New York: Routledge.

Appadurai, A., 2002. Deep democracy: urban governmentality and the horizon of politics. *Public Culture*, 14, pp. 21–47.

Bagnasco, A. and Le Galès, P., 2000. European cities: local societies and collective actors? In: A. Bagnasco and P. Le Galès, eds. *Cities in contemporary Europe*. Cambridge: Cambridge University Press. pp. 1–32.

Barber, B., 2013. *If mayors ruled the world. Dysfunctional nations, rising cities*. New Haven, CT: Yale University Press.

Bayat, A., 2000. From 'dangerous classes' to 'quiet rebels': the politics of the urban subaltern in the global south. *International Sociology*, 15, pp. 533–557.

Bloomberg, M., 2015. City century: why municipalities are the key to fighting climate change. *Foreign Affairs*, [online] September/October. Available at: www.foreignaffairs.com/articles/2015-08-18/city-century [Accessed 15 May 2018].

Boudreau, J.-A., 2017. *Global urban politics*. Cambridge: Polity Press.

Bouteligier, S., 2013. *Cities, networks and global environmental governance. Spaces of innovation, places of leadership*. London and New York: Routledge.

Brenner, N., 1998. Global cities, glocal states: global city formation and state territorial restructuring in contemporary Europe. *Review of International Political Economy*, 5, pp. 1–37.

Brenner, N., 2004. *New state spaces. Urban governance and the rescaling of statehood*. Oxford: Oxford University Press.

Brenner, N. and Schmid, C., 2015. Towards a new epistemology of the urban. *City*, 19(2–3), pp. 151–182.

Bulkeley, H., 2013. *Cities and climate change*. London and New York: Routledge.

Bunnell, T. and Maringanti, A., 2010. Practising urban and regional research beyond metrocentricity. *International Journal of Urban and Regional Research*, 34, pp. 415–420.

Castells, M., 1996. *The rise of the network society*. Oxford: Blackwell Publishers.

Chatterjee, P., 2006. *The politics of the governed: reflections on popular politics in most of the world*. New York: Columbia University Press.

Clark, G. and Moonen, T., 2016. *World cities and nation states*. Chichester: John Wiley and Sons.

Cox, K.R., 1998. Spaces of dependence, spaces of engagement and the politics of scale, or: looking for local politics. *Political Geography*, 17(1), pp. 1–23.

Curtis, S., 2016. *Global cities and global order*. Oxford: Oxford University Press.

Dahl, R.A., 1961. *Who governs? Democracy and power in an American city*. New Haven, CT: Yale University Press.

De Boeck, F., Cassiman, A. and Van Wolputte, S., 2010. Recentering the city: an anthropology of secondary cities in Africa. In: K.A. Bakker, ed. Proceedings of African perspectives 2009. *The African inner city: [re]sourced*. Pretoria: University of Pretoria. pp. 33–42.

Dehaene, M., Havik, K. and Notteboom, B., 2012. The mid-size city as a European urban condition and strategy, *OASE*, 89, pp. 2–9.

Glaeser, E., 2011. *Triumph of the city*. London: MacMillan.

Harrison, J. and Hoyler, M. eds., 2018. *Doing global urban research*. Sage: London.

Harvey, D., 1989. From managerialism to entrepreneurialism: the transformation in urban governance in late capitalism. *Geographiska Annaler B*, 71, pp. 3–17.

Hunter, F., 1953. *Community power structure*. Chapel Hill: University of North Carolina Press.

Kern, K. and Bulkeley, H., 2009. Cities, europeanization and multi-level governance: governing climate change through transnational municipal networks. *JCMS, Journal of Common Market Studies*, 47(2), pp. 309–332.

Ljungkvist, K., 2016. *The global city 2.0. From strategic site to global actor*. London: Routledge.

Logan, J.R. and Molotch, H.L., 1987. *Urban fortunes. The political economy of place*. Berkely: University of California Press.

Nicholls, W.J., 2008. The urban question revisited: the importance of cities for social movements. *International Journal of Urban and Regional Research*, 32(4), pp. 841–859.

Nijman, J., 2007. Place-particularity and 'deep analogies': a comparative essay on Miami's rise as a world city. *Urban Geography*, 28(1), pp. 92–107.

Oosterlynck, S., 2012. Path dependence: a political economy perspective. *International Journal of Urban and Regional Research*, 36(1), pp. 158–165. doi:10.1111/j.1468-2427.2011.01088.x.

Peck, J., 2015. Cities beyond compare? *Regional Studies*, 49(1), pp. 160–182.

Peterson, P.E., 1981. *City limits*. Chicago, IL and London: The University of Chicago Press.

Pieterse, E., 2008. *City futures: confronting the crisis of urban development*. London: Zed Books.

Ren, X. and Keil, R., 2018. *The globalizing cities reader*. London: Routledge.

Revi, A., 2017. Re-imagining the United Nations' response to a twenty-first-century urban world. *Urbanisation*, 2(2), [online]. Available at: http://journals.sagepub.com/doi/full/10.1177/2455747117740438 [Accessed 15 May 2018].

Robinson, J., 2002. Global and world cities: a view from off the map. *International Journal of Urban and Regional Research*, 26(3), pp. 531–554.

Robinson, J., 2006. *Ordinary cities: between modernity and development.* London: Routledge.

Roy, A., 2009. The 21st-century metropolis: new geographies of theory. *Regional Studies*, 43, pp. 819–830.

Roy, A., 2011. Slumdog cities: rethinking subaltern urbanism. *International Journal of Urban and Regional Research*, 35(2), pp. 223–238.

Roy, A. and Ong, A. eds., 2011. *Worlding cities: Asian experiments and the art of being global.* Malden, MA: Wiley-Blackwell.

Sassen, S., 1991. *The global city. New York, London, Tokyo.* Princeton, NJ: Princeton University Press.

Savage, M., Warde, A. and Ward, K., 2003. *Urban sociology, capitalism and modernity.* Houndmills: Palgrave Macmillan.

Sheppard, E., 2002. The spaces and times of globalization: place, scale, networks, and positionality. *Economic Geography*, 78, pp. 307–330.

Sheppard, E., Leitner, H. and Maringanti, A., 2013. Provincializing global urbanism: a manifesto. *Urban Geography*, 34, pp. 893–900.

Smith, M.P., 2001. *Transnational urbanism: locating globalization.* Malden, MA: Wiley-Blackwell.

Spivak, G.C., 1988. Can the subaltern speak? In: C. Nelson and L. Grossberg, eds. *Marxism and the interpretation of culture.* Urbana: University of Illinois Press. pp. 271–313.

Stone, C.N., 1993. Urban regimes and the capacity to govern: a political economy approach. *Journal of Urban Affairs*, 15(1), pp. 1–28.

Swyngedouw, E., 1997. Neither global nor local: 'glocalization' and the politics of scale. In: K.R. Cox, ed. *Spaces of globalisation.* New York: The Guilford Press. pp. 137–166.

Tavares, R., 2016. *Paradiplomacy: cities and states as global players.* Oxford: Oxford University Press.

Taylor, P.J., 2013. *Extraordinary cities: millennia of moral syndromes, world-systems and city/state relations.* Cheltenham: Edward Elgar.

Taylor, P.J. and Derudder, B., 2016. *World city network.* 2nd ed. London and New York: Routledge.

Toly, N., 2016. In the future, cities may finally solve problems that have stumped the world's biggest nations. *Quartz*, [online] 13 October. Available at: https://qz.com/807733/in-the-future-cities-may-finally-solve-problems-that-have-stumped-the-worlds-biggest-nations/ [Accessed 15 May 2018].

Van Meeteren, M., Derudder, B. and Bassens, D., 2016. Can the straw man speak? An engagement with postcolonial critiques of 'global cities research'. *Dialogues in Human Geography,* 6(3), pp. 247–267.

Vasudevan, A., 2015. The makeshift city. Towards a global geography of squatting. *Progress in Human Geography,* 39(3), pp. 338–359.

# Part I

# The city as a site for political innovation

# 2 Reflecting on the 'Global Parliament of Mayors' project

*Eric Corijn*

In memory of Benjamin Barber.
Scholar and fighter for democracy in an interdependent world,
campaigner for the Global Parliament of Mayors.
He died of pancreatic cancer at age 77 on April 24, 2017.

Benjamin Barber (1939–2017) develops the idea of a Global Parliament of Mayors in Chapter 12 of his book: *If Mayors Ruled the World: Dysfunctional Nations, Rising Cities*. Barber's book is written in the context of interconnected cities and interdependent citizens. Globalization is also urbanization. These societal developments are already represented in many city networks and in the rise of a global civil society (Barber, 2013, pp. 106–140). These networks have a long record of exchange, benchmarking, multi- and bilateral cooperation and for the more thematic networks of focused agenda setting. I argue with Barber that there is no need for yet another such network, but there is a need for more integration, more inclusive networking, of a network of the networks. These networks all represent but a section of the global urbanity and but a specific framing. There is hence room for some thinking on how a better integration can be reached. Above all, these networks do not reach a global political agenda. The challenge is to put 'urbanity' on the global agenda in an efficient way.

The central argument in Barber's book is a search for democratic regulation, increased democratic citizenship, and legitimate power in a world left to the market, private interests and corporate competition. It is an argument Barber also develops in *Jihad versus McWorld*, the book in which globalism and tribalism are analysed as two sides of the same coin, with both phenomena threatening democracy (Barber, 1995). It is exactly in that field that nations are dysfunctional, as Barber describes in *If Mayors Ruled the World*.

The challenge of democracy in the modern world has been how to join participation, which is local, with power, which is central. The nation state once did the job, but recently it has become too large to allow meaningful participation even as it remains too small to address centralized global power (Barber, 2013, p. 5).

In the first part of the book, Barber delivers the analysis (Why cities should govern globally), whereas the second part focuses on action and intervention (How it can be done). Besides the historic argument on the intimate relationship between cities, citizenship and democracy, the main argument derives from the central position of cities in the actual globalization process. Chapter 3 of the book gives a historic overview from the independent polis to the interdependent cosmopolis. Citizenship and democracy originated in place-bound territories, in walled cities opposing feudal rule. Later, regions and estates were unified in territorialized nation states, searching for independence. Today the world system and the globalization process favour connectivity on a larger scale. That connectivity is not so much obtained through incorporation of ever-larger territories within shared borders – making bigger countries – but through a growing importance of nodal connection of networks in a space of flows. The territorialized political independence of countries is thus deconstructed by transborder networking and growing interdependence. In Chapter 5, Barber explains the deeper reasons for this interdependence. Globalization has created a transnational space of flows with cities or metropolitan regions as interconnected nodes. Barber refers to the work of Castells or to the author's own to indicate how 'cities undermine national solidarities and favour "glocal" growth strategies' (Corijn, 2009).

Based on this analysis, in 2001, after 9–11, Benjamin Barber created a global Interdependence movement. He proposed September 12th as Interdependence day in contrast with the very American 4th of July Independence Day commemorating the adoption of the Declaration of Independence in 1776. This initiative has been a platform of yearly gatherings discussing democratic governance in a globalizing world. In 2008 the Brussels programme developed the theme 'The city as commons in a divided world'[1]. The meeting discussed the rescaling of the world and the remodelling of the city. It is there that the growing importance of cities as (potential) global political actors has come to the fore, under the condition that urban politics also creates self-awareness of this strategic position and the responsibility that comes with that position.

In short: according to Barber, international politics do not deliver a democratic and transparent regulation of the world system and do not seem to be able to deliver an inclusive society overcoming communitarianism and tribalism. That is why cities should 'rule the world': because the complexity of the urban system is closer to the complexity of the world system and the pragmatism, the proximity and the sense of urgency in urban governance can lead to more adequate management of cities and its problems. Barber shows that the existing state-order, the nation state and its political order, stands in the way of addressing societal problems. Both because nation states lost control over so-called internal affairs and are unable to govern global matters. So, the focal question is: how to build democratic political power at the level of the global agenda? How to democratically regulate the world system? It is here that the call to mayors comes in:

> It is only to understand that to govern their cities effectively, they may have to play some role in governing the world in which their cities fight to survive. In governing their cities cooperatively to give their pragmatism global effect, mayors need not await the cooperation of the disunited United Nations, the special interest permeated international financial institutions, private-market multinational corporations, or centuries old dysfunctional nations.
>
> (Barber, 2013, p. 338)

The project of a Global Parliament of Mayors is about creating a legitimate political authority that can add some governance capacity and regulatory power beyond and besides existing global institutions, which are all legitimized by national member states. It is about filling up the democratic deficit that has been created by the uneven globalization process, led by inter-national political and economic decisions that are made without reference to a global constituency, civil society, citizenship or global public opinion. I now want to elaborate on this argument from my own point of view and work. More specifically I will deal with two questions: (1) Why cities? (2) What is the aforementioned global agenda?

## Why should mayors become global politicians?

I see basically four strong arguments that I share with Benjamin Barber for an increased political role through a 'Global Parliament of Mayors' or a 'World Assembly of Cities'.

a   First, there is the undeniable fact that globalization takes the form of urbanization. The majority (54%) of the world population is now living in cities[2]. In one century the urban population has increased from some 220 million (1900) to near to 5 billion (2030). In 1950 we had 83 cities with more than 1 million inhabitants, now there are more than 400 urban areas of more than 1 million. In 1950 only New York exceeded 10 million inhabitants, whereas today there are 28 megacities, with Tokyo at the top with its 39.4 million inhabitants[3].

b   Secondly, the global economy operates as a space of flows with cities as nodal points (Castells, 2000). Nation states are particularly dysfunctional in this regard because they treat all problems as strictly domestic and contained within national borders; they cannot thus sufficiently regulate the transborder flows and interactions. That is why the maintenance of the territorial authority of the nation state also produces an increased nationalism, selectively closing the borders for immigration, commerce or exchange. As long as the bulk of the interactions are territorially contained, a territorial political authority can regulate them; but when they substantially exceed the borders, there are only two options for regulation. On the one hand, political territories can be integrated in supranational organisms heading towards new forms of supranational states. On the other hand, gatekeeping of flows can be organized through collaboration of the (urban) nodal points in the networks. As Barber (2013, p. 147) formulates it: 'Sovereignty is not in decline, but its exercise on the global scene is increasingly counterproductive. States are not necessarily dysfunctional as national political systems (though some are), but they are dysfunctional in their inability to cooperate across borders[4]'.

c    Thirdly, governing the complexity of an urban system leads to forms of pragmatic regulation that are mostly not contained in the ideological representation of national political parties. The urban question is hardly at the core of national politics and is mostly considered as a question only of local administration. 'Globalization' literally takes place and becomes 'glocalization'. It is happening locally and is both transforming and being fuelled by metropolitan regions (Massey, 1991), it is not structuring the national political debates. Press and media, education, culture, sociocultural mediation, political parties and even religion remain domains of the national 'imagined community' (Anderson, 1983). Many 'local' metropolitan regions are confronted with the impossible regulations and representations of the nation state and do pragmatically rearrange governance. In today's world in which more and more national governments take a protectionist turn, cities have to oppose this in favour of an open and connected world.

d    Finally, cities are directly confronted with the global challenges and feel a greater sense of urgency than that shown by heads of state and governments at their negotiating tables. Increased urban densities, the transition to post-industrial economies, the mixed forms and functions, super-diversity, and increased mobilities cannot be dealt with through a classical modern territorial approach of zoning and segmentation. Modern government has dealt with complex problems separating them in different zones, institutions or competences. Separation is a dominant modus operandi, but the urban is characterized by density, mixity and transversal complexities. Pragmatic urban governance needs to confront many interconnected challenges and is therefore more used to mobilizing more transversal and collaborative processes.

Cities are directly confronted with planetary challenges because they concentrate a significant part of the world population, because they are the nodes in the space of flows, because the complexity of such a metropolis generates a political pragmatism nearer to the world system and its planetary mechanisms and finally because the global challenges are more immediately felt in that urban context. That is why cities and their governments should take action on a global level. A Global Parliament of Mayors could be the political expression of urbanity on a global scale, just as the General Assembly of the UN is expressing nationality on the global scale. The GPM could develop inter-urban agreements, just as nations work on international treaties. The World Assembly of Cities should be composed by what cities consider their legitimate representation, just as the UN states that governments represent nations. I will come back to that question, but let us first deal with the second question: for which global agenda?

## Dealing with the global challenges

To justify the added value of a Global Parliament of Mayors compared to the existing networks that organize exchange and horizontal cooperation, we have to define its focus, its political agenda. Today, there is no elected instance dealing with the interdependence of cities and metropolitan regions and the regulation of the space of flows. Furthermore, there is no instance dealing with the global agenda affecting urbanity. Operations like UN Habitat remain international collaborations between member states with cities as mere observers, subject to their domestic competence structures.

If the world system and global markets are increasingly structured by networked cities, one can immediately see the democratic deficit created by a one-sided focus on territorialized gatekeeping. A Global Parliament of Mayors should build on the empirical evidence of specific flows and networks through which global activities are organized and regroup cities in thematic commissions to discuss democratic regulation and gatekeeping for the flows in which they are involved. Whereas national parliaments organize their work following the rather static thematic organization of the state administration, a World Assembly of Cities could be thematically organized exactly on the truly existing interdependent and transversal glocal challenges.

The whole idea of a World Assembly of Cities has been critically received, also within this book and the conference discussions preceding it (e.g. see chapters by Ward and Pinson in this collection). The arguments differ. Some argue that the 'pro-cities' literature is too essayistic and does not refer sufficiently to academic references and debates. Fair enough, but given that academic debates are often reduced to what is available in English, the inclusion in Barber's book of references to press articles or vignettes on some mayors should not be set aside as merely anecdotal. They show how local circumstances and the personality of mayors can make a difference, which is in line with other academic research showing how 'the local matters' (Moulaert, 2000; Moulaert et al., 2011).

Most of the scepticism about the possibility of turning cities into global actors is, however, related to the assessment of the relationship with the central state. Pinson (this volume) deals with the question of the conditions under which cities can be more than mere local administrations of a territorialized state. My argument here is of course not that urbanity does not need a state-form and is therefore not confronted with the question of how to territorialize legitimate power. The crux of the 'city-actor'-argument is that the national form of state has become dysfunctional because its scale is ill-adapted to the global nature of the problems, because of its incapacity to handle good multi-scalar governance, and because of its basic foundations in the specific cultural format of a national identity. The hypothesis of Pinson, that 'city-actors' need strong central states, exemplified by the case study of Lyon, needs further examination if one is looking at cases like Barcelona or even more so Brussels, which operate within a framework of a weak central state. The Brussels Capital Region with its constitutional and logistic competences is the product of the disintegration of the unitary Belgian state and the impossibility of including Brussels in the two other ethnic projects – the mono-cultural Flemish and Walloon regions. As a city-region Brussels is not merely a place in Belgium – its identity cannot even be limited to being the capital of the country; Brussels needs to develop its own project beyond the nation state as a small world city at the core of the EU project.

What seems absolutely necessary to any discussion about local sustainability is that the instruments of the local state can take a spatial form that includes (most of) the urban ecosystem. In most cases urbanity as a local system is larger than the city or the municipality in its administrative form and needs a metropolitan level of governance (see the contribution of Fricke in this volume). What is also necessary is that the central state allows for the local democracy and government to regulate at that level. There has to be either sound forms of decentralization or an equilibrium with the central state. And finally, a city is not a country and is in many cases at odds with the cultural identity of the nation state, especially when confronted with restricted identity politics against migration and multiculture. Overall, becoming a 'glocal' actor depends on the ways (global) cities relate to global challenges and on the willingness not to undergo globalization, but rather to develop democratic instruments to orient it. What is at stake is not the disappearance of the state as such, but the redistribution of sovereignty within a multi-scalar organization from glocal to national to continental and planetary. And that architecture will not only be a superposition of territories, but also of networks with more or less strong or weak ties for each locality. A(n) (eco) systemic analysis looks for the specific combination and scales of the space of places with the space of flows.

What then are the global challenges that cities can handle in a better way than dysfunctional nations? I see three main systemic challenges in an urbanized world that do not fit the scaling of international regulation.

a    *Our relationship with nature.* The combined effects of a certain functionalist modernism and industrial productivism has produced destructive effects on the environment and is now putting in danger the planet as human habitat. The ecological footprint of most developed countries is far beyond what the earth annually produces in terms of biocapacity and is only partially tempered by the extreme poverty of some peripheral territories. With an overall economic growth of 3%, the gross world product doubles every 23.4 years. In the last 500 years, the gross world product has grown 150 times. At this rate the next 150 times growth will be reached by 2200 (Maddison, 2001, pp. 639–640; Corijn and Saey, 2014). There simply is no biocapacity to support such a future. We can only conclude that the dominant discourses on economic development maintaining the frame of growth, free enterprise and market economy are irresponsible, but the urgency to propose an alternative is suppressed because of the structuring of the political agendas in a world of competing countries. Still, from a global and urban perspective, the necessity of discussing the fundamental organization of the economy is undeniable. Cities are confronted with the disastrous effects of the current economic growth model: extreme weather events, the rapid decline of biodiversity, exhaustion of resources, a structural mobility crisis, air pollution and water scarcity, an energy and food crisis, and climate change and the warming of the earth.

Some 40 billion tons of $CO_2$ have been emitted in 2013. That was 2% more than in 2012 and 60% more than in 1990, notwithstanding many international treaties. The systemic challenge is clear. Without allowing a discussion about the currently dominant economic model and a post-carbon future, an irreversible climate change will be inevitable. More than an ideological debate, we need the development of models of sustainable urban ecosystems and politics of climate justice, of just transition. Such an agenda exists in many scattered ways, but is not yet part of the core geopolitical stakes. In the meantime, cities and metropolitan regions are immediately confronted with planning sustainable land use, green and blue infrastructures, recovering biodiversity, air and water pollution, managing resources, energy production and supply, regulating food production and supply, combining urban industries and services with habitat and housing, mobility, ... in short, designing a post-industrial, post-carbon sustainable urban ecosystem.

b   The second global challenge is related to the first one: the *growing social inequality*. Worldwide, the gap between the haves and have-nots has greatly widened. A conservative estimate for 2010 finds that the world's richest 91,000 people, just 0.001% of the world's population, own at least one-third of all private financial wealth, and nearly half of all offshore wealth. The next 51% of all wealth is owned by the next 8.4 million – just 0.14% of the world's population. The richest 20% accounts for three-quarters of the world income. Tax Justice Network calculated that the super-rich are currently hiding away wealth estimated between US$21 trillion and US$32 trillion in tax havens such as Switzerland and the Cayman Islands. And the most recent Oxfam Davos report shows that 62 people own the same as half of the world population[5].

At the other end, the poorest 40% of the world's population accounts for 5% of global income. Almost half the world, over three billion people, live on less than US$2.50 a day. At least 80% of humanity lives on less than US$10 a day. Nearly a billion people entered the 21st century unable to read a book or sign their names. In 2005, one out of three urban dwellers (approximately one billion people) was living in slum conditions. In 2005, the wealthiest 20% of the world accounted for 76.6% of total private consumption, the poorest fifth just 1.5%. 1.6 billion people – a quarter of humanity – live without electricity[6].

The reality behind these global statistics can be easily seen in the urban fabric. Global poverty creates massive slums. Urban social inequality organizes a social geography in neighbourhoods and districts, in centres and peripheries, defines the housing market, structures the labour market and creates the education and health challenges of the city. In this context, creating urban social cohesion and integration is a big challenge. Many cities are thus confronted with the redistributive mechanisms and taxation systems that are still mainly in the hands of state governments. There is a growing consciousness that, even if redistribution can be improved, it will not suffice without simultaneously developing an urban commons – in the first place the provision and accessibility of basic services such as housing, education, health care, culture and mobility. Such provisions can be taught in terms of enforceable (universal) human rights and the right to the city for all. This implies a call to civil society and urban citizenry to develop new forms of production and consumption, different models of exchange, sharing and reuse, forms of cohousing, cooperatives and association, collective responsibility for public space and infrastructure, etc. Furthermore, it raises the question of successful schemes for upward social mobility. How to increase opportunities and equal chances? Social mixing is not necessarily the leverage for upgrading the poor. And finally, the social inequalities do also exist between cities in north and south, between rich and poor countries. How to organize intercity solidarities and development schemes and how bilateral and multilateral cooperation can be enhanced?

c     The third global challenge that informs the urban agenda is that of *superdiversity*, a product of migration and of sociocultural diversification. Three per cent of the world population, some 214 million people, lived outside their country in 2010. One hundred and 28 million of them moved to a developed region, 86 million migrated within developing regions, but the most important movement is that between the rural and the urban areas[7]. In 2010, in China alone, some 221 million people migrated, of which about 150 million came from rural areas. Internal migration counts for up to 40% of the urban growth in developing countries.

Cities are made by migration and the arrival of newcomers, which creates a big diversity of multiple juxtaposed lifestyles, religions, languages and cultures. A city is not a country. If the nation tries to offer inclusion through forms of national identity, the city needs to find peaceful social cohesion on the basis of great diversity and difference. Nations build on the idea of a common history that informs a shared tradition and identity. Cities cannot build on common roots, but have to project a common destiny, a futures project. Social integration and civic solidarity have to be combined with respect for multicultural and multireligious realities. These communitarian realities are enlarged with gender and social inequalities. Combining cultural diversity with social inclusion and cohesion necessitates the development of a more urban cosmopolitan culture, with respect for difference and a place for hybridity and intercultural cooperation.

Urban society cannot be based on but one community and is always trans-communitarian, interreligious and intercultural. This is captured by the often-quoted slogan: 'city air liberates'. In that sense urban culture and art are of a central concern to social integration. The creative city explores the possibilities for a post-national or post-communitarian society. It develops urbanity as a way of life.

These three global challenges are at the core of any urban project and have to be met via a process of urban transition. The resulting sustainable urban ecosystem also necessitates the development of an urban commons and a model for multicultural citizenship, but at the same time, such urban models are the only solution for the global systemic threats and are also the starting points for the necessary multi-scalar governance. This forms a strategic agenda for a Global Parliament of Mayors, an agenda that the existing international bodies are not likely to take up.

## Be pragmatic

What can a Global Parliament of Mayors or a World Assembly of Cities add to the existing networks of cities or to the existing world institutions? Again, we'll summarize some ideas in four points.

a   The major contribution lies in turning the above-developed arguments into world politics. A new agenda for world politics has to be developed by giving global challenges a new political framing, which involves developing a collective urban mission statement and common norms and policies. A GPM could translate human rights into rights to the city. The universal and abstract individual rights have to be translated into very concrete policies of provision and access that are central for an urban-grounded economy. In that sense a GPM will sometimes be an alternative to the framing of bodies like the UN or the World Economic Forum.

b   A Global Parliament of Mayors can take concrete steps in regulating the world system through measures of gatekeeping and norm setting for the spaces of flows in which the metropolitan areas are involved. The global division of labour extends the production and distribution chains and calls for transborder regulations. Combining existing local or regional competences in common agreements for interdependent processes might complement deficient international regulations.

c   A GPM can also favour very practical cooperation. Overall the urban transition needs a restructured economy compatible with a social and an ecosystemic sustainable development. Manufacturing a new type of urban commons, e.g. related to mobility, recycling, education or housing, might be enhanced through coproducing or collective orders in the same way that international agreements sustain parts of the global market.

d   The main innovation a GPM should consider is representing urban governance rather than urban government and administration. Delegations to the assembly should represent the urban system, rather than the mere political expression of (a part of) it. Besides the mayor and his/her staff, two other groups should be included. On the one hand, a metropolis contains more than the core city and must be represented by a council including the suburban and rural surroundings. But it should also be representative for the inner structures like districts and neighbourhoods. On the other hand, the urban governance depends on a development coalition including civil society. All cities are confronted with developing participatory democracy and governance models including stakeholders. Sessions of the GPM should be leveraged for expressing the urban project through mayors with their metropolitan partners and with their civil society and citizens.

Again, a combination of these four registers – global urban agenda setting, intercity regulation, economic cooperation and urban governance – will produce added value over the existing bodies. It will surely inspire existing networks to think about a network of the networks, about increased capacity building, about training, analysis and research programmes. A Global Parliament of Mayors will not replace any of the existing dynamics. By focusing on its main objective – developing (democratic) urban politics on a global level – it will inspire other world associations, for the global civil society and also for further development of local democratic governance. The GPM initiative is meant to empower the urban transition and increase democratic regulation. It seems to be a way to global citizenship beyond the national frames.

## From proposal to reality

The inaugural session of the Global Parliament of Mayors convened in the city of The Hague on 9–11 September 2016, the weekend of the 15th anniversary of 9/11. Following three years of planning, and three preparatory meetings in Seoul, Amsterdam and Washington, the effective launch of a global platform for common urban action took place in the 'city of peace and justice'. Mayor Jozias van Aartsen hosted the two and a half day convention, which drew nearly 70 cities from across the world. In addition, several urban networks and NGO's participated actively, including Eurocities, the US Conference of Mayors, OECD, the C40 Cities Climate Leadership Group, Habitat III, ICORN, EFUS and the Council of Europe. Several hundred invited observers, young people, experts, activists and representatives of civil society were present at the plenary sessions and enjoyed a special parallel programme with presentations of Advisory Board members. The opening plenary listened to short statements of the mayors of Amman, Palermo, Cape Town, Delhi and Oklahoma City.

The agenda focused on two substantive challenges for which concerted action by cities can make a crucial difference and where the urban approach is different from most national policies. The first was climate change and the specific urban role in combating it. The second was immigration and refugees, with also specific urban approaches to citizenship, integration and access to services. A third theme, the challenge of urban governance, was discussed with special focus on the relationship with civil society and the agenda of participatory democracy.

The discussions led to a draft Statement of Shared Principles on the three topics – a Mission Statement for the Global Parliament of Mayors, the instalment of a first list of members, a Steering Committee, a Consultative Committee, a Call to Action in the context of the upcoming deliberations of Habitat III, OECD gatherings, COP21 and COP22. These declarations of principles, the commitment of the participating cities and the instalment of a Steering Committee and a Secretariat are the practical outcomes of a first meeting. The further development is now in the hands of a preliminary governance structure for the GPM. Mayor Patricia de Lille of Cape Town has taken the chair of the Steering Committee after the mayor of The Hague left office[8]. It is oriented towards developing membership, organizing and financing the structure, putting in place a virtual platform and preparing for a second venue in 2018. A first annual report has been issued and a new website has been launched: http://globalparlia mentofmayors.org/.

A second convention was organized in Stavanger (Norway) on 24–26 September 2017, where four thematic resolutions were discussed and accepted: on empowering cities, inclusive cities, resilient cities and safe cities. Furthermore, the Executive Committee was re-elected and an Advisory

Committee was also installed. The GPM is now an established association with an organization, a leadership and agenda-setting bodies. It aims at broadening its audience by the next session in 2018.

## 'The times they are a changin'

Immediately after the election of Donald Trump, on 14 November 2016 Benjamin Barber published an article raising the question whether the cities could be a rampart against the xenophobic nationalist protectionism of the new president (Barber, 2016). On 18 January 2017, he issued another text answering the question positively, explaining how the resistance is going to be localized and how cities do counteract (Barber, 2017). Since Barber's book and since the first version of this chapter, the political context seems to be rapidly changing. The neoliberal globalization process did combine the development of a global market and global trade agreements, with international integration and a hegemonic discourse of combined freedom, rights and development. Globalization has taken a turn. New forms of right wing nationalism and authoritarian statism now challenge the neoliberal discourse of free trade, open borders and markets. Democracy seems to be as weak as in the Weimar times. New forms of authoritarian regimes mark politics in Russia, Turkey, the Philippines, Hungary or Poland and now the US. Right wing parties have gained dominant positions in the UK, the Netherlands, France, Austria and are threatening in many other countries. They all tend to negate the global challenges as we have described them above – climate change, social inequality and cultural mix – and thus risk turning dysfunctional nations into active opponents to planetary solutions. They all put at risk human rights, freedom of speech and democracy. Essential constituents of cities, metropolitan nodes and urbanity will be challenged within these nation states. Developing the urban agenda against right wing nationalism will increase the need for urban global political networks. The Global Parliament of Mayors project has gained relevance in an era where democracy within nation states is under threat.

The defence of democracy and human rights was the ultimate motivation for Benjamin Barber as an academic, a citizen and an activist. It is exactly that combination between rigorous scientific research, scholarly erudition, civic engagement and organizational endurance that led to the inception of a new global institution. Benjamin Barber as a person was pivotal in combining academic interest, political networking and pragmatic projects which is why the Global Parliament of Mayors can exist. Even while fighting an invasive cancer in the last days of his life, Benjamin Barber increased his contacts, organizing meetings with mayors, conference calls with his advisory board and mobilizing sanctuary cities against Trump's closing borders. The enormous energy and inspiration that he put into this project surely explains its manifestation. A more detailed analysis of how this individual scholar and activist may have changed politics on a global scale still needs to be written. His death on 24 April 2017 in his home in Manhattan, New York will surely determine the further developments of a Global Parliament of Mayors, considering that the project is now fully in the hands of mayors and thus dependent on their political will to become planetary spokespersons. Building the GPM is a political project; academics and activists now are mere consultants, but that does not reduce their responsibility, as there is a lot needing to be documented and researched. Finally, the whole project needs an active civil society. Contrary to nation states in the UN, executive powers alone cannot represent them. Urbanity is based on co-productive governance, meaning a city is not a country and urbanity as a political project is based on other forms of representation.

## Notes

1 www.kaaitheater.be/en/agenda/the-city-as-commons-in-a-divided-world. See also my key-note: 'Can the city save the world'.
2 www.who.int/gho/urban_health/situation_trends/urban_population_growth_text/en/.
3 www.citypopulation.de/world/Agglomerations.html.
4 Sassen (2006) provides a more nuanced and complex vision, that is more adequate for scholarly analysis, in her encompassing historical overview of the relationships between territory, authority and rights.
5 www.oxfam.org/en/pressroom/pressreleases/2016-01-18/62-people-own-same-half-world-reveals-oxfam-davos-report.
6 www.un.org/esa/socdev/documents/reports/InequalityMatters.pdf, http://inequality.org/, www.globalissues.org/article/26/poverty-facts-and-stats.
7 www.un.org/esa/population/cpd/cpd2013/SGreport13February.v2_changes.accepted.FP_advance%20unedited%20version_converted.pdf.
8 https://globalparliamentofmayors.org/mayor-de-lille-cape-town-new-chair-steering-committee-global-parliament-mayors/.

## References

Anderson, B., 1983. *Imagined communities. Reflections on the origin and spread of nationalism.* London: Verso.
Barber, B., 1995. *Jihad vs McWorld.* New York: Crown.
Barber, B., 2013. *If mayors ruled the world. Dysfunctional nations, rising cities.* New Haven, CT: Yale University Press.
Barber, B., 2016. Can cities counter the power of president-elect Donald Trump? The Nation, [online] 14 November. Available at: www.thenation.com/article/can-cities-counter-the-power-of-president-elect-donald-trump/ [Accessed 14 January 2018].
Barber, B., 2017. In the age of Donald Trump, the resistance will be localized. The Nation, [online] 18 January. Available at: www.thenation.com/article/in-the-age-of-donald-trump-the-resistance-will-be-localized/ [Accessed 14 January 2018].
Castells, M., 2000. Materials for an exploratory theory of the network society. *British Journal of Sociology*, 51(1), pp. 5–24.
Corijn, E., 2009. Urbanity as a political project: towards postnational European cities. In: L. Kong and J. O'Connor, eds. *Creative economies, creative cities: Asian-European perspectives.* New York: Springer Publishing. pp. 197–206.

Corijn, E. and Saey, P. eds., 2014. *Wereldvreemd in Vlaanderen. Bakens voor een progressieve politiek*. Berchem: EPO.

Maddison, A., 2001. *The world economy*. Paris: OECD Publishing.

Massey, D., 1991. A global sense of place. *Marxism Today*, June 1991, pp. 24–29.

Moulaert, F., 2000. *Globalization and integrated area development in European cities*. Oxford: Oxford University Press.

Moulaert, F., Swyngedouw, E., Martinelli, F. and González, S., 2011. *Can neighbourhoods save the city? Community development and social innovation*. London: Routledge.

Sassen, S., 2006. *Territory, authority, rights: from medieval to global assemblages*. Princeton, NJ: Princeton University Press.

# 3 Social inequality and transformation of the urban economy

*Enzo Mingione*

The current transformation is interpreted through Polanyi's approach of the double movement complemented by a contribution of Marshall on democratic welfare capitalism. The golden age of welfare capitalism minimized the tensions between market opportunities and social protection and made possible a difficult equilibrium among democracy, welfare and capitalism. Globalization, financialization and industrial restructuring reactivated the tensions among the three components and made evident the unsustainability of the current model of development. Individualization and social fragmentation make it difficult for movements to build innovative responses to the deficit of social rights. The possibility to establish durable rights and opportunities in favour of the most deprived and vulnerable groups of the population is controversial. European cities are affected by experiences of social innovation to a different degree; only in some contexts is local innovation able to limit fragmentation and growing inequality.

## Introduction

The issue of the increasing inequalities in industrialized countries has captured great political and academic attention after the publication of Thomas Piketty's *Le capital au XXIe siècle* (2013). Urban scholars have debated the impact of social inequalities in cities and the diffusion of new forms of poverty and social exclusion for a long time now; however, the accelerated socio-economic transformations connected to globalization and financialization, as well as the concomitant wave of neoliberal policies in the 1980s, have increased our interest in urban social inequalities and the diffusion of new forms of segregation and exclusion. The attention for social inequalities and social exclusion in cities has produced a large number of contributions on a wide range of issues. Criticisms of neoliberal policies, studies on global cities, debates on social polarization and/or social professionalization, attention to new waves of gentrification and the diffusion of new forms of urban segregation, and the discussion of the idea of a just city have all largely focused on social inequalities in contemporary cities.[1] However, urban scholars are not interested in social inequalities per se, but rather focus on the perspective that growing inequalities entail increasing social suffering, discrimination, poverty and social exclusion.

In this contribution, I will not be able to review the vast literature that deals with social inequalities in contemporary industrialized cities. Rather, I will limit myself to expressing a view on how socio-economic trends produce increasing social inequalities in European cities today, which in turn lead to more discrimination, vulnerability and exclusion. I will also devote some attention to how agency (policies, social movements, people's initiatives and resilience, urban social innovations) responds differently to the tensions of socio-economic change in different local contexts. Even if at a preliminary stage of research, I will try to underline how at the city level actors are implementing new forms of social protection and solidarity trends of decommodification, thus producing innovative networks and social bonds in order to counterbalance the destructive social impact of global market competition.

In the first part of the chapter, the focus is on the impact of structural socio-economic change that started in the early 1970s, with the oil crisis and industrial restructuring leading to increasing heterogeneous and unstable societies. I will argue that the socio-political order based on manufacturing and consumerism growth, on strong and standardized national regulations, and on increasing public protection is now facing great difficulties. As we shall see, industrially advanced European societies are less regulated by national standards, and have become more fragmented. The local and privatized welfare provisions are expanding with a controversial impact in terms of confronting global competition, financialization and increasing social inequalities.

In order to interpret socio-economic processes of change in capitalist industrialized countries, I will draw on Polanyi's theory of the 'double movement' and an important contribution by T.H. Marshall (1972) on 'democratic-welfare-capitalism'. This interpretative approach and the consequent narrative are still thoroughly focused on the nation state level. Then I will introduce the increasing importance of the local level (and therefore cities) and discuss the difficulties of translating the 'nation state based' parameters and narratives to explain inequalities and resistance at the urban level.

In the second part of this chapter, I will touch on the impact of the financial crisis, unfolding from 2008 onwards, on social inequalities and its consequences for austerity policies in Europe. I will raise the question of the specific urban dimension of social inequalities and new forms of poverty and social exclusion in contexts of increasing territorial diversity and inequality. In the conclusion, I will briefly discuss policies and agency responses, and consider the issues of social innovation, urban commons, the diffusion of sharing economies and new practices of solidarity and alternative welfare at the local level. I will raise the question of the importance of urban social agencies responding to and contrasting the disrupting tensions raised by globalized capitalism.

In order to understand the complex realities of contemporary cities in in-dustrialized countries, I focus on the impact of two intertwined processes of social change: the destandardization and fragmentation process associated with change, accompanied by the growth of economic inequalities gener-ated by the dominance of financial assets. This creates a large and diversified vulnerable population that has increasing difficulties claiming or defending their access to social protection and social rights. On the other hand, glo-balization and growing social heterogeneity weakens the traditional monop-olistic capacity of nation states to regulate societies and economies, and the institutional responsibilities for creating social cohesion are shifting towards the local level, at least in part. More generally, the weakening of the regula-tion capacities of the nation state and of the social protection coverage of the centralized welfare state provision activates a process of rescaling where the supranational and the local scales assume increasing importance.

The process of rescaling (Brenner, 2004; Swyngedouw, 2007; Kazepov, 2010) is controversial, discontinuous, and tension ridden. However, the in-terconnection between social fragmentation and downscaling makes the city the main arena for conflicts, inequalities, exclusion and discrimination, experiences of social innovation, as well as resilience and new forms of po-litical organization. While advanced technologies of communication and mobility make spatial distance increasingly irrelevant, the life of contempo-rary cities reflects the persistent importance of face-to-face relations among heterogeneous, multicultural, and diverse populations. As I will argue in my conclusion, the increasing importance of cities does not necessarily mean that an alternative social order is emerging at this scale but only that the main tensions of the present transition are mostly happening on the local scale.

## An interpretation framework for understanding unsustainable capitalist development: Polanyi and Marshall

In order to interpret the dynamics of contemporary societies, I will use the Polanyian paradigm of the 'double movement' (Polanyi, 1944, 1957, 1977). The commodification process at the core of the development of modern industrialized/capitalist societies offers new opportunities to work and consume that are emancipating individuals from traditional, often oppres-sive, social conditions (rural communities, clans, tribes, patriarchal families, etc.). At the same time, commodification opens a deficit of social protection and necessitates actors to reconstruct social bonds compatible with market opportunities and able to support the livelihood of the actors themselves. The double movement constitutes the permanent dynamism of modern so-cieties as exposed to commodification processes within different historical and sociocultural conditions in different temporal and spatial contexts.

The double movement paradigm can be a powerful interpretative tool shifting attention from equilibrium, sustainability and the static parameters of orthodox formal analysis to the dynamic and substantial levels of capitalist development, which is always unsustainable in different and changing ways. We have to look to the new protective social settings in combination with the commodification process instead of focusing mainly on GDP and competitiveness (Marglin, 2008). However, I am also convinced that the double movement is not sufficient to explain the dynamic of capitalist modernization. I agree with Fraser (2011, 2013, 2014) regarding the importance of the emancipation and democratization movements that struggle against all forms of oppression and discrimination. The individualization process put in motion by the double movement sets the conditions for activating agencies that struggle against traditional and new forms of oppression. Nancy Fraser (2011) argues that instead of the double movement, we should think of a triple movement including the motion of emancipation.[2] However, for both heuristic and methodological reasons, I do not agree with this proposed revision. The double movement makes sense insofar as it is composed simultaneously of two parts: the disembedding motion activated by the competitive market and the re-embedding motion activated by the necessity to create new social bonds and social protections in order to keep societies alive. The simultaneous character of the double movement is a key feature of the concept. The double movement contributes to activate movements of democratization, liberation and emancipation but organized agencies that contrast oppression emerge later on and in different ways. Emancipation movements take place in various different times and contexts, and confront both the traditional forms of oppression (for instance patriarchy) and the new forms of oppression fuelled by capitalist development (like pollution and environmental destruction or increasing bureaucratic and political oppression).

Within our interpretative framework, the emancipation motion of market opportunities generates both an instantaneous reaction necessary in order to keep society alive and the mobilization of social movements confronting all forms of oppression, variously diluted in time and mediated by diversified and changing organizational forms. The features and perspectives of our societies exposed to commodification tensions depend both on how the re-embedding agency is shaping new modes of social protection, and on how the organized emancipation movements are effectively contrasting social oppression and discrimination within a more individualized society set into motion by ever new market opportunities (Welzel, 2013). Along this line of thinking, we can suggest that the emancipation activated by the disembedding dynamic of the double movement is mainly passive and not necessarily conscious. It eradicates traditional forms of social organization generating the immediate necessity to build new social bonds and protections, and offers a potential for emancipation that becomes real only when, under diversified historical and cultural conditions, emancipation movements and organizations are effectively struggling against oppression and discrimination.

Within this interpretative framework, the necessity of reshaping social relations in order to produce a livelihood compatible with commodification constitutes the core of processes of social change. It is also the main concept needed to understand individual and collective agencies within both the simultaneous forms of re-embedding and the delayed mobilization of emancipation movements. The processes of commodification always set into motion the activation of new social bonds and protective institutions instantaneously and, eventually, promote the creation of emancipation and democratization movements and organizations. Some institutions regulating social life in modern societies, like the trade unions, are rooted in both the re-embedding and instantaneous creation of new bonds of social protection as well as in the emancipation trends that fight against social oppression.

The permanent tensions between the diffusion of market opportunities and social protection and livelihood have not been visible in the capitalist industrialized countries during the 'Golden Age' of welfare capitalism (Esping-Andersen, 1990) in the 30 years after the Second World War. An exceptional and temporary institutional combination made commodification compatible with social protection, and capitalism compatible with welfare and democracy (Marshall, 1972). Three main elements formed the base of this institutional balance: (1) the expansion of the welfare state made possible by abundant resources obtained through unequal exchange and monopolistic control of industrial technologies; (2) high manufacturing growth generating an increasing number of standardized jobs for breadwinners; and (3) stable nuclear families with women dedicated to housework and care.

At the centre of the institutional assets of welfare capitalist societies, there was a strongly legitimized nation state with monopolistic regulation power maintaining control over a massive wave of urbanization paralleling industrialization. Divided industrial cities have been growing fast at the centre of the economic, financial and social dynamic of welfare capitalism but, in general, the cities themselves expressed a rather limited capacity of autonomous regulation. Welfare, housing and urban development policies are almost everywhere bound to the nation state. Cities and local authorities have some implementing responsibilities, but with limited margins of power due to rigid legal frames and particularly the lack of financial and professional resources. As we shall see, the situation is due to change considerably in the successive phase of vertical subsidiarity, downscaling and neoliberal policies.

Diversified institutional assets have matured through turbulent processes of change, reform and political confrontation in class, gender and ethnically divided societies. All of them, however, have produced the development of a standardized institutional protection accompanying high growth and an enormous expansion of consumption. The fact that high growth and commodification have been compatible with the expansion of welfare marginalized the idea of the double movement and the influence of the Polanyian vision on the permanent tensions between market and society. However, we can easily argue that unequal global exchange and the devastating impact of dependent exposure to commodification and massive exploitation of natural resources constituted an unsustainable social vulnerability part of the double movement. The commodification process removed large parts of the population of countries in the Global South from their traditional communities and left them bereft of any form of social protection both in the countryside conquered by multinational firms and in the fast growing urban slums. This process remained 'obviously' forgotten in the mainstream economic and sociological interpretations of the 'balanced' welfare capitalist societies of the Golden Age.

Two forms of agency mobilization have been important during the period of welfare capitalism. Trade unions and political representation of working and middle class interests have been crucial in order to gain and extend standardized welfare rights and social protection. Trade unions and working and middle class parties reached the peak of institutional representation of interests, implementing circuits of mobilization in favour of welfare reforms. Towards the end of the welfare capitalist phase, however, emancipation movements against imperialism, patriarchy and ecological destruction have been fundamental in accelerating the crisis of the welfare capitalism assets. In these same years, urban protest movements have also been an important part of the process of social change (Castells, 1983).

Marshall's interpretation of the democratic-welfare-capitalist societies (defined as 'hyphenated' or 'hybrid societies' as the three components are equally essential to their nature) turns in the same direction as the double movement – i.e. the idea that capitalist development is 'normally' unsustainable – with particular attention to the importance of politics, bureaucracy and democracy, as well as to the crucial role of social inequalities. He points to the fact that the constitutive logics of capitalism, democracy and welfare are different and even incompatible. The logic of the market produces strong and cumulative economic inequalities that obstruct both the working of democratic citizenship (because the unequal distribution of resources immediately implies unequal distribution of power and representation) and the protective capacity of welfare. Moreover, the egotistic character of representative democracy often obstructs the protective capacity of welfare because the majority of the population is not in favour of protecting groups of citizens considered undeserving.[3]

Marshall argues that in the capitalist industrialized countries the intrinsically conflicting tensions between the three components of the hyphenated societies have been kept under temporary control through a precarious balance built around the legitimation of inequalities, leaving wide open the controversial question: what are acceptable levels of inequality? In the welfare capitalist countries during the Golden Age, the balance has been financed by massive resources deriving from global unequal exchange, so that democratic and welfare policies have been able to alleviate the impact of inequalities (legal protection of minorities, progressive taxation, struggle against poverty, increasing social spending, etc.). However, this same balance cannot be maintained in the successive phase of capitalist development and Marshall already signalled this in the early 1970s:

> This malfunctioning of the system of legitimate inequality is probably the most deeply-rooted threat to the viability of the hybrid or hyphenated social structure. ... The trouble is that no way has been found of equating a man's value in the market (capitalist value), his value as a citizen (democratic value) and his value for himself (welfare value). ... The failure to solve economic inequality is evidence of the weakness of contemporary democracy.
>
> (Marshall, 1972, p. 30)

## The contemporary dynamics of change: inequalities versus fragmented, unprotected and unrepresented urban populations

The post-industrial social transformations since the oil crisis of the mid-1970s are making visible again the tensions of the double movement in advanced industrial societies that are more and more individualized, destandardized and fragmented. Individuals now have greater opportunities for self-realization, they are more self-aware, more able to participate in emancipation movements and to communicate through high-tech devices; but they are also increasingly isolated and vulnerable, less protected by the welfare state, and largely dispossessed of political representation, social bonds and community relations. Inequalities are growing out of control and the possibilities to legitimate them through the democratic and welfare mechanisms are extremely weak, also because the regulation power of nation states is decreasingly able to face globalization and territorial fragmentation.

Global competition is eroding the surplus of resources based on unequal exchange and control over productive technologies and knowledge that provided means to balance high growth with expanding social protection in the previous phase. The rate of economic growth of industrialized countries has dropped and the new economic dynamism shows a decreasing capacity to compensate the impact of commodification with investments on welfare support. The process of financialization has increased the difficulties of implementing economic support for social protection, as it has constantly subtracted resources from social policies and redistributed them to the very rich and to the most powerful part of the state bureaucracy and political elite. Within industrialized countries, this process is favouring a new wave of growing economic inequalities. On the one hand, there is an increasing concentration of wealth and power in the hands of a few super rich (Stiglitz, 2012; Piketty, 2013; Franzini and Porta, 2016), while on the other hand, the decline of the stable working and middle classes generates a vast heterogeneous population lacking social protection and political representation (Sassen, 2014).

Another crucial passage of the current transformation is the decreasing regulation capacity of the nation states. Both Polanyi and Marshall assume that the regulation power and the social order are strictly in the hands of the nation state. The analysis of the welfare capitalist regimes (Esping-Andersen, 1990, 2002) is realized mainly at the national level.[4] As we have already anticipated, cities have been at the centre of economic, financial and demographic growth but they had little margins of regulation autonomy within a set of centralized systems. The current transition is controversially eroding the centrality of the national level with an impact on territorial inequalities and on the importance of urban processes and agencies. As anticipated by the notion of rescaling, local welfare provision and the urban capacity to mobilize both private and third sector support resources are now confronting the decline of the regulation of the nation state, even if with controversial and discontinuous outcomes.

At the present conditions of the dynamics of capitalism, the combination between the resources of income/employment, community/family support and protection by the welfare state is unbalanced, unstable, and largely unable to respond to old and new risks. At the same time, democratic politics are knee-deep in the crisis foreseen by Marshall. Economic inequalities are altering the representation system, delegitimizing the power elite and fuelling a large public opinion hostile to welfare, minorities, migrants and other victims of the globalized and financialized market process.[5]

Industrial restructuring and the expansion of service jobs have made labour careers more heterogeneous and unstable (Riain et al., 2015). Both new self-employed workers[6] and temporary and precarious employees are on the increase, and, at the same time, they are underrepresented by unions, political parties and professional associations. Even when they are not precarious and/or poorly paid, they suffer from a dangerous deficit of social rights and social protection. The nuclear family, still the main institution regulating private life, has a decreasing capacity to offer protection to its members, due to the lengthening of life expectancy, the drop in marriages and births and due to the spread of divorce, reconstituted families and people living alone. Families and communities, like employment, are becoming increasingly heterogeneous and unstable. Moreover, the massive entry of women into the labour market has generated tensions in relation to family responsibilities and care work. Conciliation between paid employment and family-oriented activities is difficult and opens gender, ethnic and cultural tensions in various ways.

In these same societies, a new massive wave of migration is taking place. Today's migration flows are characterized by a heightened potential for mobility (due to the growth of low-cost flights, for example) and communication (Internet, mobile phones), on the one hand, but also by occupational instability and heterogeneity, on the other hand. The diffusion of low-paid unstable jobs in the service sector for migrants has given rise to a range of difficulties in relation to their occupational, social and residential inclusion.

These mobile migrants with complex transnational identities are now facing political and cultural contexts in which welfare protection is weak and discriminatory, and xenophobic practices are widespread. In addition to the current wave of economic migrants, particularly in the last years, the diffusion of wars and violent conflicts in Africa and the Middle East is producing massive waves of refugees. This phenomenon is opening a serious political crisis in Europe and creating great difficulties in the implementation of the various traditional forms of multicultural integration at a national or local scale (Amin, 2012).

The new migration waves have had a controversial impact on cities and territorial inequalities. The wave of immigration is concerning particularly large cities and metropolitan areas, but is also present selectively in medium and small cities. Moreover, the perception of being 'invaded' is strong also in cities and regions where the number of new migrants is limited. Given present conditions, the different traditional modes of integrating migrants (from assimilation strategies to multi-ethnic arrangements) are no longer (thought to be) working. The difficulties of integrating migrants and minorities are reflected everywhere in serious political tensions and social conflicts on the local scale.

When trying to tackle the demand for social protection generated by new risks, the national welfare states can only draw on limited resources, due to globalization processes, heightened international economic competition and the growing costs of the political and bureaucratic apparatus. The expansion of social protection programmes now appears incompatible with the economic priority to maintain high levels of efficiency. The idea that active local welfare with increasing private and third sector support may solve the problem of scarcity of resources is an illusion. In general, and as we will discuss later on in this chapter, the shift towards increasing local and private welfare provisions is further expanding the dangers connected to sizable territorial and social inequalities while, in any case, social expenditures remain out of control (Andreotti and Mingione, 2014; Ranci, Brandsen and Sabatinelli, 2014).

Public welfare, as well as becoming increasingly economically problematic, has also become delegitimized in political terms, at least in countries more oriented to neoliberal policies. As Marshall feared, welfare is often perceived as a costly intervention to support 'undeserving' social groups, against the interests of the majority of the population. Moreover, bureaucratization and the influence of powerful political and economic lobbies are seriously damaging the possibility to implement efficient forms of public social protection. As Weber already foresaw more than a 100 years ago, the growth of the bureaucratic and political machines fuelled by economic development and the necessity to expand social control and public policies increased the independent powers and interests of the machines themselves in contrast with the goals of producing social protection and the social inclusion of citizens. The abundant flow of resources during the

Golden Age and the impact of global financialization more recently have made this process of accumulation of autonomous power of political and bureaucratic elites increasingly oppressive. In this sense, we could argue that capitalism is now divorcing not only from welfare but also from democracy. The three components of Marshall's hyphenated social system that, as we have seen, have always been problematically interconnected (Marshall, 1972) are now in great tension and designing unsustainable forms of development.

On the front of agency, a wave of mobilization of various emancipation movements has accompanied the development of the present historical transition (Tilly and Wood, 2009; Della Porta, 2015). The increasing oppression of the bureaucratic and political machines has favoured the mobilization of movements outside the traditional political and union organizations. Further individualization and the massive resistance to change on the part of the established political and bureaucratic lobbies are making the scenario of the contemporary tensions produced by the double movement extremely complicated. On the other hand, new communication technologies have become powerful instruments of mobilization and participation,[7] but they are also showing clear limits in the establishment of permanent and solid forms of social solidarity. New social movements easily combine high virtual interactions via communication technologies and large urban gatherings and demonstrations. In some way, we are now in a new era of urban social movements. We shall come back in the conclusions of this chapter to the importance of new movements and the question of their possible fragmentation and volatility.

### The impact of the economic crisis and the importance of the urban and local contexts

The crisis further exacerbated the tensions produced by the double movement on at least two main grounds. On the one hand, the employment crisis has greatly increased the numbers of individuals and families in need of social support. On the other, all countries, in different ways and measures, have implemented austerity policies and reduced public spending on social support policies. Moreover, austerity policies are having a negative impact on the growth rate where indebtedness and inflation are necessary to feed consumption and commodification. In urban terms, the crisis has a double negative impact, making the situation worse for cities with limited resources and a large number of vulnerable individuals and for discriminated populations already in great difficulties with no political or union representation. New forms of solidarity and social innovations have emerged in order to satisfy unmet needs, but they are unstable and insufficient due to cuts in social and regulation resources of the nation states and to the suffocating control exerted on local authorities' expenditures. Even in countries that tend to be more 'virtuous' in terms of universalistic public regulation – such as the

Scandinavian ones – the combination of welfare reforms and difficulties imposed by the crisis are opening up problematic tensions and conflicts. The violent clashes that took place between immigrants and police in the spring of 2013 in the periphery of some large cities in Sweden are a good example of these conflicts and of the urban location of new tensions.

As the crisis has been extending in time, the number of people in need of protection, such as the long-term unemployed, poor families, immigrants and discriminated minorities is growing. The ways in which the European authorities, at national and European Union levels, have managed the crisis by promoting austerity and cuts in public spending have amplified the 'short circuit' leading to inequality and discrimination. Given these conditions, some forms of solidarity and of social innovation have become the last resort to defend particular groups of citizens and communities from acute poverty. Alternatively, in a more subtle way, some local practices of social support have become the alibi for dismantling public social institutions because many politicians argue that citizens are perfectly able to 'do it themselves'. In this way, the crisis is amplifying spatial and social inequalities between localities and social groups with varying resources to produce innovative forms of solidarity and social support.

The traditional institutional assets of the worlds of welfare capitalism are less and less able to confront the impact of global trends of commodification. Therefore it becomes crucial to take into consideration responses, often from below, which are more local, more focused on the active participation and obligations of those in need of support, often characterized by the involvement of voluntary and non-profit agencies but also of private for-profit firms.

The difficulties of the national welfare states to confront increasing heterogeneity and instability with standardized programmes have shifted social policies towards local (mainly city) levels (Brenner, 2004; Kazepov, 2010), and towards activation processes that are oriented to share at least part of the responsibility for social protection with the individuals requiring welfare support (Hemerijck, 2013). Both shifts have contributed to increasing the levels of social inequality, as shown by studies on local welfare and on austerity policies (Andreotti and Mingione, 2014; Cucca and Ranci, 2017). As we have already argued, some cities have more resources to implement new social policies and social innovation experiences, some have less, and some individuals or families are in a better position than others are to assume activation responsibilities and to mobilize and be represented in order to defend their livelihood.[8] The austerity policies implemented in the face of the economic crisis have refurbished both the local and the active shifts of social policies through drastic cuts in spending, recentralization and the necessity to finance passive policies in order to confront the growth of unemployment and poverty. This step is further heightening the tensions of inequalities and discrimination, leaving some cities and some social groups with no means to face increasingly difficult situations.

Current processes of transformation are arguably driven by two different, sometimes contradictory, forces: (1) the necessity to respond effectively to the commodification pressures through the establishment of new forms of social support and political representation for a population that is increasingly fragmented and heterogeneous; and (2) the necessity to reduce public spending. It is important to underline the tensions between these two forces of change, as they entail distinct strategies for institutional reforms that are not reconcilable. There is a drive to confront commodification and achieve effective protection against the new risks, which often involves increased spending, whilst seeking to cut public expenditure. This antinomy between the invention of new forms of protection and inclusion responding to a more individualized and fragmented society, on the one hand, and the reduction of public spending, on the other hand, reflects two different ways of thinking about policies to confront social inequality and in favour of the development of new forms of social inclusion and cohesion. The first interpretation reflects the idea that new experiences should enrich a system of public responsibility and regulation that should be strong in order to confront the increase of economic inequalities and the deficit of social protection activated by global markets. The second interpretation reflects the idea that, at least in part, new social support experiences shall substitute state provision often at the expense of increasing inequalities and the dismantling of a general defence of the system of Marshallian social rights.

## Social innovations in cities versus increasing social inequalities

Social innovation experiences take place in cities that are increasingly characterized by inequalities activated by the dynamic of the double movement in the directions of more individualization and financialization and the weakening of national welfare states. Contemporary experiences of new forms of solidarity and social innovations have local and particularistic dimensions that correspond partly to the overload of highly diversified citizen demands. However, they are not able to keep inequalities under control on their own, leading to a deficit with relation to social citizenship rights. Before continuing the discussion on the question of inequalities and social rights, it may be useful to briefly explore various forms of new practices of social innovation, social support, and sharing economy in European cities.

Social innovation[9] connects in different ways to the tensions produced by the double movement. It varies from the vast typology of experiences that are defined as 'new sharing economies' (Schor and Fitzmaurice, 2015) where new commodification prevails (think of Uber or Airbnb), to local forms of solidarity networks (Moulaert and others, 2010; Kostakis and Bauwens, 2014), to new forms of cooperatives and urban commons (Borch and Kornberger, 2015). Social innovation initiatives can intervene directly to alleviate the impact of commodification, as in the case of initiatives promoting social responsibility of firms or in the case of the diffusion of short chains between consumers and producers.[10] Some experiences organize the representation and empowerment of social groups that are demanding access to rights and protection and bypass the control of bureaucratic and political oppression in the direction of self-organization and participation (Della Porta and Andretta, 2002; De Weerdt and Garcia, 2016). The organizations representing minorities, migrants, new categories of workers and the cultural mediation initiatives are good examples of innovative experiences with this kind of aim (Semprebon and Vicari Haddock, 2016). In other cases innovative urban experiences can favour the diffusion of non-commoditized practices or offer opportunities to groups of citizens with limited economic resources (microfinance, time banks, barter and community exchange, cultural encounters, community events, and so on). A large number of experiences that have raised great attention at the local scale are part of the process of urban restructuring. We could mention, for example, new uses of urban space by disadvantaged or minority groups of citizens, the revitalization of degraded areas, and other experiences based on initiatives and movements from below that match institutional attention or recognition. In most of these local experiences in which people are facing social individualization and a deficit of representation and protection, the institutionalization process that grants regulation and resources is necessary in order to make these practices effective in addressing fragmentation and inequalities (Swyngedouw, 2007; Parés, Ospina and Subirats, 2017). However, institutionalization is also, at the same time, a controversial process as it incorporates citizens' practices in the heavy bureaucratic machine of the state and in the efficiency trap of the market. Many innovative experiences from below need institutional recognition and financial help in order to survive but the institutional interventions may alter the solidarity relations that are at the base of the experience or make it more difficult to reach the goals of the initiatives. Community shops or restaurants may have to increase the prices in order to pay the increasing costs, but in so doing, they become unaffordable for the group of founders and the neighbourhood community. Workshare organizations may have to exclude some workers who are temporarily unable to contribute their share in order to pay the rent or stay by the legal rules. In the Italian case, most of the experiences of institutional arrangements for self-building based on agreements between local authorities, workers contributing work hours to build their family apartments and financial institutions have failed because the bureaucracy has been too rigid and because the financial institutions have boycotted the project.

Many initiatives of social innovation are part of social economies that alter the impact of the double movement towards differently commoditized conditions.[11] Some of them are experiences that repeat in new forms the traditional cooperative experiences with solidarity and community arrangements that produce goods and services at more affordable prices. Other experiences exchange time or voluntary work (like time banks or house sharing with repayment in working hours) without monetary intervention. However, I doubt that these alterations can really go in the radical direction of a process of '*happy decroissance*'.[12] Other initiatives are oriented to produce livelihood or social inclusion for disadvantaged social groups. Both typologies are important to confront the challenge of fragmentation and individualization produced by commodification today. It is, however, important to underline that in order to be effective, expand, and empower more people, all these experiences require investments and protections from the local authorities. This is unlikely to take place in political contexts where public responsibility is weak and austerity policies are limiting state expenditures on social support and inclusion.

In many cases, informational technologies have been essential in order to establish and enforce social innovation initiatives. For this reason, it is increasingly important for there to be local public investments in order to allow for the development and maintenance of advanced informational infrastructures, to contrast the digital divide and to promote knowledge and education. Here again the different levels of professional, cultural, political and financial resources are important and reproduce spatial inequalities.

The transformation of current societies towards more localized, active, mixed and diversified social bonds and social support is everywhere reflecting and sometimes magnifying the growth of social and geographical inequalities and eroding the system of divided but standardized rights. Within this historical transition, some conditions that may encourage initiatives of the solidarity, social innovation and political mobilization of citizens may also keep inequalities under control. In order to achieve this goal, four conditions appear crucial. First, it is important to preserve and update a strong and articulated national and supra-national institutional regulatory framework oriented to ensure acceptable and socially compatible levels of livelihood for the population. Second, a balanced system for the redistribution of resources and responsibilities from central authorities towards local bodies and organizations is crucial in order to grant support to social initiatives in favour of disadvantaged populations. Third, there should be a strong political determination to combat discrimination against minorities and other vulnerable groups at all levels.[13] Fourth, in order to promote the diffusion of non-divisive and discriminatory social policies at the local level, it is also important to favour the diffusion of knowledge, professional capacities, and solidarity and intermediation cultures. It may be possible to achieve these conditions only through the mobilization and participation of diverse agencies (local and national, public and private), associations and social movements. It is precisely the mobilization and empowerment of different actors that can generate long-lasting forms of solidarity, inclusion and social support and, at the same time, contrast the impact of the rise of social and spatial inequalities. It is on this ground that the urban movements of emancipation play a crucial role (Nicholls, 2008). The public responsibility to tackle discrimination and defend public goods is the indispensable ingredient for keeping the transformation of fragmented and individualized societies under control and only the mobilization of emancipation movements can produce it (Wilson and Swyngedouw, 2016).

## The perspectives of untenable development: inclusive cities against social fragmentation and inequalities

National welfare capitalist assets, more than being a luxury that we can no longer afford, are systems of social protection increasingly less effective in heterogeneous, unstable and individualized contexts. However, the forms of transition towards the 'new welfare' and, in particular, local welfare systems, also when filled by important experiences of social innovation, tend to accentuate inequalities and discrimination against vulnerable groups which are underrepresented in political terms. As Marshall (1972) would put it, both democracy and welfare are unable to face the current level and rise of inequalities that are totally out of control and impossible to legitimize. Capitalism is openly returning to its normal Polanyi unsustainable status under conditions where both the welfare (re-embeddedness) and the democratic responses are too weak to generate new social defences (Keane, 2009). The current unsustainable status is more evident in the cities of advanced industrial countries. Innovative forms of solidarity, community organizations, and neighbourhood mobilizations are confronting the tensions of the global transition, the decline of both national welfare and representative democracy (Swyngedouw, 2007; Sassen, 2014; Wilson and Swyngedouw, 2016; Geiselberger, 2017).

There are no blueprints for mobilization of the social economy and voluntary bodies, and 'activation' of service users, all of which are effective and equitable enough to satisfy the growing demands for protection, representation and participation. The current recession is not likely to have the regenerative impact on democracy and welfare that the great crisis of 1929 and the Second World War had, leading ultimately to the democratic Welfare capitalism in the western industrialized countries. Evers and Guillemard (2013, p. 384) provide the following scenario:

> Despite the gloomy future of the European social model, since the current recession might do away with social rights and citizenship enshrined in law, a more optimistic scenario might be played out whereby our current tribulations would revive a sense of responsibility both in national political cultures and in terms of transnational solidarity within the EU ... Active citizenship and a more civic culture might then help to find new ways for combining social cohesion and economic growth and the respective kind of social investment strategy.

This scenario lends us to ask the question of how it may be possible to mobilize the necessary powerful social movements in individualized, fragmented and localized societies. In order for this to take place, a revived active citizenship needs a massive political investment to defend citizens' livelihood and to combat inequality and discrimination. We have already anticipated that new communication technologies may be temporarily important to connect and mobilize people but have proved too volatile to consolidate strong institutional organizational bonds. It is also worth mentioning that many social movements at the local level express the 'not in my backyard' positions or other kinds of sectarian or particularistic interests that oppose rather than favour the spread of an open non-discriminative new civic culture (Hay and Payne, 2015).

In a recent book, Streeck (2016, p. 59) takes a pessimistic position on the possibility to activate mobilization and social movements able to regenerate social cohesion and a new socially inclusive society:

> The demise of capitalism … is unlikely to follow anyone's blueprint. As the decay progresses, it is bound to provoke political protests and manifold attempts at collective intervention. But for a long time, these are likely to remain of the Luddite sort: local, dispersed, uncoordinated, 'primitive'—adding to the disorder while unable to create a new order, at best unintentionally helping it to come about.

Streeck's position seems realistic in our present situation, but we know well that, for better or worse, human agencies are largely unpredictable. Moreover, as we have seen in this chapter, it is important now to consider both the global logics of contemporary capitalism and the local/city features of innovative actions. Globally, the expansion of emerging economies in the Global South is still sustaining high levels of growth and the emancipation of the conditions of life of hundreds of millions of new industrial workers (Wallerstein and others, 2013; Streeck and others, 2016).[14] Locally, we have seen how city innovative experiences are reshaping markets and communities and altering the commodification and social protection trends in various different ways.

To conclude this exploration we should underline two points of analysis that contribute to the understanding of the present transition. The first is the fact that the agencies' responses to the market processes are more important than the levels of growth and competition. The idea of unsustainable development is more useful to understand the present conditions of capitalist development than the orthodox visions of equilibrium and growth. Unsustainable development does not necessarily reflect a pessimistic vision but rather the idea that the perspectives of our societies are built by social agencies that contrast the impact of inequality and discrimination with the construction of new forms of solidarity and inclusion.

The second point is that the traditional visions exclusively centred on the western model of modernization based on a specific historical combination of the diffusion of commodification and the consolidation of the nation state are inaccurate. The double movement of capitalism happens on a global scale where different social, economic, political and cultural conditions are interconnected. Here we have only mentioned this aspect in passing, but it is important in any case to underline that the features of the current transformation of European societies are not isolated from what is happening in the rest of the world.

The western model of modernization has been built on the centrality of the nation state as the exclusive producer of political, social and economic regulations, the main base of social identity and, more recently, the institution developing welfare and social protection. As we have seen, this is no longer the case in western industrialized countries, and a controversial process of rescaling is shifting part of the regulation power towards the supranational and the local scale. We have insisted here on the fact that the urban scale is increasingly important. The specific urban features of being exposed to the market tensions, of creating new social bonds and protective relations, of mobilizing movements against discrimination in favour of the collective public good are indispensable levels of understanding of the mechanisms of change in current societies both in the Global North and in the Global South (James, 2015; Van Meeteren and Bassens, 2016). I do not share Streeck's vision that 'local' is necessarily a limit in the construction of an alternative perspective to the current crisis of capitalism. In fact, the capitalist order itself was born in European cities as a very peculiar and limited urban phenomenon. The innovative urban experiences and the city agencies that are trying to contrast inequalities and social discrimination and exclusion are not embedded in a powerful motor of diffusion, as it has been the case for the bourgeois revolution sustained and diffused by the commodification process, however historical change remain largely unpredictable.

## Notes

1  The urban debate on segregation and inequalities is well synthesized in Tammaru, Marcinczak, Van Ham and Musterd (2016). See also Wacquant (2008), Harvey (2012 and 2014), Sassen (2014), Brenner and Schmidt (2015) and Oberti and Préteceille (2016).
2  Fraser suggests '…to broaden Polanyi's problematic to encompass a third project that crosscuts his central conflict between marketization and social protection. This third project, which I call *emancipation*, aims to overcome forms of domination rooted in both economy and society. …struggles for emancipation constitute the missing third that mediates every conflict between marketization and social protection. The effect of introducing this missing third will *be to* transform the double movement into a *triple movement*.' (Fraser, 2011, p. 140)
3  'The relation between majority voting and policy decisions in welfare is very equivocal. Democratic voting is egotistic; most voters voice what they believe to be their own interests' (Marshall, 1972, p. 19). As we shall see later, this consideration explains well both the neoliberal attack on welfare expenditure supported by a large part of the public opinion – particularly in the US and UK – and the impact of the rise of populist xenophobic movements in some European countries today.
4  Max Weber underlined that the construction of the capitalist modernity has been realized by a joint venture of political and economic elites at the national level. The nationalist 'revolutions' in Europe in the 19th century have been fuelled by the

convergence of the interests of the political elites to erode the power of local autonomous entities with the interests of the economic capitalist elite to constitute a market sufficiently large and homogeneous and protect it from foreign competition.

5 See among others: Stiglitz (2012); Wallerstein, et al. (2013); Harvey (2014); Streeck and others (2016); Wilson and Swyngedouw (2016); Geiselberger (2017).

6 As mentioned in the following quotation, new self-employed workers (referred to as iPros) are the fastest growing segment of employment in the EU. Moreover, these workers are highly concentrated in large and economically dynamic cities. 'iPros are highly skilled self-employed individuals who work for themselves but do not employ others. They range from journalists and designers to ICT specialists and consultants. ... The growth in iPros in the EU since 2004 has been remarkable. Numbers have increased by 45% from just under 6.2 million to 8.9 million in 2013, making them the fastest growing group in the EU labour market' (Leighton, 2014, p. 1).

7 The Arab spring movements or the experiences of Occupy Wall Street and the Spanish indignados are good examples of both the importance and of the limits of communication technologies.

8 Think, for example, of the difference in Italy between a rich and dynamic city such as Milan and a trouble-ridden city like Naples, where both the public sector and the private and non-profit sector have little financial and professional resources to confront a large concentration of long-term unemployed, precarious workers and families in poverty.

9 There is now a vast literature on social innovation and in particular on new urban experience of various kinds. Among others see: Mac Callum, et al. (2009); Murray, Caulier-Grice and Mulgan (2010); Moulaert, et al. (2013).

10 In Italy, the GAS (Gruppi di acquisto solidale/Solidarity buying groups) can be an example of this kind of experience.

11 I am convinced that there is always one single economy where activities for profit and not for profit, formal and informal organizations mix and match in complex dynamic patchworks. Within an approach centred on the double movement and on unsustainable development, it is important to look at how, in different contexts, individuals are protected and included and find chances for self-realization, participation and representation. From our point of view the GDP, the rate of growth, and the measures of competitiveness give a completely distorted vision of how contemporary societies work and face the impact of increasing inequalities (Stiglitz, 2012).

12 Starting from the contributions of the French economist Serge Latouche (2009) there is now a movement and a debate arguing that the only way out of the contradictions of the capitalist process is to favour a controlled decrease of commodification. However, it is not clear how this can be realized under the present political and economic conditions of strong market interconnections.

13 This condition is politically difficult to achieve as often the most vulnerable groups are also stigmatized and politically underrepresented, which means that their defence may have serious electoral costs. For example, this has always been the case for inclusive policies in favour of the Roma populations (Picker, 2017).

14 The capitalist process is now, at the time, generating increasing inequalities in the life conditions of citizens of industrialized countries while dramatically changing the life chances of billions of citizens in the Global South. It is difficult to say how this controversial process will go on. As noticed by Wallerstein, et al. (2013, p. 186): 'Only after 1945 were the former peasants and working classes of the West and Soviet bloc factored into social security and prosperity by their national states. In total, this amounted to several hundred millions people. But are there now resources, let alone political will, to factor in several billion people in the global South?'.

# References

Amin, A., 2012. *Land of strangers.* Cambridge: Polity Press.

Andreotti, A. and Mingione, E., 2014. Local welfare systems in Europe and the economic crisis. *European Urban and Regional Studies,* 27 November 2014, pp. 1–15.

Borch, C. and Kornberger, M. eds., 2015. *Urban commons. Rethinking the city.* Oxon and New York: Routledge.

Brenner, N., 2004. *New state spaces: urban governance and the rescaling of statehood.* Oxford: Oxford University Press.

Brenner, N. and Schmidt, C., 2015. Towards a new epistemology of the urban? *City,* 19(2–3), pp. 151–182.

Castells, M., 1983. *The city and the grassroots.* London: Arnold.

Cucca, R. and Ranci, C. eds., 2017. *Unequal cities. The challenge of post-industrial transition in times of austerity.* London and New York: Routledge.

De Weerdt, J. and Garcia, M., 2016. Housing crisis: the platform of mortgage victims (PAH) movement in Barcelona and innovations in governance. *Journal of Housing and the Built Environment,* 31(3), pp. 471–493.

Della Porta, D., 2015. *Social movements in times of austerity.* Cambridge: Polity Press.

Della Porta, D. and Andretta, M., 2002. Social movements and public administration: spontaneous citizens' committees in Florence. *International Journal of Urban and Regional Research,* 26(2), pp. 244–265.

Esping-Andersen, G., 1990. *The three worlds of welfare capitalism.* Cambridge: Polity Press.

Esping-Andersen, G., 2002. *Social foundations of post-industrial economies.* Oxford: Blackwell.

Evers, A. and Guillemard, A.M., 2013. Introduction: Marshall's concept of citizenship and contemporary welfare reconfiguration. In: A. Evers and A.M. Guillemard, eds. *Social policy and citizenship.* Oxford: Oxford University Press. pp. 3–34.

Franzini, M. and Porta, M., 2016. *Explaining inequality.* Oxon and New York: Routledge.

Fraser, N., 2011. Marketization, social protection, emancipation: towards a neo-Polanyan conception of the capitalist crisis. In: C. Calhoun and G. Derluguian, eds. *Business as usual: the roots of the global financial meltdown.* New York: NYU Press. pp. 138–159.

Fraser, N., 2013. A triple movement? Parsing the politics of crisis after Polanyi. *New Left Review,* 81, pp. 119–132.

Fraser, N., 2014. Can society be commodities all the way down? Post-Polanyian reflections on capitalist crisis. *Economy and Society,* 43(4), pp. 541–558.

Geiselberger, H., 2017. *The great regression.* Cambridge and Malden, MA: Polity Press.

Harvey, D., 2012. *Rebel cities: from the right to the city to the urban revolution.* London: Verso.

Harvey, D., 2014. *The seventeen contradictions and the end of capitalism.* London: Profile Books.

Hay, C. and Payne, A. eds., 2015. *Civic capitalism.* Cambridge: Polity Press.

Hemerijck, A., 2013. *Changing welfare states.* Oxford: Oxford University Press.

James, P., 2015. *Urban sustainability in theory and practice.* London and New York: Routledge.

Kazepov, Y. ed., 2010. *Rescaling social policies: towards multilevel governance in Europe*. European Centre Vienna: Ashgate.

Keane, J., 2009. *The life and death of democracy*. New York: Norton.

Kostakis, V. and Bauwens, M., 2014. *Network society and future scenarios for a collaborative economy*. Basingstoke: Palgrave and Macmillan.

Latouche, S., 2009. *Farewell to growth*. Bristol: Polity Press.

Leighton, P., 2014. *Future working: the rise of Europe's independent professionals (iPros)*. EFIP: European Forum of Independent Professionals.

Mac Callum, D., Moulaert, F., Hillier, J. and Vicari Haddock, S. eds., 2009. *Social innovation and territorial development*. Burlington: Ashgate.

Marglin, S.A., 2008. *The dismal science. How thinking like an economist undermines community*. Cambridge, MA and London: Harvard University Press.

Marshall, T.H., 1972. Value problems of welfare-capitalism. *Journal of Social Policy*, 1(01), pp. 15–32.

Moulaert, F., MacCallum, D., Mehmood, A. and Hamdouch, A. eds., 2013. *International handbook of social innovation: collective action, social learning and transdisciplinary research*. Cheltenham: Edward Elgar.

Moulaert, F., Martinelli, F., Swyngedouw, E. and González, S. eds., 2010. *Can neighbourhoods save the city? Community development and social innovation*. London: Routledge.

Murray, R., Caulier-Grice, J. and Mulgan, G., 2010. *The open book on social innovation*. London: The Young Foundation.

Nicholls, W., 2008. The urban question revisited: the importance of cities for social movements. *International Journal of Urban and Regional Research*, 32(4), pp. 841–859.

Oberti, M. and Préteceille, E., 2016. *La ségrégation urbaine*. Paris: La Découverte.

Parés, M., Ospina, S.M. and Subirats, J., 2017. *Social innovation and democratic leadership. Communities and social change from below*. Cheltenham: Edward Elgar.

Picker, G., 2017. *Racial cities. Governance and the segregation of Romani people in urban Europe*. London and New York: Routledge.

Piketty, T., 2013. *Le capital au XXIe siècle*. Paris: Seuil. (English version: 2014). *Capital in the twenty-first century*. Harvard: Belknap.

Polanyi, K., 1944. *The great transformation: the political and economic origins of our times*. Boston, MA: Beacon Press.

Polanyi, K., 1957. The economy as instituted process. In: K. Polanyi, C.M. Arensberg and H.W. Pearson, eds. *Trade and market in the early empires: economies in history and theory*. New York: Free Press. pp. 243–270.

Polanyi, K., 1977. *The livelihood of man*. ed. H. Pearson. New York: Academic Press.

Ranci, C., Brandsen, T. and Sabatinelli, S., eds., 2014. *Social vulnerability in European cities. The role of local welfare in times of crisis*. Basingstoke: Palgrave and Macmillan.

Riain, S., Behling, F., Ciccia, R. and Flaherty, E. eds., 2015. *The changing worlds and workplaces of capitalism*. Basingstoke: Palgrave and Macmillan.

Sassen, S., 2014. *Expulsions. Brutality and complexity in the global economy*. Cambridge, MA and London: Harvard University Press.

Schor, J.B. and Fitzmaurice, C.J., 2015. Collaborating and connecting: the emergence of the sharing economy. In: L. Reisch and J. Thorgersen, eds. *Handbook of research on sustainable consumption*. Cheltenham: Edward Elgar.

Semprebon, M. and Vicari Haddock, S., 2016. Innovative housing practices involving immigrants: the case of self-building in Italy. *Journal of Housing and the Built Environment*, 31(3), pp. 439–455.

Stiglitz, J., 2012. *The price of inequality*. London: Allen Lane.

Streeck, W., 2016. *How will capitalism end?* London and New York: Verso.

Streeck, W., Calhoun, C., Toynbee, P. and Etzioni, A., 2016. Discussion forum. Does capitalism have a future? *Socio-Economic Review*, 14(1), pp. 163–183.

Swyngedouw, E., 2007. The post-political city. In: BAVO, ed. *Urban politics now: re-imagining democracy in the neo-liberal city*. Rotterdam: Netherlands Architecture Institute (NAI)-Publishers. pp. 58–76.

Tammaru, T., Marcinczak, S., Van Ham, M. and Musterd, S. eds., 2016. *Socio-economic segregation in European capital cities*. London and New York: Routledge.

Tilly, C. and Wood, L.J., 2009. *Social movements, 1768–2008*. New York: Paradigm.

Van Meeteren, M. and Bassens, D., 2016. World cities and the uneven geographies of financialization: unveiling stratification and hierarchy in the world city archipelago. *International Journal of Urban and Regional Research*, 40(1), pp. 62–81.

Wacquant, L., 2008, *Urban outcasts*. Hoboken, NJ: Wiley.

Wallerstein I., Collins, R., Mann, M., Derluguian, G. and Calhoun, C., 2013. *Does capitalism have a future?* Oxford: Oxford University Press.

Welzel, C., 2013. *Freedom rising: human empowerment and the quest for emancipation*. New York: Cambridge University Press.

Wilson, J. and Swyngedouw, E. eds., 2016. *The post-political and its discontents: spaces of depoliticization, spectres of radical politics*. Edinburgh: Edinburgh University Press. pp. 244–260.

# 4  Voracious cities and obstructing states?

*Gilles Pinson*

## Introduction

In recent years, we have seen a proliferation of a half-academic/half-essayist 'pro-urban' literature describing the triumph of cities. Benjamin Barber's *If Mayors Ruled the World* (2013) is just one example among many others of this literature. Somehow, this literature reverses the famous proposition made by Tilly and Blockmans about 'voracious states' and 'obstructing cities' (Blockmans, 1989; Tilly and Blockmans, 1994). In our time, reversing the historical process that saw territorial states taking precedence over cities from the 15th century on, 'voracious cities' would be threatening the hegemony of 'obstructing states' and would be able to behave as proper autonomous actors on a transnational scene. We can also quote the work of Edward Glaeser on the *Triumph of the City* (2011) or those of Katz and Bradley on *The Metropolitan Revolution* (2013). Adopting a slightly different point of view mainly inspired by the examples of North American cities, Ehrenhalt (2012) or Gallagher (2013) evoke the reversal of the demographic, symbolic and political power relationship between inner cities and suburbs and announce the reinforcement of the former and the decline of the latter. This 'great inversion' would be bringing social and political elites back into the central parts of the metropolis giving the latter a political influence and capacity that they had lost with the 'white flight' (Dilworth, 2009). Europe has not remained on the sidelines of this movement of urban economic, social, cultural and political revival. In Europe, however, the fuel of urban renaissance and city activism is not so much the recovery from economic and social decline but rather a new stage in the very long and conflictual history of the relations between central states and peripheries (Tarrow, 1977; Page and Goldsmith, 1987; Le Galès, 2002; Pinson, 2009).

Why talk about a pro-urban literature? First, because, in this literature, we find traces of a prophetic narrative about the comeback of large and dense cities, in particular of their inner areas, in a context of globalization and transformation of capitalism. Second, because we can also identify some elements of an advocacy for the upgrading of cities and mayors'

political role on the international stage in a context of generalized interdependence and erosion of the sovereignty of territorial states (Acuto, 2013;
Barber, 2013; see also Chapter 2). One might think that these writings are
only meaningful in the context of 'cultural wars' that have agitated the US
for several decades now. Recent elections have demonstrated that these
'cultural wars' had taken a spatial dimension, with the opposition between
highly urbanized and progressive coastal areas on one hand, and the inner
part of the countries composed of conservative small towns and rural areas
on the other. Thus, this pro-urban literature could be seen as an expression
of a modernist left fighting against the symbolic and political weight of conservative suburbs, small towns and rural areas. This modernist left would
have been reinvigorated by the new economic fortune of inner cities, which
allows it to demonstrate that the higher degree of diversity and tolerance
that characterize dense metropolitan centres not only give them a moral
superiority but also greater economic performance. In France, one can observe similar debates about the ongoing recomposition of the socio-spatial
basis of the Left. With the rooting of the vote in favour of National Front in
suburban and exurban areas, a harsh debate started about what should be
the constituency of the Left. Should it be the groups that embrace the values
of cultural and economic values openness and that are mainly located in the
cities (Lévy, 2003; Lussault, 2009)? Or should the Left rather be sticking to
an 'old school' constituency of native blue collar workers located in suburbs
and rural areas (Guilluy, 2013)?

Even if the urban renaissance literature is inevitably embedded in these
societal debates, I will not engage with them in this chapter. Rather, I intend
to discuss one specific underlying assumption of this literature that sometimes becomes explicit in the writing of authors like Barber or Le Galès.
This assumption is that cities are not only spaces favoured by the new organization of global capitalism. They also tend to become fully-fledged political actors evolving on the national and international stages, while so far
they have been constrained to a status of mere administrative divisions of
territorial states. I hereby position myself as part of an emerging school of
thought that tries to theorize the 'return of the cities', their revaluation as
spaces and political actors. Therefore, my point is not only to try to defend
such a vision of cities, but also to refine the argument by providing an 'internal' critique of the 'city-actor' thesis. In particular, I would like to address a
specific issue: the relationship between the rise of cities as political actors on
one hand, and the fate of territorial states on the other. All too often, publications and discourses on the political comeback of cities tend to establish a
zero-sum relationship between the political promotion of cities and the assumed decline of territorial states. I would like to test a different hypothesis.
This hypothesis is that the relationship between cities and territorial states
is not a zero sum game, as we are encouraged to think by pro-city scholars
and essayists and research in historical sociology of state building. My hypothesis is that, in the current period, those cities that manage to emerge

politically and behave as actors pushing forward their identity, agenda and interest on national and international stages are those located in national contexts in which the central states themselves have managed to reposition themselves in a context of globalization and affirmation of subnational and supranational political levels.

To this end, I will proceed as follows. I will first return to the political comeback of cities. I will discuss the elements that give credit to this hypothesis, elaborate on the objections that have been levelled against it, and discuss how these can be overcome. Then I return to what I consider one of the main limitations of the body of literature announcing the comeback of cities as political actors and spaces, namely the assumption that this comeback proceeds or benefits from a weakening of territorial states. This is a programmatic and exploratory chapter. My purpose is to articulate a hypothesis rather than to present systematized research results. I have not yet validated this hypothesis through systematic comparative work, however, it is based on over 10 years of research on urban governance and the state-city relations in France, Italy and the UK.

## 'Can cities act?' The 'city-actor' hypothesis

Can cities act? Can they be actors? Can they be anything other than spaces crossed by diverse fluxes that shape them? Administrative circumscriptions for the action of other scales, like the territorial states? This is the question recently asked by Alan Harding and Talja Blokland (2014). For the authors, until recently, most of the analytical streams of urban scholarship did 'not attach a great deal of conceptual importance to the notion of "agency"' (Harding and Blokland, 2014, p. 132). In most traditions of urban studies, the city is never really considered an actor. In urban sociology, long dominated by the Chicago School approach, scholars consider the city as a neutral space where social and ethnic groups struggle in the absence of any overarching power able to regulate these struggles. In the Marxist tradition, the urban space is the spatial projection of power relations in the productive and consumption sphere, and, in the last instance, cities are governed by dynamics that are out of reach for urban political actors and movements. None of these traditions gives much credit to the idea of mechanisms allowing the arbitration between conflicting interests at the city level and the emergence of an urban collective will.

The conceptualization of the city as a political actor dates back roughly to the 1980s and has a wide variety of sources. Eager to point out the localist shortcomings of the scholarly debate about power in the city, the provocations of Paul Peterson (1981) helped put this issue on the agenda. For him, the crucial question was not how the power is distributed among social groups in the city, which was the central focus of the tradition of scholarship on the 'power in the community' and of the debate between elitist and pluralist approaches. For Peterson, the most interesting question was how cities can act and have one voice in an inter-urban competition to capture capital and populations that have both become more mobile. In a rather different vein, David Harvey's work on the shift from urban managerialism to urban entrepreneurialism (1989) provides a picture in which the city is not seen as a mere space of services and utilities production in a context of abundant resources, but rather as an actor forced to define comparative advantages in a competitive environment where resources are scarce and scattered. American urban political economy, encapsulated by theoretical elaborations such as 'urban regimes' (Stone, 1989) or 'growth machines' (Logan and Molotch, 1987), is probably the body of literature in which the issue of the agency of cities was the most systematically addressed. Urban political economy approaches depart both from the actor-less structuralism of the Marxist approach of urban governance and from the apolitical vision of Peterson in which cities have a single set of interests. They admit that there is a pressure on cities in a competitive environment to display strategies, projects and some sort of collective will, but also consider that these strategies and projects are the outcomes of struggles and bargains, and serve the interests of some groups at the expense of others. Cities can act – there is an urban agency; but it is very likely that some elements of the citizenry will gain precedence over others to voice and embody this agency.

However, the theoretical stream that provided the most systematic formulation of the 'city-actor' hypothesis is probably the one that could be described as 'Neo-Weberian'. It is represented by researchers such as Alan Harding (1994, 1997), Patrick Le Galès (2002); Le Galès and Harding (1998), Arnaldo Bagnasco (Bagnasco and Le Galès, 2000), Angelo Pichierri (1997, 2002), and the author of this chapter (Pinson, 2009, 2010). These authors build on sections of Max Weber's *Economy and Society* devoted to the historical interlude that saw medieval cities constituting autonomous spaces of political organization and regulation between the empires' era and the rise of territorial states. Starting from this inspiration, these authors hypothesize that, in the current era characterized by globalization, Europeanization and decentralization, cities could regain a part of the share of sovereignty and part of the ability to govern societies they had largely yielded to territorial states during the previous period. The revival of urban policies, the extension of urban agendas, the greater visibility of urban leaders, the construction of metropolitan institutions and the development of urban foreign policies are not only the signs of a so-called neoliberalization of urban governance. They can also be considered as signs of a redistribution between different levels of government of political authority and of the capacity to integrate societies. Social sciences have long been, and still are, blind to this process. Born in a period of strengthening of territorial states, they tend to err on the side of 'methodological nationalism' (Beck and Sznaider, 2006). Yet, cities are back and able to affirm themselves as possible sites of government and regulation in a context characterized by globalization and the erosion of state sovereignty.

To some extent, the hypothesis formulated by Neo-Weberians are congruent with those of the proponents of the rescaling theory. The latter, who are mainly to be found in subdisciplines such as critical geography and political economy and whose main figures are Neil Brenner (2004) and Bob Jessop (2002), consider that the State was for too long considered as an entity characterized by fixed geographical coordinates, mainly those of the 'national' scale. On the contrary, scholars should analyse the State through its various scalar architectures and the way the working and reworking of these architectures define and constitute the State. 'No longer then', writes Brenner, should 'the scales of statehood [be] conceived as stable platforms of institutional organization (...) Analytical attention [should be] fixed primarily on the political strategies that operate through them.' More concretely, that means that

> state scalar structures are now understood to be historically malleable; they may be ruptured and rewoven through the very political strategies they enable. Furthermore, the rescaling of institutions and policies is now conceptualized as a key means through which social forces may attempt to rejig the balance of class power and manage the contradictory social relations of capitalism'.
>
> (Brenner, 2009, p. 126)

Thus, from the rescaling perspective, the promotion of the metropolitan scale and institutions and the transfer of functions and resources from the central States to these new scales is just another episode of the reworking of the 'statehood' architectures closely tied to the changing logics of capital accumulation. There is a common ground between Neo-Weberians and the exponents of the rescaling theory in that they both consider that the geographical coordinates of the *politique*, the State, public regulations, the level at which they can express, should not be reified. However, there is also a major difference. While the latter amalgamate the various scales of the State into a single *agency*, behind which the ruling class or the forces of capital are easily identifiable, the former consider that the reworking of state scales is a matter of struggles between different and opposing *agencies*. Inspired by a long scholarly tradition of study of the centre-peripheries relations (Tarrow, 1977; Rokkan and Urwin, 1982; Page and Goldsmith, 1987; Hesse and Sharpe, 1991), Neo-Weberians make the hypothesis that behind cities, regions and nations there are different social groups, political elites, institutions and sometimes economic interests that defend different 'state scalar projects'. For this reason, statehood cannot be identified as a single integrated entity, as it is in Brenner's work. Rather, it is highly disputed between different agencies mobilizing their own resources, building alliances and seizing the opportunity provided by the weakening of other scales. From this perspective, statehood is not determined by a single project – that of Capital; the State not only enjoys a degree of autonomy but is also internally diverse. Thus, sometimes, the rivalries between different segments of the State, corresponding to different scales or sectors, are a more robust *explanans* of scalar changes than the '*ruse*' of the capital.

More precisely, in Neo-Weberians' formulation, the city is defined as a 'collective actor' (Pichierri, 1997; Le Galès, 2002), but a collective actor that is always in the making, always confronted with legitimacy challenges, whose frontiers and internal hierarchies are always contested. To paraphrase Bourdieu comparing social sciences to physical and life sciences, we could say that cities can be seen as collective actors, just as territorial states, with the sole difference that the former might have more difficulties being identified and acknowledged as full-fledged collective actors than the latter. For Pichierri, there are five elements that can be used to identify a collective actor: a system of collective decision-making; common interests perceived as such; integration mechanisms; internal external mechanisms of representation; and a capacity for innovation. According to Neo-Weberians, the possession of these elements by cities depends on historical and local contexts, however, many clues indicate that nowadays cities display these elements more obviously than they did during the *Trente Glorieuses*. This period was characterized by an expansion of the welfare state that went with a limitation of urban governments' own resources and discretion.

## Objections

Obviously, the 'city-actor' hypothesis has raised criticism, both in its Neo-Weberian formulation as well as the urban political economy version. Here I discuss the most pertinent of these critiques.

### *Anthropomorphism*

The first criticism targets the tendency of the city-actor hypothesis's proponents to picture cities as individuals endowed with will and interests and able to express and defend them. This line of criticism is similar to that regularly made with respect to the work on international relations that tend to have a monolithic, 'gladiatorial' vision of the interests of states. If we narrowly define a state as a set of institutions, ministries, administrative organizations, and officials, it is well known that those elements constitute a 'field of forces' crossed by conflicts and competition between administrative sectors, and between elected officials, etc. It is therefore highly unlikely that that this loose and internally divided ensemble will be able to express itself with a single voice, will and strategy.

The problem arises in the same way, and perhaps even in more acute form, for cities. If we narrowly conceive the city as a set of administrations, agencies, elected officials, etc., it is hard to imagine this ensemble of actors and organizations 'acting as one'. Moreover, if we conceive cities more broadly, as a broader sets of social groups, economic actors, organized interests, service providers of all kinds, or like Amin and Thrift, as 'assemblages of more or less distanciated economic relations, which will have different intensities at different locations' (Amin and Thrift, 2002, p. 52), it is even more difficult to envisage a city acting cohesively. The fact that cities are often administrative subdivisions of territorial States and are the site for the implementation of State's and sometimes regions' services and policies makes it even harder to imagine that urban policies are the outcomes of a single actor's will.

### Localism

The second objection to the 'city-actor' hypothesis is that of localism (Ward, 1996; Cox, 1997; Jessop, Peck and Tickell, 1999). It points to the fact that talking of a city as an actor consists of postulating that the city, its space, economic base and social structure are determined by an intentionality and networks of interactions that have local bases, or are determined by dominant local forces, combining their forces to maintain a status quo or to instigate a change. It also consists of postulating that these dominant local forces are able, through their action in the city and in its environment, to determine the position of the city in the urban hierarchy or in national or international division of labour and consumption, and to resist global pressures.

Some would say that there are commonalities between the Neo-Weberian approach to the city as a collective actor and what has been called the 'new urban economy' (Florida, 2005; Glaeser, 2011; Scott and Storper, 2015). New urban economists present larger cities as wealth production machines, and for them, this productive capacity is largely due to agglomeration effects, i.e. to horizontal and localized interactions between production agents. Other economists and geographers have criticized this other expression of localism. Engelen et al. contest the 'new urban economist concept of the city as a discrete, unitary object within limits which can be understood in terms of one internal principle like agglomeration' (Engelen et al., 2016, p. 3). Building on Massey's (2007) emphasis on the relationality of the urban, on the irreducible embeddedness of cities' societies and economies in wider scales and networks, they contest the idea that the fate of a city might be in the outcome of the sole local actors' activism. Drawing on Braudel and Tilly, they insist on the role of 'external governors' and on the necessity to integrate 'hinterlands' and the protection provided at higher political scales (states, empires) in the analysis of cities' fortunes.

Thus, cities have always been, and are today perhaps more than ever before, the material and social result of the many flows and decisions that deploy on other scales. Therefore, isolating this intentionality of an urban actor is a vain quest. The agitation of local players weigh little compared to the flows and structural forces. Somehow, the 'city-actor' visions face the same criticism the scholarship on the 'power in the community' once faced. Authors involved in this body of literature provide a vision of city governance limited by 'urban blinkers' (Wood, 2005, p. 203). They tend to give prominence to local independent variables to determine the power distribution arrangements in the city, while it is highly probable that forces originating at other scales are more determining.

### Incompleteness

The third objection is the incompleteness of urban societies. If territorial states have always been more spontaneously considered as actors, it is mainly because they had the ability to articulate on the same spatial mechanisms of legitimization, of political identification, of conflicts regulation on the one hand, and on processes of public policymaking on the other hand. At their apogee, territorial states were able to make spaces of politics and spaces of policies coincide on the same territorial basis, and to enclose them within this territorial basis, limiting the influence of other spaces and scales. If territorial states have always been more spontaneously considered as actors, it is also because they had, and still have, the ability to aggregate interests (including those of capital and labour), to regulate conflicts between them, and to represent them outside. Moreover, if territorial states were able to aggregate those interests, it is because they controlled most of the parameters determining the relationships between those interests, through labour law, tax system, redistributive policies and so on.

Most of the time, these levers are not available to cities and urban governments. Urban governments often rely on financial resources transferred by other levels and local taxes will often provide a limited share of these resources. The services provided in the city, public policies implemented in the city, are not always designed and implemented by the same urban governments. This leads some scholars to question the consistency of urban policies and even the relevance of the concept of 'urban policy' itself (Le Galès, 2005). Urban governments do not have their hands on the levers that determine the relationships between social groups, including those involved in the productive process. These interests do not spontaneously identify the city as the space where the conflicts opposing them have to be regulated, nor do they spontaneously recognize urban institutions as key mediators in this regulation.

Nevertheless, in the words of Bagnasco and Le Galès (2000), cities constitute societies. However, they are incomplete societies. Their integration, after a long and conflictual process, in broader political, economic, social and administrative systems, made them lose the capacity to confine the mechanisms of identification, mobilization and regulation which they were able to control in medieval times. In contrast, territorial states have managed to organize mechanisms of political identification, representation and conflict regulation at the national scale, depriving subnational or supranational levels of the ability to organize these mechanisms. National societies have become complete societies, main loci of the construction of the mechanisms of identification and mobilization. This 'completeness' of national societies helped territorial states to become actors, i.e. to be identified as the main representative of domestic groups outside. That is precisely what cities and urban governments find hard to become.

### Elitism

Finally, talking about the city as an actor exposes the risk of identifying the city with individuals and groups who are in a position to speak on its behalf. These individuals and groups are usually part of the political, but also economic and social elites, and possess resources and access to networks and forums putting them in position to represent the city. To put it bluntly, the city-actor speaks through its elites. There is a good reason to think that the city-actor promotes strategies that are not the reflection of the interests of the urban community as a whole – which already implies supposing that such a community exists – but of the specific groups that can access forums and places of decision.

The criticism here is both normative and theoretical. First, portraying cities as actors leads researchers to confuse the interests of the city with those of its elites, and therefore to reinforce an elitist discourse that considers the city through the eyes and interests of elites (see Bourdieu, 1989 for the 'intrinsic elitist bias of the elites' social science theory'). Second, the risk is also to neglect the conflicts, censorship and processes of domination that paved the way to the building and promotion of '*a*' city strategy. As suggested by Bachrach and Baratz (1970), decisions and strategies of the city-actor often mask the processes of non-decision through which the concerns and interests of the deprived and dominated groups are excluded from urban agendas.

## Answers to objections

Most of these objections are relevant. Moreover, they are very important to keep in mind when it comes to clarifying and even reinforcing the thesis of the return of cities as political actors. In this section, I therefore consider how each of these objections can be answered and even used to strengthen the initial argument.

### Anthropomorphism

The counter-critique on the issue of anthropomorphism is probably the most established one and invites us to keep away from mechanistic, unanimist, depoliticized and 'actor-less' visions conveyed by urban policymakers' narratives when presenting one city's aim, projects, strategies and so forth. At the same time, it would be a shame if urban research got rid of any concern for the analysis of the foreign policies of Lyon or Manchester, just as it would be a pity to be deprived of an analysis of the foreign policy of Italy or of the British Labour Party strategies. Behind each of these entities there are real actors, bearing circumscribed interests that cannot be equated with the interests of all those whom they are supposed to represent. Obviously, conflicts, rivalries and differences of opinion cross any collective actor, however, there are also good reasons to think that these conflicts are dealt with through trade-offs, compromises or political arbitration that allow some specific actors, groups or segments within these collective actors to prevail, leading to an institutionalization of their interests and visions in the decisions, programmes and rules. In this case, these few actors are entitled to bear the purposes and strategies of the collective actors within which they prevailed.

Consequently, the best research strategy is to maintain the hypothesis of the city-actor and observe the struggles and legitimation mechanisms that allow some political leaders, interest groups, bureaucratic segments, and even social movements to be recognized as bearers of a strategy in the name of the city. The task of the scholars is thus to observe the mechanisms of totalization – made of enrolment and mobilization, but also of marginalization mechanisms – that lead specific actors to claim the legitimacy to build up and express a collective voice. Their task is also to determine under which conditions these actors favour the urban scale to organize these mechanisms of totalization, and why other actors or groups agree to get involved in these mechanisms.

Specifically, it can be useful to go beyond the debate over whether complex systems like cities can have economic strategies or purposes. Some do, some do not (because they are too small or too big), and in between, a vast majority have vague strategies that are not really voiced and defended and are not heard and recognized by the actors inside and outside the city. What is more interesting is to understand the mechanisms that allow or prevent us to clarify or even determine the existence of such strategies, the factors that favour or disfavour their processes of construction, the conflicts that pave the way towards to their development, and which actors and groups are legitimized to act as designers and spokesmen of these strategies. If Lyon's (France) economic development strategy is visible and heard outside the city, it is mainly because mobilizations among local business milieus and around the Chamber of Commerce to foster the city's economic competitiveness have aggregated and eventually merged with the initiatives of the top-level bureaucrats of the metropolitan authority in charge of economic development. It is also because top political leaders, foremost among them the successive mayors of Lyon, decided to redeploy their political enterprise at the metropolitan level and to use economic development strategies to enhance both the status of the city and their own political influence (Galimberti et al., 2014).

*Localism*

There are various ways of addressing the localism objection. First, by stating that if extra-local streams have always been crossing and structuring cities, this has never prevented local decisions and action of having a role in structuring urban spaces and societies. The Neo-Weberian approach to the city-as-an-actor is compatible with a sound articulation of the logics of place and of relationality, as proposed by Massey (2004, 2007). Understanding urban spaces in a relational way by considering the logic of flows and relocating urban policies in a broader context does not prevent us from acknowledging room for local agency, for mobilizations at the urban or metropolitan around shared projects (Pinson, 2009). The economic fate of cities is the result of both the vagaries of economic flows – decisions taken at other levels – but also of local choice, made by those entitled to speak in the name of the collective actor. Even theories emphasizing convergence between cities and forms of urban governance, like the neoliberalization template (Brenner and Theodore, 2002) or the body of literature on policy mobilities and fast transfer (McCann and Ward, 2011; Peck and Theodore, 2015), acknowledge the role of local agency and struggles in shaping flows. The critics of new urban economics are not questioning so much the possibility for urban actors to have an influence on the fate of their cities but the focus on developmental and supply-side policies and the spirit of competition that these theories are supposed to encourage (Peck, 2005).

One can then well imagine that the mobilizations aiming at endowing the city with an agenda and identified strategies, and at promoting them, can involve actors and institutions beyond the borders of the city in question. Romain Pasquier's research on Breton interests' mobilization shows that if this region is one of the few in France to be recognized as a scale for aggregation of interests around a common project, that is also thanks to the ability of Bretons to gain support and resources at other levels, including the French State or the EU (Pasquier, 2015). Thus, it is important to avoid defining the city-actor as a collective actor mobilizing individuals and groups only within the city itself. The city as a collective/political actor can better be defined as a wide network including external supports, relays and gatekeepers allowing access to other scales.

Thus, hypothesizing the rise of cities as actors does not imply presuming that this process occurs only through the coalition of local actors, groups and interests mobilizing only local resources. To put it bluntly, the rise of the city-actor is not a chauvinistic, autarchic and self-centred process. The demise of federal policies in the US and central states' territorial policies in Europe has certainly put pressure on cities to rely on their own resources, to mobilize their own forces. This pressure has given rise to an increased sense of locality – or rather the sense of the territory as is said in Southern Europe – among urban policymakers, which manifests in concerns for 'local development' or the development of a 'project approach' in planning (Pinson, 2009).

This rediscovery of 'locality' and 'territory', however, has been accompanied by the development of intergovernmental activities involving cities. In fact, the development of an introverted policy activism aimed at mobilizing local resources was congruent with the development of an extroverted policy activism aimed at forging alliances, learning from other cities and promoting a city's own achievement (McCann and Ward, 2011; Béal, Epstein and Pinson, 2015). All of this took place even while the actors that had been bringing resources and expertise to enhance the welfare of cities, namely the territorial states, were being seen as increasingly unreliable providers. Recent research on cities' international policies (Van der Heiden, 2010; Béal and Pinson, 2014) has shown that these policies were targeting both external objectives – promoting best practices vis-à-vis places and policy, building alliances with other cities to obtain funds, changes in legislation or the creation of new policy programmes – and internal objectives – consolidating a political leadership, building up coalitions. In sum, the recent rise of localism and urban diplomacy proceeds from the same logic as the promotion of the political profile of cities.

## Incompleteness

Cities are incomplete societies, and '(c)ity trajectories are not internally controllable' (Engelen et al., 2016, p. 13). The institutions governing urban settings do not master all control levers or the resources that could enable them to influence the behaviour of actors and groups that inhabit them, not to mention those passing by or exerting an influence from a distance. The mechanisms of identification, mobilization, or political allegiance certainly do not privilege the urban scale. Nowadays, 'incompleteness' might be however the fate of most political scales, including the national one. National societies have reached their 'peak' of completeness with the golden age of the welfare states. However, since then, territorial states have lost control of a number of these levers, and it is difficult to know precisely which other scales benefited from the process of dispersal of the political authority, and from the ability to produce public policies and integrate societies.

As the French political scientist Jean Leca (1996) wrote, governing is a twofold activity consisting of integrating social ensembles and organizing social change through public policies. Until recently, cities could hardly claim to be complete societies, governing scales and therefore fully-fledged political actors since both of these activities escaped them. Today, we can assume that cities have taken control of many public policies, however, the capacity to be a public space where conflicts, cleavages and collective identities consolidate still escapes them. Cities have arguably become political *actors* of public policymaking, involved with other actors and scales in complex games of competition and cooperation, but they have not yet succeeded in becoming political *spaces*, where legitimation and identification mechanisms are made manifest. Cities' role as political *actors* is illustrated by their increasing capacity to develop agendas, strategies, policies, and diplomacy, whereas their inability to be *political spaces* as such is made manifest in the logic of political demobilization that is especially acute in cities, and more generally in the inability of cities to be credible alternative political spaces to national representative democracy (Crouch, 2004; Hoffmann-Martinot and Sellers, 2005; Braconnier and Dormagen, 2007). In intergovernmental relations, large cities tend to move beyond their traditional passive or underdog position. They are now able to assert their interests, priorities, strategies through their own institutions, representatives, and spokespersons. They have become key players in the field of *policies*, however, for the time being, they do not seem to be able to become a space of redeployment of *politics*. As such, they remain incomplete societies, but in a context where no other scale seems to be able to establish or preserve this 'completeness'.

### *Elitism*

We can answer the objection of elitism by an argument inspired by elitist theory itself (Hunter, 1953). Granted, there is a good chance that the actors and groups able to influence the agenda of urban policy and to promote this agenda beyond the city limits belong to the urban elite, or at least to the groups that have the most voice and representation. Within cities there tends to be a concentration of resources of legitimacy, predisposing a limited number of actors to embody the city-actor.

It is, however, useful to qualify this elitist vision of the city-actor by evoking two elements. First, as shown by Dahl's theory of pluralism (1961) or Stone's regime theory (1989), accepting the existence of elites does not imply envisioning the elite as a monolithic bloc. On the contrary, the voice and action of the city-actor might actually be the result of arrangements, compromises, trade-offs or balances of power between different elements of a variegated urban elite. Second, it is therefore useful when studying the processes and mechanisms enabling the city to act as actor to explore both arrangements and conflicts between elites, but also the process through which these elites build up their dominant position and legitimacy vis-à-vis dominated groups.

## Portrait of the city as an actor

Based on these observations, we can try to recast what would be the main features of a 'city-actor' from a Neo-Weberian perspective. These features are conceived as an add-on to the propositions of Pichierri to define a collective actor (see above).

### *A strategy or a governing agenda*

Presented in official documents, programmes and speeches emanating from a variety of organizations (municipality, metropolitan government, but also chambers of commerce, employers' associations, unions, agencies, etc.), this strategy embodies the vision that has reached a position of hegemony and can be presented as the agenda of the city. It has to be the outcome of a lengthy and conflictual process of urban reflexivity that can be traced through the analysis of debates and controversies. In order to be considered as the purpose of the city-actor, it should neither be the simple standardized variation of other levels of government strategies nor the *ex post* 'collage' of sectoral routines of different segments of the urban bureaucracies.

### Mechanisms of decision and a proper bureaucracy

To act as an actor, a city has to be capable of implementing its strategy and agenda through concrete public policies. As said, the making of a strategy requires the existence of informal processes of debate and controversies, however, it also requires formal mechanisms of decision-making that allow the legitimization of the strategy and the mobilization and channelling of resources to break the strategy down into concrete policies. It also requires an administrative capacity to implement policies. One should not take the latter condition for granted, since many local administrative systems are well known for being able to design plans and schemes but also for their incapacity to operationalize these plans and schemes.

### Mechanisms of representation

Mechanisms of representation are necessary to allow the appointment and legitimization of actors able to take part in the process of the city agenda and strategy building. This is true for various spheres – political, economic, cultural, social, etc. – from which 'leaders' can emerge and participate in the debates, controversies and processes of coalition building. Mechanisms of representation are also useful for the city-actor to get access to other levels of government in order to defend the city strategy, to find institutional relays and tax resources. They also provide instruments of external representation of the city, promoting its comparative advantages and its strategies (marketing, branding, etc.).

### Mechanisms of consultation and concertation

In the city-actor 'model', formal and informal mechanisms of consultation and concertation are necessary to allow the flourishment and the regulation of conflicts between the different components of urban society, and the 'territorialization' of the expression of these interests. The latter aspect of 'territorialization' is important because, in the absence of territorialized mechanisms of consultation and concertation, other political scales can provide scenes for the mobilization of interests and the regulation of the conflicts in which they are involved. This would be detrimental to the capacity of the urban scale to totalize interests and represent them.

These basic elements necessary to bring the city-actor into existence can be supplemented with additional elements:

- **Legitimate and effective metropolitan governance arrangements** allowing the mobilization of a plurality of institutional territories constituting the metropolis around the city/metropolitan strategy;
- **Institutions providing cognitive mobilization of urban populations** around the debates on urban strategies (local media, local structures of political parties, etc.).

However, again, our hypothesis is that the building of the city as a political actor may not be accompanied by the erection of the city as a political space, i.e. as a scale where this cognitive mobilization of populations occurs. The city-actor might emerge as an essential place of public *policies* design and implementation without being a place of revitalization of *politics*. In other words, the city-actor might be a place where political mobilization only involves individuals and groups bearing resources required by the policy process rather than the entire citizenry (Pinson, 2006, 2009).

## Bringing the state back in, again

The research exploring the hypothesis of the city-actor, in particular the research conducted in the Neo-Weberian perspective, has many merits. Above all, it allows us to highlight the fact that the revival of urban policies that can be observed since the 1980s was not a mere product of the neo-liberalization of urban governance, but was also due to the activism of cities and urban elites and to the transformations of the relations between various scales of government (Pinson and Morel Journel, 2016, 2017). It reminds us that the new 'urban moment' we are experiencing should be relocated in the *'longue durée'* of the relations between cities and other scales, in particular territorial states.

In doing so, however, this research tradition has perhaps tended to regard these state-city relations as a zero sum game, in which what cities obtained in terms of agency, room for manoeuvre, functions, resources, expertise and capacity to act, was simply that which was lost by the territorial state. The case of France in the 1980s and 1990s led me to adopt such a reading. From 1982 onwards, France launched a decentralization process that led to the transfer of an increasing number of functions to municipalities, the consolidation of local government fiscal autonomy, and strengthened inter-municipal cooperation structures. Meanwhile, during these two decades, the state central and field services seemed to go through a period of disarray. The rise of the EU, globalization, neoliberal transition and the emergence of subnational authorities seem to signal the end of the modernizing and planning mission of the State. The institutions embodying this State – ministries, *grand corps* of civil servants, prestigious schools training these elites, etc. – appeared to be weakening. Conversely, metropolitan and regional institutions seemed to be on the rise. Urban and regional elected offices were looking increasingly attractive to prominent national political figures. Urban and metropolitan administrations were attracting more and more administrative elites of the State. France thus seemed to undergo a widespread 'rescaling' process in which the national space seemed to become a shadow of itself, reduced to a deserted theatre where political struggles increasingly disconnected from genuine policy issues were taking place. The same could be said for Italy, Spain or Belgium.

However, things changed considerably in the following two decades. Taking advantage of neo-managerial reforms, the central government has been able to reinvent the system of relationships linking it to local governments. In particular, it has gradually abandoned a policymaking approach based on contracts with local government (labelled '*contractualisation*' in French) in favour of new forms of interaction allowing it to reassert its control. Indeed, '*contractualisation*' was introduced during the 1980s and 1990s as a way to adjust the projects of the State and those of local government (Gaudin, 1999). *Contractualisation* embodied the difficulties that the State was facing to renew its policy agenda. The State was becoming a partner of local government, adjusting its contributions to the local projects. In contrast, in the 2000s and 2010s, the French State was able to reinvent its policy doctrine in many fields and to implement new instruments such as the competitive calls for projects. In particular, the State devised these new instruments in areas such as urban renewal or the digital economy. With competitive calls, state funds are no longer allocated to territories taking objective circumstances of deprivation into consideration. Their attribution is rather contingent upon the capacity of local communities to elaborate projects, and on the evaluation of the conformity of these projects with standards set by the State. Renaud Epstein (2013) distinguishes three phases in the central-local relations in France since the post-war period. The first, called the 'centralized government of the local' was typical of the action of the dirigiste state during the '*Trente Glorieuses*'. The 'negotiated government of territories' labels the period of the development of '*contractualisation*' that follows decentralization reforms. Since the early 2000s, a new era has opened, called '*gouvernement à distance*' by Epstein (2013). In this era, the State seeks to reassert a higher degree of control of urban policies but without turning back to the dirigiste model of the first phase. It rather seeks to preserve the new activism of cities, to encourage the formation of an autonomous political and technical capacity of cities to produce strategies and public policies, and reward those whose initiatives are consistent with its own objectives.

Recent research has shown that these new devices of competitive allocation of resources are not only a way for the State to govern and standardize the behaviours of urban governments. They are also meant to feed the State agenda with new ideas and best practices in certain areas of public policy, including those relating to territorial policies (Béal, Epstein and Pinson, 2015). Indeed, the calls launched by the central ministries generally implement selection criteria implicitly or explicitly inspired by the policies and strategies previously implemented by cities considered as trendsetters in a particular sector (Béal, 2014; a similar trend has been identified by Perulli for the Italian case: cf. Perulli, 2013). Thus, even though the State seemed at a loss to define a policy doctrine in certain areas, its services managed to reconstruct the elements of doctrine based on the census and standardization of urban governments' best practices. Leveraging the essential resources that remain in its possession such as legitimacy in defining the common good, the ability to frame the law and organize the redistribution of resources, the French State has thus repositioned itself as an actor capable of governing from a distance the behaviours of urban governments by rewarding good local practices, turning them into labels of quality.

In the end, the logic of mobilization and policy innovation that we can observe at the city level, the increasing autonomies of cities in the making of their agenda and the implementation of policy programs, i.e. the ability of cities to become actors – all this does not conflict with the ability of the State to reposition itself in the multi-level governance.

## Strong cities within strong states and conflictual intergovernmental relations

This leads us to the hypothesis with which we would like to conclude this chapter. This hypothesis is twofold and relies on two different bodies of literature. First, I hypothesize that the emergence of strong city-actors occurs not necessarily in institutional contexts where the territorial states' influence and profile has weakened, as some proponents of the Neo-Weberian approach of cities, including myself, have postulated before. Quite conversely, my present hypothesis is that the emergence of city-actors takes place in national contexts characterized by a strong state. I would thus reverse Tilly and Blockmans' title (1994) and say that voracious cities have been supported by strongly resilient states. The second part of the hypothesis is that there is another factor fostering the city-actor phenomenon: this factor is the competitive and even conflictual nature of state-city relationships, and more broadly of centreperiphery relationships.

I borrow the definition of a strong state from the old and contested, yet robust, work of Badie and Birnbaum (Badie and Birnbaum, 1983; Birnbaum, 2011). In a quite classical Weberian vein, the two political scientists defined the strong state as a differentiated and institutionalized state. In their approach, the State's strength relies on its will, capacity and legitimacy to organize society and economy, and its ability to put society at a distance in order to transform it thanks to its own administrative apparatus and policymaking capacity. The strong state model is embodied by the French State but also by the Prussian, Spanish or Italian cases, even if the two last represent incomplete or unachieved versions. By contrast, societies with a weak state (US, UK) are characterized by a weak state bureaucracy, processes of policymaking in which the intervention of organized interests is legitimate, and political institutions that are poorly differentiated and captured by social and economic elites.

In our hypothesis, the most dynamic city-actors have emerged in contexts where various parts of the state apparatus – central administrations, local and regional government – have been struggling to maintain a high degree of regulation of social and economic and mechanisms, in a changing context characterized by urbanization and globalization, among others. The numerous reforms which state administrations have gone through should not be taken lightly. They may have taken the form of decentralization, Europeanization and even privatization and deregulation reforms. Nevertheless, we should not interpret these reforms as the result of a straightforward will to disempower the State, or deprive it from its governing capacity. Rather, they embody what Philippe Bezès, inspired by Foucault, called the *'souci de soi de l'Etat'* (2009), awkwardly translatable as 'State self-concern'. Somehow, these reforms were the outcome of a reflexive State seeking to restore its grip on society and economy at various scales.

Thus, the emergence of the city-actor does not proceed from a simple 'rescaling' of statehood (Brenner, 2004), more or less orchestrated by a ruse of capitalism or by neo-liberal reforms. Nor does it proceed from a revenge of newly voracious cities against a weakened but obstructive territorial state, leading to a form of federalization of political regimes, or a return of city-states. Rather, it proceeds from a logic of adaptation of policymaking systems to a new context, characterized by urbanization, metropolitanization and the territorialization of economic development mechanisms. In this adaptation process, the territorialization of policymaking at the urban and metropolitan scales, and the promotion of local government, have been identified as potential levers to regain governing capacity. Urban governments have been one of the beneficiaries of this process, but central state administrations have also seen how it could help them to implement their own policies and strategies. Thus, the emergence of city-actors would not lead to a situation in which the city/metropolis scale would take precedence over the territorial state scale, but in which each of the two levels seek to benefit from the other's capacity to act to achieve its own ends. In quite a similar approach, Crouch and Le Galès (2012) formulated the hypothesis according to which, in a context of heightened global competition, large cities and their urban or metropolitan governments would be the new 'champions' of territorial states. These new champions would replace the old industrial champions of yesteryear, which have become much less reliable and controllable. We agree with this hypothesis insofar as it helps to understand why states are no longer seeking to confine the policymaking capacity of cities, but rather to strengthen them, namely, to build up a powerful network of cities able to take part in the international competition (Béal, Epstein and Pinson, 2018).

The second part of my hypothesis might look contradictory to the first part, but the two are actually dialectically linked. The city-actor phenomenon has blossomed in national contexts characterized by conflictual, or at least competitive, relations between centres and peripheries, in particular between states and cities. One should refer here to the tradition of comparative studies of European systems of local government (Page and Goldsmith, 1987; Hesse and Sharpe, 1991; Bobbio, 2002; Goldsmith and Page, 2010; Pinson, 2016) and centre-peripheries relations, a body of research that has shown how different the role of local government and the nature of the relations between central state and local government are between northern and southern parts of Europe. Smith (1985) distinguishes countries where 'efficiency' is the main focus as far as local government is concerned, and those where the 'community' dimension dominates. In the first category, which corresponds largely to the northern part of the continent where the process of national integration went quite smoothly, the representation of local communities absorbed by a centre was not a big issue and local government 'naturally' became endowed with mostly technical and managerial functions. By contrast, in the South, the process of national integration and state building was a highly conflictual one. As a result, in the early periods of the state-building process, local authorities in Southern Europe have been deprived of many functions by state administrations, but local elected officials have had a prominent role in defending local interests against what is seen as the voracious central state (Grémion, 1976; Tarrow, 1977). Thus, in Southern Europe, the nature of the relations between the state and local government is a permanent concern and issue of intergovernmental rivalries. Local politicians and officials are permanently seeking to expand their room for manoeuvre against the state, which is almost inconceivable for, say, English or Swedish urban leaders. Hence, the recent emergence of strong boosterist and competitive urban agendas in southern European cities must not be interpreted solely as the result of the imposition of the neoliberal paradigm from above. Here, urban entrepreneurialism is also, and above all, a manifestation of a change in state-city relations, and of the activism of urban leaders struggling to gain more autonomy in a context where the central state has released its grip upon them (Le Galès, 2002; Le Galès and Pinson, 2005). Thus, somehow, central-local latent and permanent conflict provides the fuel for urban activism and agency.

From this vantage point, the city-actor phenomenon is more likely to emerge in national contexts combining a tradition of (achieved) strong state with a history of conflictual central-local relationships. France displays this kind of combination; the UK, on the contrary, offers a context with almost no room for the city-actor. A tale of two cities, in which we briefly compare Lyon and Manchester, will illustrate this.

Lyon, France's second largest city, has clearly benefited from the deep transformations of intergovernmental relations. France has long been known for its system of very asymmetric centre-periphery relations in which the design and implementation of public policies were monopolized by the state central and field services. In return, local officials had access to the centre, which endowed them with a significant political weight and capacity to influence state policies. As indicated by Page and Goldsmith (1987), this compromise is the product of a history of very conflictual centre-periphery relations, typical of southern European countries. With decentralization reforms and the building of powerful intercommunal structures in cities, things have changed dramatically (Borraz and Le Galès, 2005). Competition between levels has not disappeared, but has become much more balanced. Central state administrations have become accustomed to dealing with increasingly competent urban and metropolitan bureaucracies. Some state officials have even migrated to these structures that tend to be identified as innovation cradles. Central state administrations gradually realized that reinforced urban governments might be beneficial to them. They also learned how to (re)construct a more subtle form of control. For their part, urban and metropolitan governments have learned to 'do without the state' by developing their own capacity for action and addressing other partners such as foreign cities or the European Union. By comparing themselves to other cities in other countries, however, they also understood that being able to rely on a robust central state administration with financial capacity and technical expertise and guaranteeing some form of fiscal redistribution could be useful. A game of complementarities has thus progressively been installed between the state and major French cities, probably to the detriment of other scales and spaces such as regions, *départements* and rural spaces.

The case of Lyon is interesting in how it illustrates this complementarity between a still robust central state and increasingly powerful metropolitan powers (Galimberti et al., 2014; Galimberti, 2015). The 'Métropole de Lyon', a metropolitan authority with special status, has been the governing body of the urban area of Lyon since 1 January 2015. This new institution governs an area with 1.3 million inhabitants and exerts a very broad set of functions ranging from urban planning, water, transport, to social policy, higher education or international relations. On its territory, the new institution now exercises the functions so far exerted by the *département du Rhône*, which no longer has authority over the metropolis. The budget of the metropolitan authority is about €2.5 billion, the same as the Rhône-Alpes region. The creation of a Lyon metropolitan government (also called Grand Lyon) was driven by Gérard Collomb, mayor of Lyon and president of the Urban Community of Lyon, the inter-municipal cooperation authority that existed before the creation of the metropolis. This creation has largely inspired the law on the affirmation of cities passed by Parliament in 2015 that created metropolitan governments in a dozen other French city-regions including Paris and Marseilles. It is said that the services of Grand Lyon wrote the sections of the Act, relating to the Lyon case. The city of Lyon is probably now one of the most powerful metropolitan governments in Europe and its creation was encouraged, even sponsored by the central government, even though in recent years the Lyon metropolitan structure had the ability to marginalize state field services. This voracious nature of the *Métropole de Lyon* does not prevent state services from promoting within and beyond the French borders the good practices of Lyon metropolitan government in areas as diverse as economic development, research and higher education or the smart city. What we have here is an example of this positive-sum game in which a state, repositioned into a steering posture, is sponsoring the emergence of a powerful metropolitan government.

The case of Manchester is quite different. In my view, it is the result of the combination of a traditional 'weak state' on one hand, and the absence of an institutionalized conflict between the state and local government on the other hand. The absence of conflict between the monarchy and the landed nobility, due to their common origin in the Norman invasion, led to a situation in which there were no strong conflicts between the king and the aristocracy, between the centre and the periphery. There was thus no need to build up a strong state bureaucracy to control either society or peripheries, which Badie and Birnbaum considered as being at the origin of the strong state. The absence of a strong alliance between the executive power and the bureaucracy paved the way for the rise of the Parliament, traditionally considered as the principal holder of the political legitimacy and power in the UK. The British paradox is that the weak nature of the state, i.e. the executive-bureaucracy nexus, did not prevent some form of centralization, just like the strong state did not prevent the rise of a genuine local power in France (Ashford, 1982). In Britain, the centralization trend, which became evident with the construction of the welfare state, marginalized local government to the benefit of the Parliament, not of the bureaucracy. Indeed, the other peculiarity of the British political system is the smooth nature of the central-local relationships that deprived local societies of any strong sense of identity and any antibodies against centralization. As a result, 'local government has few friends at the court of Westminster' (Chandler, 2001, p. 85). The weak nature of the British State did not prepare it to efficiently resist the Thatcherite reforms aiming at downsizing the state regulation capacity. It did not incite it to organize the rescaling of state authorities at the regional or metropolitan levels. Efforts to build up proper scales of regulation at these levels were actually timid and short-lived. Besides, local government and politicians were never in a position to obtain the tools and resources allowing them to build up new institutional capacities at the subnational level. A weak state tradition and a history of non-contentious central-local relationships allow a tiny political space for local government and paradoxically pave the way to a situation detrimental to the rise of the city-actor.

At first sight, Manchester would appear to be the sole example – alongside London – of a genuine city-actor. After all, it is one of the rare examples of British cities where a public-private coalition rose up to rebuild the city after the 1996 IRA bombing and to endow the city with a clear redevelopment agenda (Harding, 2000; Peck and Ward, 2002; Pinson, 2009). The participation in bids to organize great events, in particular the Olympic Games, and the actual organization of events, like the 2002 Commonwealth Games, provide the processual framework for this coalition building, uniting elected officials, top municipal civil servants, construction and development firms, banks and cultural institutions. These mobilizations and physical transformation processes allowed Manchester to emerge as one of the rare islands of prosperity outside South-East England, and as a centre of tertiary and knowledge economy. Did it make Manchester into a city-actor? It is far from certain. Again, of all the British cities, it is probably closest to the ideal type we drew up above. Nevertheless, the agenda of the ruling regime is limited to real estate boosterism, while many dimensions of urban policies (social services, housing, metropolitan planning) are not really integrated in this agenda. As in many other British municipalities, many urban services and bureaucratic segments are functioning on automatic pilot, with elected officials having only a loose grip. Manchester City Council is still governed by a committee system with a council leader who cannot benefit from the political status French mayors have. Hence, the capacity for external representation is quite low. Still, on representation issues, Manchester is not that present in European urban networks and Brexit will not make anything easier. Since the abolition of the Greater Manchester County Council in 1986, the city is deprived of any serious metropolitan political authority. Even if the range of single purpose joint boards and quangos that were set up after 1986 have been replaced by a 'combined authority' (GMCA), it cannot compare with the *Métropole de Lyon*'s scope, resources, technical expertise, political authority and innovation capacity. Finally, yet importantly, the tradition of low turnout in local elections in Britain is not of any help to get local populations enrolled in Manchester's strategy.

To be sure, the model I have presented in this chapter needs refinement and should be tested on a wider array of national and local situations with various combinations of the weak/strong state and conflictual/smooth central-local relations independent variables. Other cases in Europe might prove trickier to enter into the model. Italy, for instance, is kind of a 'failed' strong state with conflictual central-local relations. A form of city-actor phenomenon appeared in the 1990s but was quickly disarmed by the competition of increasingly strong regions and the return of parties. Indeed, in the early 1990s, a movement of newly directly elected urban mayors emerged. These mayors claimed that cities might be the cradle of the reconstruction of politics and policies in Italy, just as they were during the medieval communal movement. These independent mayors were depicting cities as places where the domination of parties could give way to a political practice more attentive to the provision of services and the delivery of policy programs. Nevertheless, the following decades saw a slowdown in institutional reforms and the return of particracy. The inability of the Italian State administration to pursue reforms seems to have hampered the consolidation of this

process of urban policies revival: the shadow of powerful regional governments was detrimental to the rise of the city-actor, which gives us an insight
about a possible third factor.

## References

Acuto, M., 2013. City leadership in global governance. *Global Governance*, 19(3),
pp. 481–498.
Amin, A. and Thrift, N., 2002. *Cities: reimagining the urban*. Cambridge: Polity Press.
Ashford, D.E., 1982. *British dogmatism and French pragmatism: central-local policy-
making in the welfare state*. London: Allen & Unwin.
Bachrach, P. and Baratz, M.S., 1970. *Power and poverty: theory and practice*. Oxford:
Oxford University Press.
Badie, B. and Birnbaum, P., 1983. *The sociology of the state*. Chicago, IL: University
of Chicago Press.
Bagnasco, A. and Le Galès, P. eds., 2000. *Cities in contemporary Europe*. Cambridge:
Cambridge University Press.
Barber, B.R., 2013. *If mayors ruled the world: dysfunctional nations, rising cities*. New
Haven, CT: Yale University Press.
Béal, V., 2014. Trendsetting cities': les modèles à l'heure des politiques urbaines
néolibérales. *Métropolitiques*, [online]. Available at: www.metropolitiques.eu/
Trendsetting-cities-les-modeles-a.html [Accessed 25 January 2018].
Béal, V. and Pinson, G., 2014. When mayors go global: international strategies,
urban governance and leadership. *International Journal of Urban and Regional
Research*, 38(1), pp. 302–317.
Béal, V., Epstein, R. and Pinson, G., 2015. La circulation croisée. Modèles, labels
et bonnes pratiques dans les rapports centre-périphérie. *Gouvernement et action
publique*, (3), pp. 103–127.
Béal, V., Epstein, R. and Pinson, G., 2018. Networked cities and steering states. Urban
policy circulations and the reshaping of state-cities relationships. *Environment and
Planning C: Environment and Planning C: Politics and Space*, 4 January, online first.
Beck, U. and Sznaider, N., 2006. Unpacking cosmopolitanism for the social sciences:
a research agenda. *The British Journal of Sociology*, 57(1), pp. 1–23.
Bezès, P., 2009. *Réinventer l'Etat. Les réformes de l'administration française
(1962–2008)*. Paris: PUF.
Birnbaum, P., 2011. Défense de l'État 'fort'. Réflexions sur la place du religieux en
France et aux États-Unis. *Revue française de sociologie*, 3(52), pp. 559–578.
Blockmans, W., 1989. Voracious states and obstructing cities. *Theory and Society*,
18(5), pp. 733–755.
Bobbio, L., 2002. *I governi locali nelle democrazie contemporanee*. Bari: Laterza.
Borraz, O. and Le Galès, P., 2005. France: the intermunicipal revolution. In:
B. Denters and L.E. Rose, eds. *Comparing local governance. Trends and develop-
ments*. New York: Taylor & Francis. pp. 12–28.
Bourdieu, P., 1989. *La noblesse d'État. Grandes écoles et esprits de corps*. Paris: Minuit.
Braconnier, C. and Dormagen, J.Y., 2007. *La démocratie de l'abstention: aux origi-
nes de la démobilisation électorale en milieu populaire*. Paris: Gallimard.
Brenner, N., 2004. *New states spaces. Urban governance and the rescaling of state-
hood*. Oxford: Oxford University Press.

Brenner, N., 2009. Open questions on state rescaling. *Cambridge Journal of Regions, Economy and Society*, 2(1), pp. 123–139.

Brenner, N. and Theodore, N., 2002. Cities and the geographies of 'actually existing neoliberalism'. *Antipode*, 34(3), pp. 349–379.

Chandler, J.A., 2001. *Local government today*. 3rd edition. Manchester: Manchester University Press.

Cox, K., 1997. Governance, urban regime analysis, and the politics of local economic development. In: M. Lauria, ed. *Reconstructing urban regime theory: regulating urban politics in a global economy*. Thousand Oaks, CA: Sage. pp. 99–121.

Crouch, C., 2004. *Post-democracy*. Cambridge: Polity.

Crouch, C. and Le Galès, P., 2012. Cities as national champions? *Journal of European Public Policy*, 19(3), pp. 405–419.

Dahl, R.A., 1961. *Who governs? Democracy and power in an American city*. New Haven, CT: Yale University Press.

Dilworth, R. ed., 2009. *The city in the American development*. New York: Routledge.

Ehrenhalt, A., 2012. *The great inversion and the future of the American city*. New York: Alfred A. Knopf, Inc.

Engelen, E., Froud, J., Johal, S., Salento, A. and Williams, K., 2016. How cities work: a policy agenda for the grounded city. *CRESC Working Paper Series*, 141, pp. 1–31.

Epstein, R., 2013. *La rénovation urbaine: démolition-reconstruction de l'État*. Paris: Presses de Sciences Po.

Florida, R., 2005. *Cities and the creative class*. London: Routledge.

Galimberti, D., 2015. Gouverner le développement économique des territoires: entre politique et société. Une comparaison des régions de Lyon et Milan (1970–2011). PhD in political science. Université Jean Monnet de Saint-Étienne / Università degli Studi di Milano.

Galimberti, D., Lobry, S., Pinson, G. and Rio, N., 2014. La métropole de Lyon. Splendeurs et fragilités d'une machine intercommunale. *Hérodote*, (3), pp. 191–209.

Gallagher, L., 2013. *The end of the suburbs: where the American Dream is moving*. New York: Penguin.

Gaudin, J.P., 1999. *Gouverner par contrat: l'action publique en question*. Paris: Presses de la Fondation nationale des sciences politiques.

Glaeser, E., 2011. *Triumph of the city: how our greatest invention makes us richer, smarter, greener, healthier, and happier*. New York: Penguin.

Goldsmith, M. and Page, E. eds., 2010. *Changing government relations in Europe: from localism to intergovernmentalism*. London: Routledge.

Grémion, P., 1976. *Le pouvoir périphérique*. Paris: Editions du Seuil.

Guilluy, C., 2013. *Fractures françaises*. Paris: Flammarion.

Harding, A., 1994. Urban regimes and growth machines: toward a cross-national research agenda. *Urban Affairs Quarterly*, 29(3), pp. 356–356.

Harding, A., 1997. Urban regimes in a Europe of the cities? *European Urban and Regional Studies*, 4(4), pp. 291–314.

Harding, A., 2000. Regime formation in Manchester and Edinburgh. In: G. Stoker, ed. *The new politics of British local governance*. London: Palgrave-Macmillan. pp. 54–71.

Harding, A. and Blokland, T., 2014. *Urban theory: a critical introduction to power, cities and urbanism in the 21st century*. London: Sage.

Harvey, D., 1989. From managerialism to entrepreneurialism: the transformation in urban governance in late capitalism. *Geografiska Annaler. Series B. Human Geography*, 71(1), pp. 3–17.

Hesse, J. and Sharpe, L. eds., 1991. *Local government and urban affairs in international perspective*. Baden: Nomos.

Hoffmann-Martinot, V. and Sellers, J. eds., 2005. *Metropolitanization and political change*. Wiesbaden: VS Verlag für Sozialwissenschaften.

Hunter, F., 1953. *Community power structure. A study of decision makers*. Chapel Hill: The University of North Carolina Press.

Jessop, B., 2002. *The future of the capitalist state*. Cambridge: Polity.

Jessop, B., Peck, J. and Tickell, A., 1999. Retooling the machine: economic crisis, state restructuring and urban politics. In: A. Jonas and D. Wilson, eds. *The urban growth machine*. Albany, NY: State University of New York Press. pp. 141–159.

Katz, B. and Bradley, J., 2013. *The metropolitan revolution: how cities and metros are fixing our broken politics and fragile economy*. Washington, DC: Brookings Institution Press.

Le Galès, P., 2002. *European cities: social conflicts and governance*. Oxford: Oxford University Press.

Le Galès, P., 2005. Elusive urban policies in Europe. In: Y. Kazepov, ed. *Cities of Europe: changing contexts, local arrangements, and the challenge to social cohesion*, London: Blackwell. pp. 235–254.

Le Galès, P. and Harding, A., 1998. Cities and states in Europe. *West European Politics*, 21(3), pp. 120–145.

Le Galès, P. and Pinson, G., 2005. State restructuring and decentralization dynamics in France: politics is the driving force. *Working papers du Centre d'études européennes de Sciences Po*, 7, p. 5.

Leca, J., 1996. La 'gouvernance' de la France sous la Cinquième République. Une perspective de sociologie comparative. In: J.-L. Quermonne, ed. *De la Ve République à l'Europe*. Paris: Presses de la FNSP. pp. 329–365.

Lévy, J., 2003. Vote et gradient d'urbanité. L'autre surprise du 21 avril. *Espacestemps.net*, [online]. Available at: www.espacestemps.net/articles/vote-et-gradient-urbanite/https://www.espacestemps.net/articles/vote-et-gradient-urbanite/ [Accessed 25 January 2018].

Logan, J. and Molotch, H., 1987/2007. *Urban fortunes: the political economy of place*. Berkeley: University of California Press.

Lussault, M., 2009. *De la lutte des classes à la lutte des places*. Paris: Grasset.

Massey, D., 2004. The political challenge of relational space: introduction to the Vega Symposium. *Geografiska Annaler: Series B, Human Geography*, 86(1), p. 3.

Massey, D., 2007. *World city*. Cambridge: Polity.

McCann, E. and Ward, K. eds., 2011. *Mobile urbanism: cities and policymaking in the global age*. Minneapolis: University of Minnesota Press.

Page, E. and Goldsmith, M. eds., 1987. *Central and local government relations: a comparative analysis of West European unitary states*. London: Sage.

Pasquier, R., 2015. *Regional governance and power in France. The dynamics of political space*. London: Palgrave Macmillan.

Peck, J., 2005. Struggling with the creative class. *International Journal of Urban and Regional Research*, 29(4), pp. 740–770.

Peck, J. and Theodore, N., 2015. *Fast policy*. Minneapolis: University of Minnesota Press.

Peck, J. and Ward, K., 2002. *City of revolution: restructuring Manchester*. Manchester: Manchester University Press.

Perulli, P., 2013. Return of the State and attempts of centralisation in Italy. *Métropoles*, (12), [online]. Available at: https://metropoles.revues.org/4665 [Accessed 25 January 2018].

Peterson, P.E., 1981. *City limits*. Chicago, IL: University of Chicago Press.

Pichierri, A., 1997. *Città stato: economia e politica del modello anseatico*. Venice: Marsilio.

Pichierri, A., 2002. Concertation and local development. *International Journal of Urban and Regional Research*, 26(4), pp. 689–706.

Pinson, G., 2006. Projets de ville et gouvernance urbaine. Pluralisation des espaces politiques et recomposition d'une capacité d'action collective dans les villes européennes. *Revue française de science politique*, 56(4), pp. 619–651.

Pinson, G., 2009. *Gouverner la ville par projet: urbanisme et gouvernance des villes européennes*. Paris: Presses de Sciences Po.

Pinson, G., 2010. France. In: M. Goldsmith and E. Page, eds. *Changing government relations in Europe. From localism to intergovernmentalism*. London: Routledge. pp. 68–87.

Pinson, G., 2016. The French way to multi-level governance. In: R. Elgie, E. Grossman and A. Mazur, eds. *The Oxford Handbook of French Politics*. Oxford: Oxford University Press. pp. 102–127.

Pinson, G. and Morel Journel, C., 2016. The neoliberal city. Theory, evidence, debates. *Territory, Politics, Governance*, 4(2), pp. 137–153.

Pinson, G. and Morel Journel, C., 2017. *Debating the neoliberal city*. London: Routledge.

Rokkan, S. and Urwin, D., 1982. Introduction: centres and peripheries in Western Europe. In: S. Rokkan and D. Urwin, eds. 1982. *The politics of territorial identity. Studies in European regionalism*. London: Sage. pp. 1–18.

Scott, A.J. and Storper, M., 2015. The nature of cities: the scope and limits of urban theory. *International Journal of Urban and Regional Research*, 39(1), pp. 1–15.

Smith, B., 1985. *Decentralisation: the territorial dimension of the state*. London: Allen & Unwin.

Stone, C., 1989. *Regime politics: governing Atlanta 1946–1988*. Lawrence: University Press of Kansas.

Tarrow, S., 1977. *Between center and periphery. Grassroots politicians in Italy and France*. New Haven, CT: Yale University Press.

Tilly, C. and Blockmans, W.P., 1994. *Cities and the rise of states in Europe, AD 1000 to 1800*. Boulder, CO: Westview Press.

Van der Heiden, N., 2010. *Urban foreign policy and domestic dilemmas: insights from Swiss and EU city-regions*. Colchester: ECPR Press.

Ward, K., 1996. Rereading urban regime theory: a sympathetic critique, *Geoforum*, 27, pp. 427–438.

Wood, A., 2005. Comparative urban politics and the question of scale. *Space and Polity*, 9(3), pp. 201–215.

# 5   Global-urban policymaking[1]

*Kevin Ward*

## Introduction

There is a long tradition of academic work on city mayors: from anthropology to economics, human geography to international relations, planning to political science, sociology to urban studies – research in each of these disciplines has explored the emergence of mayors, the governance systems of which they are a part, and their failures and successes, however broadly understood and represented (Ferman, 1985; Svara, 1990; Hambleton, Savitch and Stewart, 2002; Jayne, 2012). Many of these contributions have studied the work of city mayors in the US, where the capacities afforded to them are greater than in more centralized governmental systems, such as for example the UK (Hambleton and Sweeting, 2004). Much of this work predates Barber's (2013) *If Mayors Ruled the World*, the publication of which was the initial inspiration for this volume. The academic work on mayors is, of course, continuing to multiply after the publication of Barber's oeuvre. A large, multidisciplinary and sprawling field continues to expand as nation states restructure themselves in such a way as to empower their city mayors, asking more of them while simultaneously affording them greater financial and political responsibility (Curtis, 2014; Ljungkvist, 2016). In some cases, such as my own city, Manchester, this has meant the establishment of mayors for the first time. As one of 10 boroughs in Greater Manchester, it has recently voted for its first 'metro mayor'. This came about not from some local public campaign but rather as a wider restructuring of the nation state that entailed various budgets and responsibilities being delegated to city regions, which are in turn required to elect a mayor.

*If Mayors Ruled the World* featured in *City Lab*, the *Financial Times*, *Foreign Affairs*, the *New York Times* and *Prospect*, supported by various seminars by Barber, including a *Ted Talk*. This mediation and translation (including literally into other languages) constructed a wide audience for the book and its central arguments, consisting largely of those with an interest in urban policymaking and politics. Replete with anecdotes and accounts from the mayors of 11 cities around the world, all of whom Barber (2013, n.p.) has been 'lucky enough to know', in general terms the book joins the ranks of a number of other books for whom the primary target audience is the informed general public and a range of policymakers, such as consultants, government officials and others with a stake in the governance of cities. Most noteworthy of these books that make the case for an investigation into the generative role of cities are Richard Florida's (2004) *Cities and the Creative Class* and Edward Glaeser's (2011) *Triumph of the City*, but we can also think of Anthony Townsend's (2014) *Smart Cities* and Charles Montgomery's (2015) *Happy City*. Presented as a response to the ideological and political failings of the nation state as too bureaucratic, too corrupt, too undemocratic and too unwieldy, the focus on cities offers them up as the new urban political frontier, to paraphrase various contributors but particularly Neil Smith (1996). Here is an opportunity for political renewal; for starting afresh, we are told. Keep those institutions and networks that have emerged, connecting cities across the world, while augmenting, enhancing and formalizing their capacities. Writing in *The Nation*, Goldberg (2014, n.p.) argues that in the case of the US, these books – and others – reflect a growing conviction amongst those on both the Left and the Right that the city is now the primary site for policy experiments, a laboratory for innovations, whether in the introduction of living wages or the downsizing/rightsizing of government. This rendering of the urban as the primary site for policymaking across a range of fields, from ageing to climate change, sits squarely alongside another persuasive 'truth' which presents the global urban population as reflecting a new 'urban age' (Brenner and Schmid, 2014). This argument underpinned Barber's (2017) *Cool Cities: Urban Sovereignty and the Fix for Global Warming*, his last book before his untimely death.

Accompanying these works – and creating the conditions for how they have been received by the media, policymakers and think tanks – is the informational infrastructure that has been carefully crafted and assembled – blogs, co-eds, interviews, presentations, press releases, podcasts and WordPress sites (McCann, 2011). While there is much that divides these books, such as ideological orientation and geographical focus, what they constitute together is a renewed focus on the agency and capacity of something called 'the city' to act and to be a site for progressive environmental, economic, political and social transformation. In the interest of limiting the scope of this chapter, I choose to ignore the risk of spatial fetishism that accompanies claims of local agency (Cox and Mair, 1989; Duncan and Savage, 1989), even if extended through notions of international political agency (Ljungkvist, 2016). I will instead draw upon Barber's (2013) contribution regarding 'cities governing globally' (Barber, 2013, p. 2): to consider the role cities might play in the future governance of the world.

> As a starting point for giving institutional expression to the need to realize some form of constructive democratic interdependence, I propose then the convening of a global parliament of mayors – call it a World Assembly of Cities. To begin, such an assembly would represent

a modest first step towards formalizing the myriad network of cities already actively cooperating across borders around issues as mundane an [as?] express bus lanes, bike-share programs, and web-based information collection, and as momentous as climate change, nuclear security, and intelligence gathering.

(Barber, 2013, p. 340)

This notion of a 'World Assembly of Cities' has been picked up and discussed by various commentators (Florida, 2013; Jenkins, 2013; Rogers, 2013; Goldberg, 2014; Lechtenberg, 2014). Leaving aside its virtues or otherwise, which as Glaeser (in Lechtenberg, 2014, n.p.) notes would transform executives into parliamentarians, in this chapter I want to pick up on another aspect to Barber's argument. On the one hand, this is a statement about the current set of relations between cities and their mayors. As he puts it (2013, p. 302, original italics):

The intercity civic infrastructure already in place comprises in its present form an informal approximation of the kinds of collaboration and confederal partnership that a prospective mayors' parliament will represent.

Specifically, he identifies a number of features of the current global urban system that point to the generative capacity of city-to-city collaboration, comparison, exchange and learning. Many of these are formal or 'official' relations between city governments – the 'myriad network of cities' of which Barber (2013, p. 340) writes. On the other hand, he also points to more 'informal' means through which ways of doing things emerge in particular urban locations and then find themselves travelling, rendered potentially appropriate for introduction elsewhere. These practices tend to emerge in and around city governments, involving them in some cases but not in others, in less organized and structured arrangements. Hence the emphasis on the 'informal':

There's already a great deal of what you might call best-practice sharing – as with bike sharing programmes that started in Latin America in the 1970s. That's a viral best practice that has spread across the world. Beyond that, there's room for ratcheting that up because the dysfunctions of nation state means that the kinds of cooperative, bilateral, multilateral agreements that we have looked for simply aren't happening.

(Barber in Derbyshire, 2013, n.p.)

It is on this particular aspect – the informal global-urban policymaking – upon which I want to focus in this chapter.[2] I want to highlight how we might understand the contemporary extrospective nature of urban policymaking that locates the city in the world (and, of course, the world in the city). Or, put another way, in the current institutional and political climate, in the context of 'dysfunctional nations', how are we to understand those 'intercity networks' (Barber, 2013, p. xx) that are held to be the governance building blocks for a new geo-political order? In focusing on this aspect of Barber's (2013) wider argument, the chapter draws upon some foundational work in urban policy mobility studies (Ward, 2006; McCann 2008, 2011; McCann and Ward 2011; Peck 2011), which is concerned with the relational and territorial geographies and histories behind the arriving at and making up of contemporary urban policies. This research involves actors of various geographical backgrounds and reach, and includes mayors and others with a political stake in the future of cities and of their role in the world. It questions what work is being done to bring the formal structures alive and what is happening between and beyond those structures? What are the more ad-hoc, ephemeral, fleeting, incremental and informal ways in which cities perform governing, and how/why do they become global practices? The next section turns to the various ways in which those who arrive at and make up policies look outwards, and in doing so appear to be exercising a form of political agency similar to that in the arguments made by Barber (2013). The chapter's third section discusses how this sort of working across and within cities might be understood, engaging directly with what Barber writes in his *If Mayors Ruled the World*. Finally, the conclusion takes up Barber's (2013) argument on cities as sites for policy innovation and outlines a case for a more global and relational understanding of the policymaking process.

## Cities in the world, the world in motion

> … increasingly it seems that what propels mayors to adopt a more global outlook is their new focus on urban policymaking. Study tours, involvement in transnational city networks and organization of international events have gained a new importance within cities' international strategies.
>
> (Beal and Pinson, 2014, p. 305)

The urban policymaking world seems to be in almost constant motion. In a figurative sense, policymakers – understood as a broad category of political actors in differing geographical reach – seem to be under increasing pressure to make things happen, to keep up with what is in and what is out in their own particular field of policy. As wave after wave of experimentation and innovation moves across the urban policymaking landscape, restructuring cities along the way, and as in turn policies themselves are restructured, so has an accompanying 'churning' been identified in urban policy. New ideas and initiatives replace older – though not necessarily 'old' policies – with increasing regularity. Contemporary urban policymaking, at all scales, therefore involves the constant scanning of the policymaking landscape via blogs, professional publications, the media, webinars, websites, and word of mouth for ready-made, off-the-shelf policies and best practices that can be quickly applied locally. Less instrumentally, but sometimes as importantly, are those ephemeral and fleeting engagements – between policymakers and their work – which can lead to the making up and arriving at policies (Ward, 2006; Robinson, 2015). However they are arrived at, it is increasingly common for those who govern and plan cities to look elsewhere, even if the production comes from within:

> [a]n extrospective, reflexive, and aggressive posture on the part of local elites and states ... Today cities must actively – and responsively – scan the horizon for investment and promotion opportunities, monitoring 'competitors' and emulating 'best practice', lest they be left behind in this intensifying competitive struggle for the kinds of resources (public and private) that neoliberalism has helped make (more) mobile.
>
> (Peck and Tickell, 2002, p. 399)

In this context figurative mobility in the policymaking world becomes literal motion. Policy actors act as 'transfer agents' in the words of Stone (2004, p. 550). Critiquing existing work, she argues:

> Key actors in the mechanics of policy transfer are international organisations and non-state actors such as interest groups and NGOS, think tanks, consultant firms, law firms and banks ... Recognition of non-state and international organisation roles complicates understanding of policy transfer processes beyond that of simple bilateral relationships between importing and exporting jurisdictions to a more complex multilateral environment.

While this is a welcome intellectual development, there is still room to widen the scope of who gets included as 'transfer agents': a diverse and large group that should include academics, activists, consultants, politicians, policy professionals and practitioners who have found themselves in the political process of making policy, even if at times their work is not understood as such. Each can do different types of work in the shuttling of policies and knowledge about policies between cities – in some cases through the networks identified by Barber (2013). This occurs inter alia through attending conferences, doing consultancy work, and engaging in fact-finding study trips or 'policy tourism' (Gonzalez, 2011; Ward, 2011). These activities then set the conditions for the potential making up and arriving at of urban policy and involve policymakers of various geographical reach in networks that extend globally, bringing certain cities into interaction with each other (while pushing others further apart). They create mental maps of 'best cities' for policy that in turn informs future strategies: Amsterdam for cycling; Austin for quality of life and creativity; Barcelona and Manchester for urban planning and regeneration; Bogotá for Bus Rapid Transit

(BRT); Curitiba for environmental planning; Detroit for urban farming and gardening; Freiburg for sustainability; Portland for growth management; Porto Alegre for participatory budgeting and direct democracy and so on. Thus, in a policymaking sense as in many other ways, cities are constituted through their relations with other places and scales.

Yet, while mobility and relationality define some aspects of contemporary urban policymaking (and there is evidence of numerous historical precedents, even if the conceptual language that has been used differs), this is, of course, only half the picture. Urban policies and the policymaking process are also intensely and fundamentally embedded, grounded, proximate and territorial. The examples just outlined confirm this argument. Our ability to refer to complex approaches to the vexing problems of urban policy through the use of a shorthand consisting of city names indicates how tied to specific places certain policies are. There is a 'Barcelona model' of urban regeneration, for example, which is contingent on the historical-geographical circumstances of that city and its relationship with other regional and national forms of decision-making. While other cities might be encouraged to compare, imitate or learn, it is generally understood that, in so doing, adjustments, modifications and translations will need to be made. So, it is understood that the success of participatory budgeting in Porto Alegre, for example, will not necessarily guarantee its successful adoption elsewhere as policy is fundamentally territorial in that it is tied up with a whole set of locally dependent interests, with those involved in growth coalitions being the most obvious examples; although activists NGOs and social movements have also historically often been involved in the making up and arriving at urban policies.

As such, urban policymaking must be understood as both relational and territorial, as both in motion and simultaneously fixed, or embedded in place. The contradictory nature of policy should not, however, be seen as detrimental to its operation. Rather, the tension between urban policy as both dynamic and relational on the one hand and fixed and territorial, on the other, is generative. It is necessary to the production of policy and places, as a number of human geographers have argued over numerous decades (Massey, 1993). And it is this aspect, where cities are seen to be central to how nations are responding to global challenges such as climate change or terrorism, which Acuto (2013) and Ljungkvist (2016) argue lies behind what might be labelled as the urban turn in international relations.

## Theorizing 'intercity civic infrastructure'

Barber (2013, p. 340) highlights the collaborations between cities when he writes about a 'Global Parliament of Mayors'. He identifies this as a 'civic infrastructure', and envisages the Parliament formalizing it as one example of the various already existing informal inter-city networks. How might we understand all the ways of generating, maintaining and nourishing this infrastructure? Or, put another way, in what ways might we think through all the big and small acts of assembling policy from a relational-territorial perspective? Of course, this is not an entirely new question. With this in mind, an appropriate point of departure is to consider the already existing political science literature on policy transfer. This explores how policies – including urban policies – are learnt from one context and moved to another with the hope for similar results. In one sense, this is a literature that is all about global relations and territories. It has a relatively long history, dating back to the 1960s, but it was really in the early 1990s that the field expanded, as 'the scope and intensity of policy transfer activity... increased significantly' (Evans, 2004, p. 1). This now voluminous literature shares some common characteristics, even while being internally differentiated and heterogeneous in many regards. It focuses on modelling how transfer works, on creating typologies of 'transfer agents' and on identifying conditions under which transfer leads to successful or unsuccessful policy outcomes in the new location. Contemporary urban political scientists continue to generate fruitful insights in these regards (Evans, 2010).

Yet, while this literature is certainly about global relations and territories, it has not consisted of an engagement with the full range of social territoriality. It is limited in its definition of the agents involved in transfer, focusing largely on national and international elites, and usually working in formal institutions. In many regards its focus on formal structures mirrors Barber's (2013) emphasis on the existing policymaking landscape of inter-city networks. It focuses almost exclusively on national territories – the transfer from A to B among nations or among localities within single nations. It neglects to consider inter-city transfers that transcend national boundaries, connecting cities globally. Furthermore, it tends not to consider transfer as a socio-spatial process in which urban policies, for example, are apt to morph and mutate as they travel. It also fails to theorize the non-linear movement from A to B, which might involve a range of other locations and where movement might consist of various stops and starts as opposed to a seamless journey from one location to another. It is focused on the transfer of a policy rather than on the practices through which a policy is transferred and what happens on its travels.

These limits to the 'traditional' policy transfer literature offer a series of opportunities for further theorization from a number of perspectives that understand transfer – albeit often in different ways – as a global-relational, social and spatial process which interconnects and even constitutes cities. For Wacquant (1999, p. 321), the aim should be 'to constitute, link by link, the long chain of institutions, agents and discursive supports' that constitute the current historical period. Larner (2003, p. 510), taking a governmentality approach, also advocates a move in the same intellectual direction, towards a 'more careful tracing of the intellectual, policy, and practitioner networks that underpin the global expansion of ... ideas, and their subsequent manifestation in government policies and programmes.' Explicitly interested in understanding both how and why governing practices and expertise are moved from one place to another, she advocates the 'detailed tracings' of social practices, relations, and embeddings. More recently, work in this vein has developed a sensibility to the ways in which mobility and mutation are mutually constitutive processes. That is, to the ways in which policies are constituted and reconstituted over the course of the various travels. For Peck (2011, p. 2), 'the movement of policy is more than merely a transaction or transfer, but entails the relational interpretation of policymaking sites and activities'. Jacobs (2012, p. 418) argues that translation

> brings into view not only the work required for a thing to reach one point from another, but also the multiplicity of add-ons that contribute, often in unpredictable and varying ways, to transportation, arrival, adoption and ... non-arrival and non-adoption.

Much of the mobilities work attempts to understand the details of a particular form of mobility, or a specific infrastructure that facilitates or channels mobilities, in reference to wider processes and contexts:

> [It] problematizes both 'sedentarist' approaches in the social sciences that treat place, stability, and dwelling as a natural steady-state, and 'deterritorialized' approaches that posit a new 'grand narrative' of mobility, fluidity or liquidity as a pervasive condition of postmodernity or globalization ... It is a part of a broader theoretical project aimed at going beyond the imagery of 'terrains' as spatially fixed geographical containers for social processes, and calling into question scalar logics such as local/global as descriptors of regional extent ...
>
> (Hannam, Sheller and Urry, 2006, p. 5)

This language is useful for thinking through urban policymaking in a global context. It allows the framing of the creation of urban policy in the world in such a way as to emphasize the social and the scalar, the fixed and the mobile features of urban policies. The notion of mobilities is being marshalled in the sense that people, frequently working in and across a variety of institutional contexts and settings, draw upon and mobilize expertise, ideas and knowledge to serve particular interests – with particular material consequences.

We can then perhaps see convergences among scholars about the need to look at both the why and the how of the movement of urban policy, particularly of the sort referenced by Barber (2013). The argument made here necessitates that we pay attention not only to how – through 'ordinary' and 'extraordinary' activities – policies are made mobile (and immobile), but also to why this occurs, and to the relationship between the movement of urban policies and the socio-spatial (re)structuring of cities. This will form the cornerstone of an institutionalizing of cities and their mayors of the sort suggested by the notion of a 'Global Parliament of Mayors'. The question remains how we might best frame these sorts of empirical discussions in light of understandings of the global that are increasingly generated from the urban.

Three suggestions are made here with a focus on how to study the ways in which cities and their mayors are and will be involved in the current and future creation of global governance. The first is to refine the theorization of urban policymaking and place-making as an assemblage of 'territorial' and 'relational' geographies, recognizing the material reality of each and the productive tensions between them without affording either of them pre-eminence. Following Olds (2001, p. 8), there is a requirement for 'a relational geography that recognizes the contingent, historically specific, uneven, and dispersed nature of material and non-material flows'. That means in the case of Barber's (2013) argument exploring the various types of infrastructure that support the inter-city relations of which he writes. This includes events, people and websites, examples of elements that hold together and give meaning and shape to the cross-border cooperation to which Barber (2013) points.

Second is to work with a broad appreciation and understanding of those who fall within the category of 'transfer agents'. This requires taking seriously the movement and translation of inter-urban, transnational expertise, ideas, and knowledge, understanding this movement as a socio-spatial, power-laden process in which urban policies are subject to contestation and change as they are rendered mobile. Of course, mayors are already important but the argument is that their importance may be on the increase; however, what the limiting factors to that agency are is not yet clear. Moreover, what this agency might actually mean in different cities and under different governance arrangements is almost certainly going to be geographically variable. It is the case now that not all mayors are equal. We should not confuse formal equivalence with functional equivalence. This level of granularity regarding the spatial context of each mayor should be built into the assessment of how any Global Parliament of Mayors is organized and structured, which might usefully shift the focus away from just elected officials in liberal democracies to a wider group of political agents.

Third, the methodological and analytical approach demonstrated here has at its core an orientation to the mutual constitution of structure and agency. The high-profile mayors of which Barber (2013) writes – high-profile not least in part because of the cities over which they govern – are understood to have made a difference. *If Mayors Ruled the World* is full of such examples. This an important component of the evidence base upon which Barber (2013) makes his case. It could be argued, however, that much of this making of a difference by each mayor is not done under terms of his or her making. Rather, a set of mutually constituting macro supply and demand contexts frame the potential and realized movement of policies, empowering certain 'idea brokers' (Smith, 1991) or mediators (Osborne, 2004) and disempowering others. Some, more than others, are likely to have their ideas and policies made mobile and they are, in turn, most likely to stamp their authority on the bringing forth of urban policy. The policymaking landscape is an uneven one, produced through past, and contributing to future, unevenness. That is not to say that urban policymaking is always a closed shop. Those who draw their power from a strong resource base do not always get their own way. Oppositional or subaltern groups as well as governments with ideologies that differ from the norm are able to inhabit and utilize the same global circuits of policy knowledge to develop alternative assemblages of policy and power, even if in most examples they do so against the odds. Acknowledging the importance of the context in which the different city mayors operate and upon which they draw their power and resources, and highlighting the role of other sorts of actors in the performance of politics, would advance the position outlined by Barber (2013).

## Conclusion

It is easy to understand the popular success of Benjamin Barber's (2013) *If Mayors Ruled the World*: it appeared amid a mood of unease among the general populace. The restlessness of the geopolitical 'order' – if that is the right word – that emerged across the globe after the Second World War shows no sign of abating. The decision of the UK to exit the European Union and the election of Donald Trump as the 45th US President are the latest examples of just such destabilization. No less important, the political uncertainties in Eastern Europe and in the Middle East continue, their effects being felt in cities around the world. Contestation and tension over the bordering and re-bordering of territories evidently continues to characterize contemporary supranational, national and urban politics. This instability – in which the almost taken-for-granted centrality of the nation state is no longer what it once was (without any post-national settlement on the horizon) is one of the currents behind the projecting forward of a world in which cities become more central to the global order. Indeed, the sort of flux we have been witnessing and that lies behind some of the calls and claims about a new urban political order might be the new normal.

Thinking through contemporary ways in which cities relate to one another – through either the formal networks of which Barber (2013) writes or the less formal circuits and webs in which cities are located – raises some important questions about urban policy (Ward, 2006; Robinson, 2015). In many instances this involves the bringing together of the global and the urban. This is about the complex realities of urban governance and urban politics. It is no longer either possible or desirable to imagine the world through lenses that implicitly or explicitly place things in scalar hierarchies. The focus on relations between cities and the work involved in making urban policy acknowledges the extent to which urban politics, by its very nature, incorporates actors and interests that are often understood to be located elsewhere. Of course, this 'elsewhere' is quite literally a point on the map.

This leaves us with four fundamental insights as we think about the city in the world – the city as a global political actor. First, that urban policymaking and the politics around it cannot, and should not, be taken for granted. Or, put differently, it is difficult to define them as clearly identifiable academic or scientific objects of study. There is no formula or template that will render them fixed in place and explain their operation as flowing from some necessary set of relations. Rather, urban policymaking is an active and continuing process of production and reproduction. Cities are made and remade, generated if you like, assembled in particular ways at particular times in specific locations. This involves inter-urban relationships, such as those inter-city networks about which Barber (2013) and others write, through which occur processes of comparing and learning.

The second insight is that urban politics cannot just be approached through the lens of 'place', notwithstanding its indisputable importance; urban politics is also global in a much more nuanced sense, because of the extent to which it is based on systems of comparing, borrowing, exchanging, imitating, learning, reinterpreting and translating. Place is thus both territorial and relational. This is revealed in the work that gets done through the many infrastructures of inter-city networks that constitute the building blocks of Barber's (2013) *If Mayors Ruled the World*. Barber refers to 'best-practice sharing' (in Derbyshire, 2013, n.p.), though a more nuanced understanding of the phenomenon, as advocated in this chapter, is one way of generating a more refined appreciation of the potential of cities to perform as global political actors.

The third insight is that creation of urban policy draws upon a wide range of actors of differing geographical reach, from the local to the global. The term 'transfer agents' entails more than those involved in city government and more than those who have historically been understood to be involved in politics, although these 'official' policymakers remain important. The term encompasses others whose work is less visible: actors who may be unelected and less accountable to local communities, but whose work of interpreting, mediating and translating can shape which policies move – and to where. The agents operate in a range of economic and political geographical contexts, which shape how and to what end they do their work. From community activists to private sector consultants, from political think tanks to the representatives of professional groups, such as architects, planners and realtors, a range of interests and their representatives are involved in global urban policymaking.

The fourth and final insight offered by this chapter relates to the dangers of over-emphasizing the agency-like qualities of cities as political actors in the context of the wider transformations in economic, political and social systems. Here it is pertinent to return to some formative work of the late 1980s. Writing with a number of colleagues, Kevin Cox tackled the notion of spatial fetishism in relation to localities as agents. This has parallels with some of what has been written about cities having political agency in the context of contemporary international relations. His way out of the apparent intellectual impasse was to suggest the following:

Locality as agent defined as the sum of individual local actors is certainly problematic. But locality as agent defined as a local alliance attempting to create and realize new powers to intervene in processes of geographical restructuring has now become a vital element of those very restructuring processes.

(Cox and Mair, 1991, p. 406)

We might understand the forms of agency being referenced in the current academic/scientific work on cities and international relations in this manner, with mayors joining other economic and political actors – both local and non-local – in alliances that are both territorial and relational.

## Notes

1 I would like to acknowledge the work with Eugene McCann (Simon Fraser University) as well as the comments of the reviewers, which have informed the writing of this chapter. The usual disclaimers apply.
2 Barber (2013) uses the notion of 'formal' to refer to official, governmental relations, while using 'informal' to denote relationships beyond the governmental or 'official' sphere. My argument in this chapter is that there is a need to focus on the range of relations involving cities in one way or another, many of which cannot be called 'official' even though bits of government may be involved in them. These exhibit different combinations of 'formality' and 'informality' and in doing so, challenge both the assumed binaries at work in the use of these terms and the assumptions that the latter automatically comes 'from above' and the former 'from below'. The use of these terms by Barber (2013) differs from how the formal/informal distinction is used more generally in urban studies, including in a number of chapters in this collection.

## References

Acuto, M., 2013. City leadership in global governance. *Global Governance*, 19(3), pp. 481–498.

Barber, B., 2013. *If mayors ruled the world: dysfunctional nations, rising cities.* New Haven, CT: Yale University Press.

Barber, B., 2017. *Cool cities: urban sovereignty and the fix for global warming.* New Haven, CT: Yale University Press.

Beal, V. and Pinson, G., 2014. When mayors go global: international strategies, urban governance and leadership. *International Journal of Urban and Regional Research*, 38(1), pp. 302–317.

Brenner, N. and Schmid, C., 2014. The 'urban age' in question. *International Journal of Urban and Regional Research*, 38(4), pp. 731–755.

Cox, K.R. and Mair, A., 1989. Levels of abstraction in locality studies. *Antipode*, 21(2), pp. 121–132.

Cox, K.R. and Mair, A., 1991. From localized social structures to localities as agents. *Environment and Planning A*, 23(1), pp. 197–213.

Curtis, S. ed., 2014. *The power of cities in international relations.* London: Routledge.

Derbyshire, J., 2013. If mayors ruled the world: a conversation with Benjamin Barber. *Prospect*, [blog], 22 November. Available at: www.prospectmagazine.co.uk/blogs/ jonathan-derbyshire/if-mayors-ruled-the-world-a-conversation-with-benjamin-barber [Accessed 1 May 2016].

Duncan, S. and Savage, M., 1989. Space, scale and locality. *Antipode*, 21(3), pp. 179–206.

Evans, M. ed., 2004. *Policy transfer in global perspective.* Aldershot: Ashgate.

Evans, M. ed., 2010. *New directions in the study of policy transfer.* London: Routledge.

Ferman, B., 1985. *Governing the ungovernable city: political skill, leadership and the modern mayor.* Philadelphia, PA: Temple University Press.

Florida. R., 2004. *Cities and the creative class.* London: Routledge.

Florida, R., 2012. What if mayors ruled the world? *CityLab*, [online], 13 June. Available at: www.citylab.com/politics/2012/06/what-if-mayors-ruled-world/1505/ [Accessed 1 May 2016].

Glaeser, E., 2011. *Triumph of the city.* Basingstoke: Macmillan.

Goldberg, M., 2014. The rise of the progressive city. *The Nation*, [online], 2 April. Available at: www.thenation.com/article/rise-progressive-city/ [Accessed 1 May 2016].

Gonzalez, S., 2011. Bilbao and Barcelona 'in motion'. How urban regeneration 'models' travel and mutate in the global flows of policy tourism. *Urban Studies*, 48(7), pp. 1397–1418.

Hambleton, R. and Sweeting, D., 2004. US-style leadership for English local government. *Public Administration Review*, 64(4), pp. 474–488.

Hambleton, R., Savitch, H.V. and Stewart, M. eds., 2002. *Globalism and local democracy: challenge and change in Europe and North America.* London: Palgrave.

Hannam, K., Sheller, M. and Urry, J., 2006. Editorial: mobilities, immobilities and moorings. *Mobilities*, 1(1), pp. 1–22.

Jacobs, J., 2012. Urban geographies I: still thinking cities relationally. *Progress in Human Geography*, 36(3), pp. 412–422.

Jayne, M., 2012. Mayors and urban governance: discursive power, identity and local politics. *Social and Cultural Geography*, 13(1), pp. 29–47.

Jenkins, S., 2013. Comment: sorry, archbishop – but London is where the action is. *Evening Standard*, [online], 29 October. Available at: www.standard.co.uk/comment/simon-jenkins-sorry-archbishop-but-london-is-where-the-action-is-8910719.html [Accessed 1 May 2016].

Larner, W., 2003. Guest editorial: neoliberalism? *Environment and Planning D: Society and Space*, 21(5), pp. 508–512.

Lechtenberg, S., 2014. *If mayors ruled the world*, A New Freakonomics Radio Podcast. [podcast] 10 April. Available at: http://freakonomics.com/podcast/if-mayors-ruled-the-world-a-new-freakonomics-radio-podcast/> [Accessed 1 May 2016].

Ljungkvist, K., 2016. *The global city 2.0. From strategic site to global actor.* Routledge: London.

Massey, D., 1993. Power-geometry and a progressive sense of place. In: J. Bird, B. Curtis, T. Putman, G. Robertson and L. Tickner, eds. *Mapping the futures: local cultures, global change.* New York: Routledge. pp. 59–69.

McCann, E., 2008. Expertise, truth, and urban policy mobilities: global circuits of knowledge in the development of Vancouver, Canada's 'four pillar' drug strategy. *Environment and Planning A*, 40(4), pp. 885–904.

McCann, E., 2011. Urban policy mobilities and global circuits of knowledge: toward a research agenda. *Annals of the Association of American Geographers*, 101(1), pp. 107–130.

McCann, E. and Ward, K. eds., 2011. *Mobile urbanism. Cities and policymaking in a global age.* Minneapolis: Minnesota University Press.

Montgomery, C., 2015. *Happy city: transforming our lives through design.* London: Penguin.

Olds, K., 2001. *Globalization and urban change: capital, culture, and pacific rim mega-projects.* Oxford: Oxford University Press.

Osborne, T., 2004. On mediators: intellectuals and the ideas trade in the knowledge society. *Economy and Society*, 3(4), pp. 430–447.

Peck, J., 2003. Geography and public policy: mapping the penal state. *Progress in Human Geography*, 27(2), pp. 222–232.

Peck, J., 2011. Geographies of policy: from transfer-diffusion to mobility-mutation. *Progress in Human Geography*, 35(6), pp. 773–797.

Peck, J. and Tickell, A., 2002. Neoliberalizing space. *Antipode*, 34(3), pp. 380–404.

Robinson, J., 2015. 'Arriving at' urban policies: the topological spaces of urban policy mobility. *International Journal of Urban and Regional Research*, 39(4), pp. 831–834.

Rogers, B., 2013. 'If mayors ruled the world: dysfunctional nations, rising cities' review. *Financial Times*, [online], 8 November. Available at: www.ft.com/content/f586be20-4570-11e3-b98b-00144feabdc0> [Accessed 1 May 2016].

Smith, J.A., 1991. *The idea brokers: think tanks and the rise of the new policy elite.* New York: Free Press.

Smith, N., 1996. *The new urban frontier: gentrification and the revanchist city.* London: Routledge.

Stone, D., 2004. Transfer agents and global networks in the 'transnationalisation' of policy. *Journal of European Public Policy*, 11(3), pp. 545–566.

Svara, J., 1990. *Official leadership in the city: patterns of conflict and cooperation.* Oxford: Oxford University Press.

Townsend, A., 2014. *Smart cities: big data, civic hackers and the quest for a new utopia.* London: W W Norton & Company.

Wacquant, L., 1999. How penal common sense comes to Europeans: notes on the transatlantic diffusion of the neoliberal doxa. *European Societies*, 1(3), pp. 319–352.

Ward, K., 2006. 'Policies in motion', urban management and state restructuring: the trans-local expansion of business improvement districts. *International Journal of Urban and Regional Research*, 30(1), pp. 54–75.

Ward, K., 2011. Entrepreneurial urbanism, policy tourism, and the making mobile of policies. In: G. Bridge and S. Watson, eds. *The new Blackwell companion to the city.* Chichester: Wiley-Blackwell. pp. 726–737.

# 6 The politics of the (global) urban

## City strategies as repeated instances[1]

*Jennifer Robinson*

## Introduction

When we say that **cities have become global actors**, this is a shorthand (cities don't act as such) which indexes a desire to understand the ways in which urban actors and institutions have agency to shape globalizing processes of economic growth, politics, cultural form and environmental change. Urban political analysis has sought to explain the territorial basis for global competition amongst cities (Cox and Mair, 1988; Leitner, 1990), in which cities compete to attract circulating capital investment, to project a potent image on a global stage as destinations for tourism, or, more recently, to be recognized as successful leaders in urban management and policy innovation (González, 2011). There has also been significant concern with the ways in which circulating investments (Harvey, 2013) and policies (Peck and Theodore, 2015) both shape and are shaped by the agency of urban actors. Thus, as cities experiment with neo-liberalization, for example, they are influenced by wider global norms of urban policy, but in turn contribute, according to Peck, Theodore and Brenner (2009) to the 'syndrome' of neo-liberalization.

Both of these analytical series identify the grounds for global urban agency in the local interests and territorially circumscribed dependencies of the urban actors. In this context, the configuration of different actors into stable, 'urban regimes' (Logan and Molotch, 1987; Stone, 1989), or local regulatory contexts (Peck, Theodore and Brenner, 2009) emerges from deep historical path dependencies in specific places and in relation to specific territorialized institutions. However, even as these analyses explore the interface of localized urban politics with the transnationalization of urban processes, neither of these approaches has revisited the (territorial) grounds of urban agency in the context of highly transnationalized, even substantially exteriorized, processes of urban governance. This chapter argues that new understandings of the grounds or territorializations of 'urban agency' need to be developed, not least in response to the changing empirical realities of urban politics and globalization.

Significantly, a sense of enhanced global urban agency has emerged on the stage of international development policy and politics. The scope for agency on the part of institutions representing local government, mayors and urban policymakers (Acuto, 2013) has expanded considerably with the upsurge in international interest in urban development policy, such as the UN agreement to support an Urban Sustainable Development Goal (Acuto and Parnell, 2016; Parnell, 2016), or in the role of cities in addressing climate change (Bulkeley, 2010). Furthermore, the financialization of urban development process has led to a strong transnationalization of the actors involved in urban development in many contexts (Halbert and Rouanet, 2014; Weber, 2015; Shatkin, 2017). Here, though, the rescaling of 'urban' agency at a global scale (Acuto, 2013), and the transnationalization of urban actors (Allen and Cochrane, 2014) has arguably contributed to enhanced territorializations of agency at the level of the project (Pinson, 2009), or in emergent 'transcalar territorial networks' (Halbert and Rouanet, 2014).

Thus the spatiality of global urban political agency is being reconfigured beyond the binary of territorialized local government, locally embedded communities and firms anchored in structured coherence of localities versus circulating or 'footloose' capital (Cox and Mair, 1988; Harvey, 1989). I will suggest that '*terrain*', or territory ('locality') remains important in shaping the grounds of the global agency of the urban, but I will show how this needs to be conceptualized as emergent in the transnational and trans-scalar practices of urban governance (Allen and Cochrane, 2007), rather than imagined as an effect of the *a priori* territorialization of interests at a local 'scale' of urban politics.

A second element contributing to the need to reconceptualize global urban agency emerges from the territorial reconfiguration of urbanization processes and of urban theory. The dispersal of urbanization across sprawling city-regions/urban territories has ignited a strong theoretical debate on the complex spatiality and nature of 'the urban' (Scott, 2001; Merrifield, 2013; Brenner and Schmid, 2015). New grounds need to be established for speaking about the global agency of 'the urban', or 'cities', then, partly because the very content and spatial form of urbanization has been brought in question (Brenner and Schmid, 2015). But it is also essential to revisit analyses of global urban agency in the wake of an extensive postcolonial critique in urban studies, which has generated a heightened awareness of the great diversity of urban experiences across the globe which need to be drawn into theoretical conversation (Robinson, 2006; Roy, 2009; Parnell and Oldfield, 2014). The nature of urban agency itself needs to be interrogated across the great variety of urban political formations, stretching across contexts with different kinds of actors, different interest configurations, and different formats of the operation of power. In this chapter, I draw on the potential of a comparative imagination to inform efforts to theorize urban political agency across a wider range of urban experiences. A reformatted comparative practice (Robinson, 2011a, 2016a, 2016b) enables analysis to bring a diversity of urban experiences into conversation, thereby stretching existing concepts and inventing new repertoires for understanding global urban agency.

I use three cases of city strategy in very different urban contexts (London, Johannesburg and Lilongwe) to stretch analyses of the politics of the global urban; this will help us to draw insights from the range of political dynamics at work across different urban contexts as urban actors seek to launch themselves on the global stage, or to localize 'global' processes. Rather than being led analytically by the experiences of Euro-American cities, which have dominated theoretical reflections on how cities act globally, our insights on these issues need to also be inspired by urban processes from different regions, shaped by diverse histories of globalization, highly disparate levels of poverty and wealth, and different regulatory and institutional forms.

In this light, this chapter addresses two questions in relation to re-theorizing global urban agency: what are the grounds, or territorializations, of global 'urban' agency? And, what are the diverse forms of global political agency which might be identified across different urban situations? Thinking across the variety of global urban politics asks that we extend the theoretical registers which have helped us understand these processes. Trans-scalar and emergent territorializations of urban agency, as well as informal, fleeting or spectral forms of engagement with globalizing urban processes come into view. Starting analyses of urban political agency from 'anywhere' and including a diverse range of cities as starting points – perhaps as diverse as possible, which could be a methodological guidance for starting to theorize a problematic – enables an expansion of theoretical registers, and supports the potential to launch insights from different urban situations.

## From local economic development politics to city strategies: genetic grounds for comparison

I have found it intriguing that many of the analytical concerns of the earlier local economic development literature come into focus, but in new ways, when considering a widely circulating genre of urban practice, long-term city visions and plans, or city strategies (McCann, 2001; Healey, 2007; Pieterse, 2008; Robinson, 2011b). These have been prepared in most cities around the world, and reflect the strongly interconnected world of urban policy production. Their serial production in very different cities and their shared circuits of influence suggest an opportunity to analytically frame an interrogation of the politics of urban development starting from any city where city strategies have been prepared; thus, their ubiquity can provoke thinking across a diversity of urban outcomes. They allow insights into issues which will be familiar to researchers of the politics of urban development: how different kinds of actors with differential influence and capacity can come together across the city to define or contest paths for development, perhaps as an 'urban regime' but perhaps as an array of actors and agency more fragmented and spectral than that; how these pathways of policy formulation can shape and influence potential for different economic or collective activities; how more or less stable and shared visions of a city's future can be articulated in relation to other cities and wider global processes (Healey, 1998).

Taking such a ubiquitous governmental technology as our starting point means that while existing analyses focus largely on the US and Europe (although see Harding, 1994), we can internationalize these concerns to attend to a much wider range of actors, forms of constitution of urban agency, as well as different processes of globalization. Starting from cities like Lilongwe, Malawi, difficult questions are posed to the existing literature: How are urban political interests articulated in conditions of the strong transnationalization, even exteriorization of economy, state and civil society, the dominant role of international agents (donors, advisors, consultants, investors, foreign governments), the significance of informalized political processes and challenges of extreme inequality? In fact, I will suggest that these issues are also important to consider with respect to both London (at the top of the putative international urban hierarchy) and Johannesburg (a regional centre in a middle-income country). Bringing these cities together for analytical reflection challenges us to explore whether experiences from poorly resourced contexts can be effectively drawn into wider theoretical conversations on urban governance. This would demonstrate the potential to significantly extend the theoretical range of urban political analysis through comparative research with a wider global scope. The two questions which this chapter (and this book) addresses, could therefore be extended in ways that could be helpful to thinking about all cities.

Firstly, on what is the capacity for urban actors to engage in global processes grounded? For Cox and Mair (1988), whose analysis was concerned with US cities, local agency in urban development politics rests in the territorialized interests of the state, business and community in a locality – states have interests in generating income and securing electoral victory; residents are concerned with jobs and meeting their needs through local facilities and services; locally dependent businesses are invested in the 'structured coherence' (Harvey, 1989) of local environments that support their activities; and incoming businesses are interested in reducing the costs of doing business. The potential for an alliance around the agenda of economic growth is secured amongst these actors, with different outcomes in different contexts, dependent on the nature of formal and informal arrangements of political relationships (Logan and Molotch, 1987; Vicari and Molotch, 1990). Savitch and Kantor (2004) expand on this analysis, with eight case study cities across the US and the UK, attending to how local actors position themselves in relation to the circuits of governmental resources and the role of the strategic positioning of the local economy in global economic processes more generally. A quite rounded theorization emerges in their analyses, informed also by the variety of institutional and political dynamics that their cases bring to light, including, for example, the relative democratic responsiveness of governments to the electorate.

In contrast, and despite a sometimes celebratory approach to this, the increasingly successful representation of 'cities' and local government at the highest levels of international diplomacy and policymaking highlights the importance of taking a networked view of the capacity of urban actors. The transnationalization of urban agency at the scale of international politics has been effectively achieved in both climate change and development policy (Bulkeley, 2010; Acuto, 2013; Parnell, 2016), speaking to a networked capacity across city leaders in emergent transnational organizations. Territorially based urban actors engage in policy development and sharing, in international diplomacy and lobbying, and configure their agency on the global/international stage in strongly networked ways. However, in this context, there has been little work to interrogate exactly how and why specific territorially located urban actors participate in such networked global politics. While in this emergent networked capacity, 'cities' in general might become 'obligatory passage points for global environmental governance' (Acuto, 2013, p. 15), Bulkeley (2010) explores more directly how transnational networks also provide a platform for enhanced visibility for both individual urban actors and their cities, indicating that the territoriality (local and national) of political interests in this phase of the city as networked global actor remain vital (Söderström, 2014). Policymaking processes within these networks can 'enhance international legitimacy' and 'improve policymaking independence' for individual cities in relation to national level government (Acuto, 2013, p. 16). But some lack capacity to influence such networks, or face significant challenges in implementing networked policy innovations for lack of resources and capacity or poor fit with their circumstances.

Thus, a territorial perspective remains important for understanding global urban agency. It brings into view the multiplicity of networking opportunities and diversity of transnational connections which cross-cut any given urban territory. Thinking across and with this dense multiplicity of connections is in fact the task of a city strategy, which emerges out of a vast array of policy influences and competing agendas to frame a city-wide agenda (Robinson, 2011b). Recent studies of financialized urban development projects have documented the emergent territorializations of transnational and transcalar networks coalescing from the particular associations amongst actors involved in a specific project (see Halbert and Rouanet, 2014). The practice of preparing a long-term vision or strategic plan convenes an imaginative and practical territorialization of politics concerning a particular urban context. But this is formed in the midst of transnational agents and ideas, including sometimes strongly exteriorized actors and concerns. Here we can learn much from methodological innovations of policy mobility research (for example, Ward, 2006; McCann and Ward, 2010; McCann, 2011; Roy and Ong, 2011; Robinson, 2011b; Peck and Theodore, 2012; Söderström, 2014; Wood, 2014). Thus, while the vigorous policy circulation associated with preparing city strategies is noteworthy (Healey, 2007; Robinson, 2011b), each strategic vision or plan is also distinctive, representing both a close engagement with a specific context and a certain territorialization of urban politics (see Robinson, 2011b for a wider discussion of city strategies).

For some time I have been following three examples of city strategy, from cities not conventionally considered together: Johannesburg, London and Lilongwe. Aside from my opportunistic interest in each case, I justify bringing them into comparative conversation on the grounds of their common genesis in interconnected policy circuits: they emerge from the shared, if differentiated, global circulation of a specific governmental technology (city strategy), and shared, if differentiated, circuits of globalizing urban policies. There seem to be good reasons for exploring these 'repeated instances' (Jacobs, 2012; Robinson, 2016a, 2016b). To be clear, the 'cases' in such an analysis are not cities as such, but the 'repeated instances' of long-term city strategies which form part of an interconnected field of urban policy, in which city strategies, as well as the policies and practices involved in their production, are circulating.

Reformatting comparative tactics for 21st century cities benefits from working with the spatialities of urbanization itself (Robinson, 2011a). Thus, comparative tactics can be inspired by the ways in which the urban comes to be formed (and known) as it is assembled transnationally, across a multiplicity of specific but strongly interrelated outcomes. This supports the possibility for proliferating insights across a world of cities. I have therefore suggested 'genetic' grounds for comparison, based on the extraordinary array of interconnected processes which lead to often repeated, but distinctive, urban outcomes around the globe (Robinson, 2016a, 2016b). Here, then, we can start our thinking with the many repeated instances which make up the urban, such as residential high rises (Jacobs, 2006), mixed use 'new urbanism' developments (Moore, 2013), low-income financialized housing (Caldeira, 2016) – or indeed, city strategies. I also propose the value of 'generative' starting points for thinking the urban, which build on conceptual curiosity about shared and differentiated urban experiences. Scholars, practitioners and observers build insights across different experiences (Myers, 2014; Söderström, 2014), based on shared features that enable them to think with the variety of urban experiences, generating and extending concepts, perhaps coming to their useful limits, and needing to invent new concepts. In this way, a reformatted comparativism can contribute to developing new vocabularies and conceptualizations of the urban (Robinson, 2011a, 2016a; Brenner and Schmid, 2015; Schmid et al., 2018), including understandings of the politics of the global urban.

## Thinking across three cases of city strategies

The cases of city strategies that I have assembled here have the potential to stretch understandings of urban development politics by drawing insights from three very different urban contexts. The London Plan, a strategic 'spatial' plan for the UK capital city, is mandated in legislation. It draws together statutory thematic plans of the Mayor for the Greater London Authority, established in 2000. The Plan is reviewed and updated periodically, and usually redrafted significantly with the election of a new mayor (4-year terms of office). Its function is to set high-level planning and wider policy guidance for the administratively fragmented metropolitan area. In Johannesburg, South Africa, strategic city-wide visioning was initiated in the late 1990s alongside preparations for the creation in 2000 of unitary local government in the municipality after a period of institutional transformation with the end of apartheid. The review and revision of the plan has been episodic, linked to moments of institutional and political change, and while initially ad hoc, is now an expectation of national legislation on local government, and nominally linked to the annual and 5-year planning and electoral cycles of the city. Institutional tensions exist between operation and strategic control, and between spatial and strategic planning, and the implementation of the high-level city strategy is not very streamlined. In Lilongwe, Malawi, an ad hoc city strategy was formulated in 2009 through close cooperation with the City of Johannesburg, partly supported by the United Cities and Local Government, and partly by the Cities Alliance, but substantially self-financed by Johannesburg and Lilongwe (UCLG, 2013). The cooperation came to an end with change in personnel in Lilongwe, and while there have been some positive outcomes in terms of further funding and investment in housing, implementation was patchy.

My evidence for each of these three cases is rather different, driven partly by my lines of access to learning about the strategy in each context but, as the following section will clarify, also driven by the distinctive issues that each case raises. I have followed the Johannesburg case closely since 1999, and have conducted numerous and repeated interviews with key city government personnel and other actors involved in the city strategy process. My comments here are drawn from these interviews, as well as from the attendance of public consultation meetings, and the analysis of background documents as well as draft and final policies. Here the strategy represents the negotiation of a post-apartheid vision, balancing priorities of growth and redistribution. In London I have been fortunate to cooperate with the community-based network, Just Space, learning through their collective organization and engagement with the London Plan process, primarily through the formal consultation process and public hearings, but also through informal interactions with officials (Brown, Edwards and Lee, 2014). Here the key issue which emerges concerns the potential for a more inclusive policy agenda to take hold in a large city with very powerful property development and business groups. My knowledge of the Lilongwe case was gained from discussions with several Johannesburg officials who were involved in a city to city exchange process, as well as from the analysis of detailed background documents, correspondence and evaluation materials.

Here the question that arises concerns the implications of external actors playing a strong role in policy and urban development.

As a comparative practice, I benefit here from the observation that each 'repeated instance', while interconnected with the others through the practice of preparing city strategies, is a 'singularity' and offers a distinctive starting point for reflection; each experience of city strategy prompts analytical reflection in its own right. There is much more to be done with these three cases as well as with the potential of a genetic comparative method working with the specific connections between cases and the wider relational contexts which they bring into view (see Deville, Guggenheim and Hrdlicková, 2016, for some wider discussion; Hart, 2016; Robinson, forthcoming). Here, though, I use this particular tactic: starting with each (repeated) instance, grounded in an interconnected practice, I allow them to raise questions of each other in order to 'thicken' the interpretation of each of the singularities in turn (Akrich and Rabeharisoa, 2016). I start below with London, as it proposes some perhaps surprising answers to the first of the two questions I am addressing in this chapter: what are the grounds of global urban agency? Johannesburg signposts well some helpful understandings of the second issue, what are the forms of global political agency? And the case of Lilongwe importantly stretches understandings of both of these issues. Each case offers a different analytical starting point, and in a specifically post-colonial move, in the final section below I bring the experiences of both Lilongwe and Johannesburg back to review the understandings of the London case.

### London: trans-scalar territories of global urban agency

The London Plan itself is a circulating practice, a form of strategic planning drawn on and promoted by the UK national government. The Department of Communities and Local Government under New Labour had been influenced by US-centric practices of strategic city-wide visioning, including those of Monitor's Michael Porter, who was also influential in shaping the early analyses and processes in Johannesburg (Institute for a Competitive Inner City, 2003; Interview, former Johannesburg city official A, 2009). The use of strategic planning to cohere the work programme of the newly-established GLA also reflected the more general global and European circulation of a strategic approach to urban planning as good practice, also promoted by New Labour's DCLG (Raco, Parker and Doak, 2006; Tewdwr-Jones, Gallent and Morphet, 2010). But it also represented a long-standing strategic policy function associated with the authorities responsible for London, which shifted over time from the GLC (abolished in 1986) to an amalgam of local authorities' representatives (Local Planning Advisory Committee), under the auspices of central government's Office for London until the GLA was re-established in 2000 (Bowie, 2010).

Does London's plan itself act on this global policy stage? The London Plan office does attract international visitors concerned with observing their practices, and officers have visited other contexts to present their approach (Interview, 2011, GLA Planning Officer). Unlike Johannesburg, where the city strategy has been self-consciously positioned in relation to wider circulating ideas about strategic urban planning, the London Plan itself has had a more confined presence as a local working document and is just as strongly embedded in local planning ideas and inherited practice as in global urban policy. This despite the myriad learning practices of officers and elected representatives (as well as business and community actors) involved in preparing the London Plan who are inserted into global policy circuits, which they draw from as well as contribute to, benefitting from their relatively privileged location on conference, academic and professional circuits looped through London.

The assessment of UK global urban agency more generally has been rather muted, since for many decades local government has been strongly directed and funded by the central state, making it one of the most centrist systems of urban governance, and often a poor fit for analyses of local agency such as regime theory (Harding, 1994; Ward, 1996). Local states did not raise money through attracting business investment, so the link to promoting employment was weak (business rates were redistributed nationally through an agreed formula). Local authorities had little discretion on most of their spending, despite historically playing a strong role in delivering major welfare state services, such as education and housing. Applying urban regime theory to the UK case has been notoriously challenging, with only a few cases standing out for their more entrepreneurial approach to local development, e.g. Manchester (Harding et al., 2004).

However, this has changed significantly over time. Central government grants have been reduced, and instead, local business rates are increasingly allocated directly to local authorities, albeit with some fiscal cross-subsidy amongst authorities over time (DCLG, 2017). Thus, local authorities are increasingly dependent on urban development to generate their core budget: the generation of income for key welfare state functions has been downscaled to rely on the extraction of value from the urban environment. Reinforcing this, but in a longer-term trend, opportunities for property development provide one of the few means for discretionary spending on infrastructure, services and housing via localized developer taxes and charges. Each of the 33 sub-metropolitan local governments have strong incentives to seek close alliances with property developers. Alongside these fragmenting trends in urban development, the creation of metropolitan mayors, the London Mayor being the first, has initiated new spheres of governance at the metropolitan scale in realms of policy, strategic vision, international engagements, coordination with lower level authorities, and management of an expanding range of city-wide services (such as transport, police, fire, and economy). However, the mayor needs to operate in an urban environment characterized by significant territorial fragmentation and institutional complexity (Travers, 2004; Gordon and Travers, 2010).

In London, agency in relation to globalized processes must therefore emerge from significantly dispersed agency. The leading analyst of London politics, Tony Travers (2004), coins this an 'ungovernable city', with 33 sub-metropolitan councils, capped by a relatively weak Greater London Authority. The latter is effectively an agency of metropolitan strategic guidance and some operational responsibilities, including for transport, police, fire and broader economic development, which is effected partly through a business-led agency, a Local Enterprise Partnership. This is not to suggest that London lacks significant scope for action, these responsibilities have expanded significantly since the GLA was established. But in this city, urban political agency is notoriously complex, involving many governmental and quasi-governmental institutions across different levels of authority, with historically strong central state direction of policy formulation and service delivery, adding to the mix numerous central agencies operating locally – most notably, health care. Furthermore, many core urban services are provided by large private sector actors, making long-term planning for infrastructure renewal, for example, highly challenging (GLA, 2015b), combined with the need to navigate shared concerns with the many authorities neighbouring the GLA boundaries (GLA, 2017). Agency is dispersed across institutions, and in such a situation of messy governance, it relies strongly on interpersonal negotiations and alliances. Building a growth coalition for the city has taken adept political agency from mayors, lobby groups and central state officials (Gordon, 2004; Thornley et al., 2005; Gordon and Travers, 2010); it has also, however, relied upon informal networks cohering around specific urban development projects or consolidating through repeated interactions over numerous projects (e.g. associations that are forged at presentations, conferences and developer milieus)[2]. In its institutional dispersal, then, the 'city' as a global actor in the case of London is very much a multiplicity (Allen and Cochrane, 2007). The formulation of the London Plan, as a city strategy, establishes a vision and direction of travel across London's institutional complexity, with the GLA needing to navigate the multiplicity of actors relevant to the formulation of policy.

How then does the relative coherence of a city strategy emerge from highly 'distributed' agency? Territorial coherence and temporal consistency in the discourses and practices underpinning London's strategic visioning have been significant and I would suggest the reasons for this can be properly located in three sets of processes. Firstly, the role of policy discourses and transnational regulations (e.g. the adoption of European strategic spatial planning practices – see Nadin, 2007; Tewdwr-Jones, Gallent and Morphet, 2010). Secondly, the strongly path-dependent challenges, imaginations and practices which frame actions and interventions to understand the city on its own terms. And, thirdly, the imagination at work to competitively mobilize 'London' in global circuits of investment.

Coherence and consistency in the London Plan over time emerges to some extent from the technical nature of the discourses which subtend this particular city strategy, and which establish or constrain the terms of engagement amongst different actors shaping the London Plan (Rydin et al., 2004). Analytically, then, the case of London requires us to attend to the agency which emerges from the territorializations of 'discourse'. The London Plan is quite a technical document, unlike the other two plans considered below which are looser and more strategic. The London Plan sets both direction and specific policy in local strategic planning issues, and so must provide conformity with wider legislative and policy frameworks and expect conformity from subsidiary local planning authorities. It also incorporates high-level consequences of national and European or international policy in different areas of the GLA's responsibilities (health, environment, transport, housing, economy). In each of these policy areas, the London Plan is part of the banal but thoroughgoing transnationalization of urban policy through learning networks, best practice and national-level circuits of information. And 'London' is in turn a participant in the invention of the broader regulatory context of the global urban through these processes.

The territorial grounds for agency are therefore emergent in a strongly transnationalized field: local actors are constituted in relation to a diversity of transnational alliances and networks. Thus 'local' actors range widely in their sources of policy and strategic inspiration, not least because many Londoners, including many small and medium business operators, bring experience from elsewhere to their political engagements (Allen and Cochrane, 2014; Roman-Velazquez, 2014; cf. Theodore, 2007). The mayors of London have been closely involved in the creation of a global political stage for cities; Ken Livingstone and deputy Mayor Nicky Gavron were influential in forming and sustaining international networking amongst powerful cities and other actors, such as in the C40 network (Gavron, 2007; Interview, Nicky Gavron, 2009). This provides a strong example of the transnational nature of 'local' policymaking: 'as a political officer from the GLA put it: "the [C40] meeting offers a sweeping window to survey the state of urban planning in our global competitors, and twin global action with global competitiveness"' (Acuto, 2013, p. 16). The Mayor of London also has access to readily available platforms at key events, such as the annual gathering of business leaders at Davos. And the prolific engagement between all the different local boroughs as well as the mayor with property developers introduces a profound and quotidian agency for the city in terms of shaping the global circulation of financial investment capital.

In the face of the institutional complexity which underpins the plan, perhaps the most surprising feature is the consistency in the precise form of strategic visions for London over time. Continuities can be identified in the consistency of top-level objectives, the stability over decades of the spatial form of the plan (central activity zone; smaller town centres; 'opportunity' areas for significant and concentrated development; inner city vs outer suburbs; convergence between east and west), and the continuity of staff working on London's strategic planning across these institutions (the lead planner until 2017 having worked on the strategic plan for London in different capacities since the mid-1980s). There has also been a very long time-horizon in which the dual imagination of London as a 'global City' and a site for sustainable urban development practices has been operative. Locally embedded policy-led research on London's economic future, as a world, or 'global', city, was initiated in the late 1980s. Initially this came as a response to a devastating decline in population and economy, and later as a basis for competitive positioning vis-à-vis other cities dependent on advanced services and finance sectors (Kennedy, 1991; Gordon, 1995). Sustainable urban development discourse began to gain purchase internationally and influence strategic thinking about London around the same time.

A form of territorialization of London's policy can therefore be seen to emerge from the broader competitive eye of policymakers, businesses and politicians: to make London the 'best big city' in the world according to former Mayor Boris Johnson (GLA, 2011); to sustain its competitive positioning in certain sectors, such as finance, services and cultural industries; to promote new areas of specialization, such as green or circular economies; to facilitate London's relatively rapid growth; to construct a built environment which advertises this global status (GLA, 2015a). Certainly, responding to the implications of such globalized economic growth places enormous pressure on wider policies on housing, local economies and high streets, for example, which ostensibly claim a continuing interest in supporting London's distinctive neighbourhoods and quality of life. Global ambitions are therefore distilled, giving shape to the overall framing of the mayor's strategic spatial plan as well as informing specific elements of it, such as creating planning exceptions to protect key central office spaces against more lucrative housing development, and to intensify development in so-called 'Opportunity Areas'.

The London Plan, then, demonstrates that the basis of global urban agency is achieved through and against significant institutional fragmentation, a multiplicity of transnational circuits of learning and practice and is profoundly trans-scalar in that it is mediated by national and (for the moment) supranational institutions while also demonstrating deep local influences and continuities.

### Johannesburg: launching and localizing developmental agendas

In contrast to the muted global agency of the London Plan as such, Johannesburg has claimed strong global effects for its early post-apartheid city strategies, which emerged at the same time as the Cities Alliance (CA) was founded – the CA being a key promoter of city strategies amongst poorer countries. While some Johannesburg officials involved at this time suggest they 'invented' the city strategy idea which became so important to the CA, at the very least Johannesburg's strategic planning formed a case study for the Cities Alliance (CA, 2006; Harrison, 2006; Robinson, 2011b), and informed their development of the idea of the city strategy as a framework for supporting a city-wide and inclusive approach to urban development planning in poorer cities.

From the vantage point of Johannesburg, we can trace some shared circuits of influence across the three cases presented here. The consultants who developed the city strategy concept for the CA drew on literatures and practices related to long-standing US practices of city-wide visioning (Harris, 2002; Cities Alliance, 2006), which also informed the Monitor Group and Michael E. Porter's (1990) Competitive Advantage of Nations (and cities) methodology. As we saw above, these were also relevant to the UK city strategy initiatives of the 1990s and 2000s. There are significant shared influences, then, amongst the strategic planning practices which shaped both the formulation of the Cities Alliance and Johannesburg engagements with stakeholder-led city visioning (Robinson, 2011b), and the national legislation on regional and spatial planning governing the London Plan (Allmendinger and Haughton, 2007). As a best practice case of city strategy preparation, UCLG subsequently targeted Johannesburg to support a city-to-city learning initiative across Southern Africa and became involved in a collaborative relationship with officials in Lilongwe (Interview, former city official A, 2009). Eager to promote itself in international circuits, an international strategy has seen Johannesburg circulate its own achievements, such as the city strategy, through selected international networks (C40, UCLG, Metropolis), where targeted marketing of the city strategy work was undertaken (Interview, city official JE, 2011; Peyroux, 2016). City to city learning also led to international study visits and exploring city twinning with cities in different parts of the world (Europe, Africa, South America), on different themes, including security, strategies, youth, environment (Interview 1 and 2, city official B, 2011 and 2012; Peyroux, 2016; this volume).

With few resources available for international interaction in a middle-income resource-stretched city with many urgent delivery priorities, international networking organizations have provided valued opportunities for strong global visibility for the city (Interview 3, city official B, 2013; Peyroux, this volume). Successive Mayors have been promoted as leading figures in international networks of cities, such as UCLG and Metropolis, with the former Mayor Masondo co-President of the UCLG, and former Mayor Parks Tau currently (2017) president of UCLG, even though he lost his post in the 2016 elections which ousted the ANC from power in Johannesburg City Council (Interview 3, city official B, 2013; Harrison, 2015). While learning and policy development have clearly been important ambitions of Johannesburg's international policies, a key official commented that, 'basically, your allies are very important as well in terms of your political and international profile, so it is building your brand to some extent as well' (Interview 1, city official B, 2011). The form of global urban agency morphs here into a form of marketing, seeking to raise visibility and influence to bring reputational enhancement and a larger role on the world stage of urban policy and local government networks.

Johannesburg has prepared a sequence of city strategies, with different emphases. The early 2000s saw a broad-ranging initiative, *iGoli*[3] 2010, which sought to set a strategic agenda alongside institutional restructuring and privatization or corporatization of various functions for the newly integrated City of Johannesburg. This initiative stalled with trade union and popular protests against privatization, leading to a much-reduced stakeholder process (Parnell and Robinson, 2006). An in-house consolidation of the private sector research undertaken for the iGoli process (by Monitor, the Harvard-based company) saw a strategy – Joburg 2030 – strongly focused on marshalling all council initiatives to promote economic growth as a priority. And after city officials stalled the implementation of this strategy, a more balanced coordination of a series of strategies covering growth, human development and environmental issues was brought together in the 2006 Growth and Development Strategy (Interview 1, former City Official A, 2009). While this gained some traction discursively in the city institutions, officials in different departments were skeptical as to its influence on implementation (Interview, former city official D, 2011). A revised version in 2011 was launched with considerable publicity and stakeholder 'participation', although this was rather late in the day to influence the actual document, which was already substantially drafted. Very little time was allocated to implement any changes, sometimes overnight, as the consultation reports were coming in (Interview 1, city official B, 2011). Implementation was again compromised as the influence of the City Strategy Unit responsible for driving the strategy was undermined when the Urban Development Planning department gained control of planning the capital budget, more fundamentally shaping city priorities (Interview, former city official A, 2009; Interview former city official D, 2011; Interview, City Official G, 2016).

As in London, international policy circuits are an everyday presence for Johannesburg's officials, through their own professional training, internet and institutional connections. Policy development draws selectively and intelligently on key ideas and practices prominent in different settings around the world (Robinson, 2013), or packaged for circulation (Wood, 2014). Ideas are available easily on the internet – arriving in Johannesburg as quickly as anywhere else – and trajectories of learning include maintaining close contact with local and international academics, officials who have themselves recently returned from studies abroad (Interview, former city official A, 2009), or commissioning desk top studies of key areas (Interview 1, former city official B, 2011). Face to face discussions with peers while on trips or in international policy settings can make a strong impression and provide leads for further research or implementation. Although ideas shared may be familiar or relatively slight, chance discussions can help to focus practices (Interview 2, former city official C, 2011). A former city official from the office responsible for preparing Johannesburg's Growth and Development Strategy noted, for example, that on a trip to New York City, where they were discussing 'PlanNYC', New York's long-term plan, they discussed their public consultation process. He

> took home [that] they just simplify the stuff, even this [GDS] document you have was deliberately simplified – and still people find it dense .... They were saying to us if you want to engage with stakeholders and citizens across the board you have to take stuff in a form they can engage... we hadn't finalized this consultation process, so we got back and that was at the back of my mind....
>
> (Interview 2, former city official C, 2011)

More generally, though, apparently international learning can be embedded in ongoing recursive interactions and policy development where the exact origins are unclear (Wood, 2014). Phil Harrison, himself a former city official as well as senior scholar notes, for example, that although South Africa had long experience in *in situ* upgrading of informal settlements (van Horen, 2000), and academic knowledge about the Brazilian experience was widespread, two study tours to Brazil and Peru (1999) and to Brazil in 2008 'led more immediately to the adoption of new policy' (2015, p. 216) in this area.

However, as in the London case, the folding of Johannesburg's city strategy and wider policy development through diverse transnational circuits (Robinson, 2011b, 2013) sits alongside the strong path dependence of the strategic visions. Here we are able to highlight perhaps even more strongly than in London the importance of historical trajectories in shaping the grounds for global agency. In Johannesburg, the idea and the capacity to frame a distinctive vision for the future of the city, and to articulate the terms of the city's insertion into globalizing processes, has been strongly informed by the national political project of democratization and redistribution after apartheid (Parnell and Robinson, 2006). This has been grounded on a strong, integrated local government across the metropolitan area of Johannesburg – although the wider city-region is fragmented into three similar sized municipalities. Headline commitments to redressing balances in service delivery, and to accommodating both the poor and the large number of migrants arriving in the city from across the region, country and continent, owe their formulation to this shared national political project. A key idea for the GDS 2006, 'the proactive absorption of the poor', was influenced by emerging practices in another South African city, Durban, where the idea that people should be able to live in the city without needing to incur costs had emerged (Interview 1, former city official A, 2009). A strong political commitment to this idea saw it adopted into the strategy (Interview, city official E, 2011). Equally, the desire to promote (shared) economic growth alongside other objectives derives from both national and local level understandings of Johannesburg's dominant place in the export economy of

South Africa, and the need to secure jobs and income for residents, as well as sustain national (income) and local (property) taxes.

The electoral system plays a strong regulatory role in shaping these strategic aims. Concerns about being re-elected were high on the ANC Mayors' minds over the years, and extensive survey analysis of residents' views informed election and policy planning (Parnell and Robinson, 2006; see for example, https://www.wits.ac.za/news/latest-news/general-news/2016/elections/quality-of-life-survey-for-gauteng/). These concerns were well-placed, as the 2016 elections saw the ANC lose control of Johannesburg to a relatively unlikely coalition of the liberal, largely white Democratic Alliance (which won the most votes) and the radical youth ANC breakaway group, the Economic Freedom Fighters (who provide the required voting majority in the council).

Formulating strategic policy, then, could not simply rehearse global policy ideas, such as promoting resilience, or environmental sustainability, for example, although these played a strong role in the 2011 version of the Joburg 2040 growth and development strategy (Interview, city official E, 2011). It was also not feasible to offer a programme of urban development which sought simply to insert Johannesburg as a competitive 'global' or 'world class' city into wider global circuits. In the light of the multiple competing agendas presented by the demands of that context (Parnell and Robinson, 2006) the city strategy had to address hard questions from politicians concerned with delivering jobs and houses (Interview 1, former city official C, 2009). Indeed, these priority agendas contributed to shaping the global agency of the city globally, in terms of the policy circuits and places it sought to influence and learn from. Here, a distinct South-South shift in policy learning had been underway, in which Johannesburg officials have sought closer links with other 'southern' contexts, reflecting both a search for more relevant policy ideas (Harrison, 2015) and a revised assessment of the geopolitical positioning of Johannesburg and South Africa (Peyroux, 2016). This led to important opportunities for learning with respect to, for example, participatory budgeting, bus rapid transit and transit-oriented development, informal settlement upgrading, smart cities, security and resilience (Harrison, 2015; Peyroux, 2016). The form of policy learning was also influenced by this territorial imperative – embedded developmental learning based on long-term and substantive exchanges related to topics and sectors which mattered to Johannesburg came to dominate the city-to-city relationships (Interview 2, city official B, 2012).

Nonetheless, the city strategy navigated a range of trans-scalar dynamics. Shared responsibilities with other tiers of government in key areas relevant to urbanization ensured that political concerns and policy concepts reaching across local-national-provincial government shaped or delimited local trajectories. Close attention needed to be paid to new national government agendas and concepts – such as shared growth or human settlements policy – as well as alignment sought with ANC party concerns at local and regional levels (Lipietz, 2008). These were interwoven with widely circulating policy ideas but also with more locally developed academic concepts. This made the arrival at a specific range of policy agendas and commitments in the city strategy a complex and rather untraceable, even spatially topological, process (Robinson, 2011b, 2013; Interview city official E, 2011). Two of the key authors of the 2006 and 2011 city strategies expressed this in the following ways:

> The way the stuff works in truth is that a small team of people and, almost always, one or two individuals within that team are engaged in policy debates more generally, read incredibly widely on all sorts of issues and it just becomes part of the amorphous mass of their thinking. And then as they engage with what people are saying within the city, engage with stakeholders, engage with communities, a synthesis process happens by which the thoughts become a particular policy statement or a particular program of action. But if you were to say now where did that idea come from, you'd say well it came out of the work we were doing in this particular department; but in truth actually the idea probably came from somewhere else.
>
> (Interview 1, former city official A, Johannesburg, 2009)

> Honestly it's been very difficult; it has been very, very challenging. It would have been easy if we had agreement around the concept [of resilience] and theories of change – and then further develop it ... and then at the same time I guess the challenge has been trying to bring together the political imperatives, national and provincial priorities, and the theoretical stuff together.
>
> (Interview, city official E, 2011)

In addition, overlapping responsibilities for aspects of urban development across tiers of government led to creative rethinking about who is to be involved in strategic urban policymaking and implementation at an early stage in the Johannesburg process of city strategy preparation. One observer who had worked in national local government policy networks and in another large city, commented in relation to Joburg 2030 and the 5-year planning instrument, Integrated Development Plans,

> We started thinking that the limitations of the IDP [integrated development plan] was that the IDP is a municipal plan, it's not a plan for the municipal area ... IDPs only deal with powers and functions of municipalities and municipal budgets, cutting out safety and security, health, welfare, education, foreign policy... all sorts of things that you need to run a city. And I would give credit to Joburg for opening up that debate; they were pioneers there, I give credit to [officials] and all those people because they were doing it and at the end of the day their conclusion was

that their two main interventions had to be around crime and skills, at that stage, and then they thought, whoops, but we don't do that, so who can we influence to do that? Then they had to start constructing alliances and networks and how do they influence those people, although I don't think they necessarily succeeded in that.

(Interview, former Cape Town city official, Cape Town, 2009)

On the one hand, then, the Johannesburg city strategies evidence a capacity to work effectively across the demands of a range of different territorially embedded constituencies and to address tough redistributional and developmental questions politically and technically. But on the other hand, they have also managed to engage closely with global urban policy and developmental practice. For the purposes of this chapter we can see at work in this case a trans-scalar formation of territorial imperatives for urban development and a multidirectional process of engagement with wider global processes of knowledge circulation. However, this transnational environment is simultaneously territorialized – circulating ideas are already present and localized. Moreover, local practices are effectively launched into wider knowledge circuits and policy networks to achieve both developmental goals and global visibility. As I have argued elsewhere, a sophisticated spatial vocabulary is needed to appreciate these dynamics (Robinson, 2011b, 2013; Allen and Cochrane, 2014; Allen, 2016). In this case, the territorialized concerns and interests of the city strategy are grounded in cross-cutting trans-scalar and transnational formations. One consequence of this was the increase of Johannesburg's strategic policy team's influence in shaping strategic planning practices internationally – which opens up the story of their engagement with Lilongwe.

### Lilongwe: slow policy transfer and the exteriority of urban agency

If the wider national and supranational political contexts, as well as globally circulating ideas, shape the particular territorializations of urban political agency at a metropolitan scale in London and Johannesburg, turning to the case of Lilongwe we are reminded to attend to the precise nature of these articulations of global and localized processes, to the form of global urban agency. In Malawi, and in many African contexts, local government has historically been strongly directed by central government ministries, and still has little discretionary scope for operation. As a result, central government agencies are important players in key areas such as housing, planning and infrastructure (UN-Habitat, 2011). More than this, international non-governmental and development organizations are active partners or providers in so many of the functions of local government that significant problems of overlap and duplication of effort can emerge. But without local government engagement, the moral authority and coordinating capacity to implement policies is frequently lacking (Manda, 2014). At the time

Johannesburg officials began engaging in a close peer-to-peer mentoring relationship with Lilongwe officials with the intention of producing a city strategy (2007–2009), there was no functioning political authority there and somewhat limited administrative capacity. Local elections had not been held since 2000, the 2005 elections had been cancelled, a number of officials had been suspended because of a financial corruption scandal (including the financial oversight officials). The financial probity of the council's activities was a difficult issue to confront (Interview, Johannesburg city official and consultant, 2012). The Lilongwe city strategy was not prepared in auspicious circumstances, then, for realizing either the political capacity or the organizational resources to constitute local agency in relation to wider global processes. In this case transnational influences took the form of direct policy inputs from Johannesburg and the UCLG, as well as the search for international NGO and development aid funding to address the extreme developmental needs of this city of around 700,000 people.

Following initial engagements with UCLG's City Futures programme to promote international networking at the World Urban Forum, June 2006, in Vancouver and the Africities conference, Nairobi, September 2006, a dedicated workshop was held in Johannesburg in July 2007. Here officials and politicians from South African and Southern African cities met to discuss South Africa's experiences with city strategies (Interview 2, former city official A, 2011). Some of the officials I interviewed indicated that they had not been strongly aware of the aims of the workshop in advance, and appear to have been surprised to learn at the time that the underlying motivation was to implement mentorship arrangements amongst Southern and South African cities. Thus UCLG and Cities Alliance organizers seem to have been driving the process at this stage. Once engaged, however, Johannesburg officials, in particular a key official in the Central Strategy Unit responsible for drawing up the city strategy, saw this as an opportunity to engage in a positive developmental international relationship with a city closely connected to South Africa through labour migration and economic interdependence (Interview, city official and consultant, 2013). This opportunity was welcomed as a marker of a shift in the city's international policy towards a strategic and substantive process of city-to-city learning, rather than the more ceremonial approach to city twinning which had become customary, or the proliferation of uncoordinated international links across many different departments which predominated at the time (Harrison, 2015; Peyroux, 2016; Interview 3 City Official B, 2014).

At the commencement of the city planning exercise with Lilongwe, Johannesburg officials engaged together with UCLG officers in discussions with national and local government departments, NGOs and donor organizations. Their first visit to Lilongwe in April 2008 resulted in a confirmation of support from these bodies, however, Johannesburg participants observed that no political authority was in place to guide the officials there, and little scope existed for institutional leadership. A previously elected council had not assumed their positions, but also most of the directors of administrative units were acting appointments, and the suspension of the financial department because of extensive corruption significantly hampered both day-to-day operations and longer-term planning (Lilongwe CDS, 2009; City of Johannesburg, Central Strategy Unit, 2011). Nonetheless, in the incumbent CEO of some years, Professor Donton Mkhandawire, a motivated champion for the process was found, although unfortunately he left to serve as an MP in 2009 (*Nyasa Times*, 29 December 2011).

Interviews with Johannesburg city officials involved in the mentoring relationship revealed that their view of the state of the city government of Lilongwe was fairly dismal, and some examples they recounted seemed to indicate a low point in both local government capacity and international development practice. As documented in the city strategy document that emerged from the process, at the time, only 10% of the anticipated revenue was being collected. Recently donated used fire engines languished in the car park, needing spare parts no longer produced anywhere in the world. A recent investment in a road system (through Japanese aid) had significantly disrupted walkways for the majority of the city's pedestrians. Not one of the 17 Lilongwe city assembly health centres was functioning in the city for want of personnel and drugs (Lilongwe City Council, 2008, p. 11). And this in the face of one of the highest levels of malaria anywhere on the continent, 1,478 reported cases of cholera in 2008 that led to stunting and malnutrition in children, and significant HIV infection rates (12%) (Lilongwe CDS, 2009).

A small grant from Cities Alliance (US$72,375) offered start-up costs for city strategy preparation, financing an exchange between city officials in the two cities, and was sponsored by GTZ (German Organization for Technical Co-operation), Japan International Cooperation Agency, UCLG and USAID Malawi (GHK Consultants, 2010). The exchange continued over several months, and then years, building engagements between officials from Johannesburg and Lilongwe across a number of areas of city activity, such as security, health, GIS, planning, infrastructure/engineering, bye-laws. Over time the programme expanded beyond the initial brief Phase 1, and perhaps a dozen officials travelled from Johannesburg to Lilongwe for a week or two at a time, funded largely by Johannesburg's own budget, reviewed regularly by the mayor. It was estimated that this second phase of activity cost around US$250,000 (GHK Consultants, 2010; Interview 3, City Official C, 2013; Interview, City Official and Consultant, 2013). The agreement between Lilongwe City Assembly and Johannesburg City Council remained informal, partly because of the lack of political direction in Lilongwe. According to the Johannesburg team, this allowed flexibility for the process to evolve according to need and not to face excessive external expectations (City of Johannesburg, Central Strategy Unit, Powerpoint Presentation, Cities Alliance meeting, Mumbai, January 2010).

Some very positive features were identified by the officials who attended the mentoring events. The Planning Department had excellent paper records, with traceable cases, about which officials were knowledgeable (Interview, city official and consultant, 2013). And the bye-laws were in good shape, needing only an update of the fees charged to be effective and supply an income stream (Interview, city official F, 2013). Based on the evaluation reports and interviews with Johannesburg officials, a lively opening meeting was held with local citizen groups, including groups linked to the Slum and Shack Dwellers International, involved in local housing provision, and locally based international NGOs (Kitchin, 2009; GHK Consultants, 2010). A report back to a meeting of these constituents was presented by Lilongwe officials at the end of the process (Interview 3, city official B, 2014). Ideas for a set of priorities for a medium-term strategic vision for the city were generated through some commissioned local research and through numerous follow-up meetings with Lilongwe officials. Together the groups of officials generated an assessment of the needs of the city and its residents, and the potential to address these, with Johannesburg officials carefully thinking through some general insights from their policy knowledge, and drawing on their experience in their own city, which had some resonances with Lilongwe's more extensive urban challenges (Interview, city official H, 2013). The final document reflected a significant investment from officials in both cities. The numerous Johannesburg visitors reflect positively on the experience, the mutual learning involved, and the potential to turn around the administrative malaise. A major achievement noted in the internal review by Johannesburg was that on the basis of the CDS the development process could be demand-driven, rather than 'donor-led' or 'supply-led' (City of Johannesburg, 2011).

There were also some more challenging aspects to the programme of work. A recent Japanese (JICA) donor master planning exercise had delivered an ambitious set of plans for infrastructure development in Lilongwe. However, these were inaccessible to the officials – their computers were not powerful enough, and they did not have the software package to read the data. One colourful hard copy sufficed to frame discussion (Interview, city official and consultant, 2013). In fact, computing equipment was so poor that the Johannesburg official felt it important to inform the Cities Alliance that advertising their activities via email and internet was not adequate (City of Johannesburg, 2011). At one point, the officials from Johannesburg had to confront their Lilongwe counterparts with ongoing corrupt behaviour, extended lunch breaks and lack of attention to meeting times (Interview, city official and consultant, 2013; City of Johannesburg, 2011). A specific intervention and training on local financial management funded by the South African Development Community was poorly attended by officials enrolled, who were in their turn frequently distracted or left the meeting (South African Cities Network, 2011). More generally Johannesburg observers noted that training itself seemed to play a negative role in government capacity: officials were often abroad, not doing their allocated work, but receiving 'training' through various multilateral and bilateral agencies, which seldom translated into improved implementation (Interview, city official and consultant, 2013).

The global life of the city strategy, initiated through an international networking programme of the UCLG and CA, continued after its completion but in a rather different form. A new incumbent in the role of CDS Manager brought in to take forward the city strategy swiftly submitted the mentoring relationship behind the city strategy for an international award. The award was duly received in Ghuanzhou, China, for international cooperation. It was received by the manager, who had not been part of the process at all, and the Cities Alliance, whose early subvention of the process was rather less than the investment of money, time and commitment from Johannesburg. Johannesburg officials who had driven the process were not named in the award or invited to the award event (Interview, city official and consultant, 2013), although later publications did bring some desired publicity (Cities Alliance in Action, 2010; Thorpe, 2011).

Other outcomes were more material, including an investment of significant funds. US$2.6m was given from the Gates Foundation to implement the City Strategy, specifically related to informal settlements needs and upgrading service delivery there (GHK Consultants, 2010; Interview, city official and consultant, 2011; Bill and Melinda Gates Foundation, 2010). Cities Alliance funding of nearly US$250,000 was forthcoming for Phase 3 implementation of institutional elements of the strategy. A series of low-cost implementation engagements with Johannesburg through 2010 saw some successes, such as planning for future projects, and opening some local clinics (City of Johannesburg, 2011). However, the formal relationship with Johannesburg ended around this time amidst some distrust and ongoing concerns about good governance and the continued absence of elected local government, which was seen as a precondition for Johannesburg's continued involvement (Interview 1, city official B, 2011). They had also withdrawn, for example, from the Addis Ababa collaboration during a period of political uncertainty and diminished adherence to democratic governance (Interview 3, city official B, 2014). Notes from regular contact sessions through 2010, looking to work with Johannesburg officials on the implementation strategies for different elements of the programme, indicate concerns about lack of preparation between meetings, commitment to attend, and core project planning skills to take forward the programme of work set out in the city strategy, including in the office of the CDS Manager. Despite acknowledgement of these challenges, the significant new funding streams from Cities Alliance and the Gates Foundation were dedicated to the initiative – and Johannesburg was sidelined from future implementation plans. Johannesburg officials' concerns here are pertinent, as they observed the potential for the high expectations of donor agencies – and the large sums of money they wished to disburse – to be at odds with the mentors' own flexible and low-key approach, able to engage with capacities and programmes at a more realistic level (City of Johannesburg, 2011).

The experience of the Johannesburg-Lilongwe cooperation had attracted the attention of myself and Elisabeth Peyroux, as researchers both focused on Johannesburg and interested in global policy circulation. We were intrigued by the developmental focus of the programme, and the slow, engaged process of policy transfer – quite at odds with the 'fast policy' analyses currently dominant (Peck and Theodore, 2015; Peyroux, 2016). In some ways, though, aside from the practical investment leveraged from the Gates Foundation through the report, and some early low-key implementations, this lengthy, committed, developmental process of strategic policy formulation, with a number of feasible suggestions for effective development in a well-researched document disappeared in the face of political change and administrative challenges – although a newly elected mayor was still able to refer to the 2009 City Strategy as a valid programme of work in 2017 (The Nation, 28 January 2017). However, the city strategy was most impressively specularized in a process of international recognition of the personal and institutional collaboration. And it has been 'banked' as organizational capital, part of the ongoing achievements of international city-to-city networking programmes promoted by the CA and UCLG, leading to further investment in policymaking in Lilongwe despite some less favourable assessments of the actual dynamics of the process (Interview, city official and consultant, 2013; City of Johannesburg, 2011, p. 13).

The Johannesburg-Lilongwe cooperation to produce a city strategy again illuminates the strongly transnational grounds of urban agency – in this case, the capacity to act both locally and in the global arena was founded on a strong and direct engagement from a series of international actors – Johannesburg municipality, the Cities Alliance, UCLG, the Gates Foundation, JICA and the many other international organizations, donors and NGOs active in the city. More generally on the basis of this case, analyses of urban agency need to focus more strongly on the external and exteriorized nature of interests and capacity. This case also provides insights into the range of forms of global urban agency: specific and personal links and connections in substantial face to face engagements; fleeting but perhaps influential engagements at international conferences; formal interactions through applications and evaluations with transnational organizations; speculative presencing to bring to visibility certain representations of places, which might have little foundation in reality but which can have strong material impacts, such as enhancing reputations, strengthening institutional credibility and directing financial flows.

### *Looking again at London: spectrality and welfarism*

If we turn back to London with Johannesburg in mind, the powerful continuities of both democracy and a strong, welfare-oriented state need to be layered into any assessment of the agency of that city (the point of key commentators on the global city interpretation of London – see Buck et al., 2002).

While there is significant and growing pressure on local councils to fund their budgets from local business rates (as opposed to direct government grants which drew on the national receipts of these rates) and property development in the face of budget cuts, which significantly changes their priorities in spatial planning terms, the reorientation of funding sources does not necessarily destroy policy processes appropriate to a democracy founded on citizen rights, elections and consultative practices; the London Plan remains focussed on key policy outputs relevant to these welfare agendas, including ensuring the provision of education, urban services and being held responsible for health and the quality of urban life. Reviewers hired by the City of Johannesburg to scope international practice in city strategies observed of the London Plan that it was notable for, amongst other things, 'The progressive feel of the document; The ability to link economic growth with an emphasis on inclusivity' (Silverman et al., 2005, p. 26). The outcomes often fall far short of the expectations of a universal welfare state, and nowhere more so than in the purposeful devastation of affordable housing stock, diminished capacity to deliver social infrastructure and appropriate levels of open space, and the failure to take action to lower pollution levels (see for example, Lees, 2014). London's capacity for global action has been fundamentally re-engineered and expanded, across successive Labour and Conservative governments, directly through establishing the institutions of the GLA, and indirectly through the wider competitive ambitions fostered through discursive achievements on the basis of platforms such as the city strategy/London Plan. The terms of that engagement are widely seen to have been configured in a 'neo-liberal' or competitive idiom (Massey, 2007), but it would be inaccurate to disregard the substantive residual welfare orientation of urban policy and politics.

Returning to London with experiences of city strategy making in Lilongwe in mind, I was surprised by the resonance I observed between the 'spectrality' and informality of the state and policy in Lilongwe with the situation in London (see Hillier, 2000). Here, the occasion was the Evidence in Public hearings, a semi-judicial situation in which a Planning Inspector tests the soundness of the plan against available legislation, and the public and different interested parties (local governments, developers, business organizations, community groups) are permitted to make contributions to discussions of matters identified by the Inspector. In the UK, in 2012, a substantial body of planning legislation was discarded in favour of a new, slimmed down 'National Planning Policy Framework'. In the process of institutional and legal reform, the London Plan had become an anomaly, as the only 'regional' (supra local government) plan. At one stage in these hearings in 2014 I noted the Planning Inspector suggesting that since there was no available legislation to inform his thinking, he would take guidance from a now defunct piece of legislation. Something of the spectrality of the state, operating in a legal void, was exposed (Tuvikene, Neves Alves and Hillbrandt, 2016). In this set of hearings, too, the growth in the population of London was under the spotlight, and the uncertainty of the recent census data as a basis for predicting future growth was very apparent; the question being, had fewer people moved out of London in the recession period (because of the stagnation of the property market) and would this resume in the future, returning growth rates to lower pre-recession levels? Population projections for 2036 ranged from 9.8m to 10.4m, for example, due to this uncertainty (GLA, 2015a). This conundrum was evidenced not least in the double dealing of the planners, suggesting on the one hand that London must plan to accommodate this entire predicted population growth. Based on a mid-level of growth scenario, this would need 49,000 new homes p.a., far beyond historic levels of construction achieved although still less than the objectively identified need. This ambition would run against the failure to translate planning permission and identified sites into realized construction of homes – effectively giving London a 'ghost city' of over 200,000 homes (Mayor of London, 2012, p. 20). On the other hand, the mayor's team was writing what became a notorious letter to the relatively distant commuter town of Bedford, warning their planning officers to plan for growth despite the census figures showing population decline in those areas. The uncertainty at the core of a key driver of the London Plan and its competitive global ambitions – especially given the potential impacts of the strong steer for densifying housing development and increasing building heights across the city on residents' quality of life – was palpable. The Inspector required that a full review should begin immediately, and it is underway at the time of writing (The Planning Inspectorate, 2014).

## Conclusion: revisiting the global political agency of the urban

In the context of multiple globalizations, and with an analytical view committed to widening the range of urban contexts informing theorization, conventional assessments of the territorializations of 'urban' agency need to be revisited. This chapter has suggested that it is important to consider how global urban political agency comes to be constituted across different urban situations, and in relation to a multiplicity of transnational and trans-scalar processes (Allen and Cochrane, 2007). In thinking about the production of the global political agency of the urban through the example of city strategies, it is necessary, then, to appreciate the diversity of circuits and processes through which a territorialized urban agency is configured. In the midst of a range of globalizing processes and agendas in which the urban is implicated, there is room for global urban political agency to be assembled at specific moments of imagination, presentation, articulation and institutional coordination.

Urban agency, then, appears as an emergent, trans-scalar but territorialized possibility, articulated in many different ways in relationship to a wide range, and different configurations, of globalizing processes. Different forms of global urban agency have also been evidenced in this chapter. The capacity to scan and assess circulating policy for locally resonant agendas, for example, speaks to a strong agency for analysis and assessment within institutions and city officials in both London and Johannesburg. But the Lilongwe case, where a capable official mobilized the achievements of their collaboration with Johannesburg for maximum global visibility while disconnecting these representations from the outcomes on the ground, reveals the importance of recognizing the spectrality of some aspects of 'global urban agency'. This is especially relevant when urban actors are pressing for visibility and packaging achievements for circulation – publicizing Johannesburg's City Strategy work, for example, or making political capital from London's role in C40. More than this, the international development industry needs to look at the ways in which their own practices foster more spectral forms of urban agency, promoting and publicizing ambitious programmes, such as international networking and city to city learning, despite significant and known challenges of implementation. Thus learning, and competing, disciplining and developing, fabricating or acting on uncertain grounds, bringing to visibility precarious achievements or invisibilizing failure through managing representations, indicate a range of forms of global urban agency.

Working across three interconnected cases, based on the widely circulating practice of preparing city strategies, has brought into view the need for new interpretations of global urban agency in the midst of multiple globalizations. This chapter has shown the value of bringing the three cases into conversation with each other in order to enrich understandings of urban politics in each context. Together they have drawn attention to the wide range of globalization processes shaping urban outcomes, and to the variety of forms of urban political agency at work. This demonstrates the potential in following connections to generate comparisons across a wider variety of urban contexts, thereby stretching analyses of urban politics.

## Notes

1 Acknowledgements. I would like to thank the officials at the City of Johannesburg who so kindly shared their experiences with preparing city strategies, and their recollections of the city to city link which they developed with Lilongwe. Special thanks are due to Jan Erasmus for his enthusiastic support for the research. I also would like to thank the members of Just Space who have been open to collaborations in support of their engagements with the London Plan process. I have learned so much from their hard work, especially the inspiring commitment of Michael Edwards, who has motivated several generations of London-based urban scholars to conduct their research in support of community-based organization.

2 These observations are based on current esrc-funded research on a Mayoral Development Corporation, established 1 April 2015, in Old Oak and Park Royal.

3 The vernacular name for Johannesburg, 'place of gold'.

## References

Acuto, M., 2013. The new climate leaders? *Review of International Studies*, 39(4), pp. 835–857.

Acuto, M. and Parnell, S., 2016. Editorial: leave no city behind. *Science*, 352(6288), pp. 873–874.

Akrich, M. and Rabeharisoa, V., 2016. Pulling oneself out of the traps of comparison: an auto-ethnography of a European project. In: J. Deville, M. Guggenheim and S. Hrdličková, eds. *Practising comparison: logics, relations, collaborations.* Manchester: Mattering Press. pp. 130–165.

Allen, J. and Cochrane, A., 2007. Beyond the territorial fix: regional assemblages, politics and power. *Regional Studies*, 41(9), pp. 1161–1175.

Allen, J. and Cochrane, A., 2014. The urban unbound: London's politics and the 2012 Olympic Games. *International Journal of Urban and Regional Research*, 38, pp. 1609–1624.

Allen, J., 2016. *Topologies of power: beyond territory and networks.* London: Routledge.

Allmendinger, P. and Haughton, G., 2007. The fluid scales and scope of UK spatial planning. *Environment and Planning A*, 39, pp. 1478–1496.

Bill and Melinda Gates Foundation, 2010. *Foundation announces effort to tackle urban poverty in five African cities.* [online] Available at: https://www.gatesfoundation.org/Media-Center/Press-Releases/2010/09/Foundation-Announces-Effort-to-Tackle-Urban-Poverty-in-Five-African-Cities [Accessed 6 March 2018].

Bowie, D., 2010. *Politics, planning and homes in a world city: the mayor of London and strategic planning for housing in London 2000–2008.* London: Routledge.

Brenner, N. and Schmid, C., 2015. Towards a new epistemology of the urban? *City*, 19(2–3), pp. 151–182.

Brown, R., Edwards, M. and Lee, R., 2014. Just space: towards a just, sustainable London. In: R. Imrie and L. Lees, eds. *Sustainable London? The future of a global city.* Bristol: Policy Press. Chapter 3.

Buck, N., Gordon, I., Hall, P., Harloe, M. and Kleinman, M., 2002. *Working capital. Life and labour in contemporary London.* London: Routledge.

Bulkeley, H., 2010. Cities and the governing of climate change. *Annual Review of Environment and Resources*, 35, pp. 229–253.

Caldeira, T., 2016. Peripheral urbanization: autoconstruction, transversal logics, and politics in cities of the global south. *Society and Space*, 35(1), pp. 3–20.

Cities Alliance, 2006. Guide to city development strategies: improving urban performance. Washington, DC: The Cities Alliance.

Cities Alliance in Action, 2010. *Johannesburg-Lilongwe partnership leads to a robust city development strategy.* [online] Available at: http://www.citiesalliance.org/sites/citiesalliance.org/files/Lilongwe%20-%20FINAL.pdf [Accessed 16 April 2018].

City of Johannesburg, 2011. Growth and development strategy (GDS2040). [online] Available at: https://www.joburg.org.za/about_/Documents/joburg2040.pdf [Accessed 15 April 2018].

City of Johannesburg, Central Strategy Unit, 2011. *Mentoring in city government. Evaluating the Mentorship Programme between the City of Johannesburg and the Lilongwe City Council.* June 2011. Central Strategy Unit, internal report.

Cox, K. and Mair, A., 1988. Locality and community in the politics of local economic development. *Annals of the Association of American Geographers*, 78(2), pp. 307–325.

DCLG, 2017. *100% business rates retention. Further consultation on the design of the reformed system.* UK Government: Department of Communities and Local Government.

Deville, J., Guggenheim, M. and Hrdličková, Z., eds., 2016. *Practising comparison: logics, relations, collaborations.* Manchester: Mattering Press.

Gavron, N., 2007. Towards a carbon neutral London. In: R. Burdett and D. Sudjic, eds. *The endless city.* London: Phaidon. pp. 372–385.

GHK Consultants, 2010. Evaluation of project implementation modalities of the cities alliance (Excerpt 3.2.2 Malawi). Made available by Johannesburg City Council Officials.

González, S., 2011. Bilbao and Barcelona 'in motion'. How urban regeneration 'models' travel and mutate in the global flows of policy tourism. *Urban Studies*, 48, pp. 1397–1418.

Gordon, I., 1995. London: world city: political and organizational constraints on territorial competition. In: P. Cheshire and I.R. Gordon, eds. *Territorial competition in an integrating Europe.* Aldershot: Avebury. pp. 295–311.

Gordon, I., 2004. *Capital needs, capital demands and global city rhetoric in mayor Livingstone's London Plan.* GaWC Research Bulletin 145, University of Loughborough.

Gordon, I. and Travers, T., 2010. London: planning the ungovernable city. *City, Culture and Society*, 1, pp. 49–55.

GLA, 2011. *The London Plan. London: Greater London Authority.* [online] Available at: https://www.london.gov.uk/what-we-do/planning/london-plan/past-versions-and-alterations-london-plan/london-plan-2011 [Accessed 31 October 2017].

GLA, 2015a. *Further alterations to the London Plan. London: Greater London Authority.* [online] Available at: https://www.london.gov.uk/what-we-do/planning/london-plan/current-london-plan [Accessed 31 October 2017].

GLA, 2015b. London Infrastructure Plan 2050. London: Greater London Authority. https://www.london.gov.uk/what-we-do/business-and-economy/better-infrastructure/london-infrastructure-plan-2050. [Last accessed 31 October 2017].

GLA, 2017. Policy and infrastructure collaboration across wider south east; https://www.london.gov.uk/about-us/organisations-we-work/policy-and-infrastructure-collaboration-across-wider-south-east. [Last accessed 31 October 2017].

Halbert, L. and Rouanet, H., 2014. Filtering risk away: global finance capital, transcalar territorial networks and the (un)making of city-regions: an analysis of business property development in Bangalore, India. *Regional Studies*, 48(3), pp. 471–484.

Harding, A., 1994. Urban regimes and growth machines: towards a cross-national research agenda. *Urban Affairs Quarterly*, 29(3), pp. 356–382.

Harding, A., Deas, I., Evans, R. and Wilks-Heeg, S., 2004. Reinventing cities in a restructuring region? The rhetoric and reality of renaissance in Liverpool and Manchester. In: M. Boddy and M. Parkinson, eds. *City matters: competitiveness, cohesion and urban governance.* Bristol: Policy Press. pp. 33–50.

Harris, N., 2002. Cities as economic development tools. *Urban brief*, December 2002. [pdf] Washington: Woodrow Wilson International Center for Scholars. Available at: http://www.citiesalliance.org/sites/citiesalliance.org/files/cities-economic-development-tools%5B1%5D.pdf [Accessed 6 March 2018].

Harrison, P., 2006. Integrated development plans and third way politics. In: U. Pillay, R. Tomlinson and J. du Toit, eds. *Democracy and delivery: urban policy in South Africa.* Cape Town: HSRC Press. pp. 186–207.

Harrison, P., 2015. South-south relationships and the transfer of 'best practice': the case of Johannesburg, South Africa. *International Development Planning Review*, 37(2), pp. 205–223.

Hart, G., 2016. Relational comparison revisited: Marxist postcolonial geographies in practice. Progress in Human Geography, First Published 16 December 2016, https://doi.org/10.1177/0309132516681388.

Harvey, D., 1989. From managerialism to entrepreneurialism: the transformation in urban governance in late capitalism. *Geografiska Annaler B*, 71(1), pp. 3–17.

Harvey, D., 2013. *Rebel cities: from the right to the city to the urban revolution.* London: Verso.

Healey, P., 1998. Building institutional capacity through collaborative approaches to urban planning. *Environment and planning A*, 30(9), pp. 1531–1546.

Healey, P., 2007. *Urban complexity and spatial strategies: towards a relational planning for our times.* London: Routledge.

Hillier, J., 2000. Going round the back? Complex networks and informal action in local planning processes. *Environment and planning A*, 32(1), pp. 33–54.

Institute for a Competitive Inner City (ICIC), 2003. *City growth strategy: a new agenda for business-led urban regeneration.* Boston, CA: ICIC.

Jacobs, J., 2006. A geography of big things. *Cultural Geographies*, 13(1), pp. 1–27.

Jacobs, J., 2012. Commentary: comparing comparative urbanisms. *Urban Geography*, 33(6), pp. 904–914.

Kennedy, R., 1991. *London: world city moving into the twenty-first century.* London: HMSO.

Kitchin, F., 2009. Strategic planning mentoring: case studies from Southern Africa. Prepared for UCLG.

Leitner, H., 1990. Cities in pursuit of economic growth: the local state as entrepreneur. *Political Geography Quarterly*, 9(2), pp. 146–170.

Lilongwe City Council, 2008. *Lilongwe city assembly annual report.* Lilongwe: Lilongwe City Council (made available by Johannesburg City Officials).

Lilongwe City Council, 2009. *City development strategy: 'a shared future'.* Lilongwe: Lilongwe City Council.

Lipietz, B., 2008. Building a vision for the post-apartheid city: what role for participation in Johannesburg's city development strategy. *International Journal of Urban and Regional Research*, 32, pp. 135–163.

Lees, L., 2014. The urban injustices of new labour's 'new urban renewal': the case of the Aylesbury Estate in London. *Antipode*, 46(4), pp. 921–947.

Logan, J. and Molotch, H., 1987. *Urban fortunes: the political economy of place.* Berkeley: University of California Press.

Manda, M.Z., 2014. Where there is no local government: addressing disaster risk reduction in a small town in Malawi. *Environment and Urbanization*, 26(2), pp. 586–599.

Massey, D., 2007. *World city.* Cambridge: Polity.

Mayor of London, 2012. *Barriers to housing delivery: what are the market perceived barriers to residential development in London.* London: GLA.

McCann, E., 2001. Collaborative visioning or urban planning as therapy: the politics of public-private urban policy making. *Professional Geographer*, 53(2), pp. 207–218.

McCann, E., 2011. Urban policy mobilities and global circuits of knowledge: toward a research agenda. *Annals of the Association of American Geographers*, 101(1), pp. 107–130.

McCann, E. and Ward, K., 2010. *Mobile urbanisms*. Minneapolis: University of Minnesota Press.

Merrifield, A., 2013. The urban question under planetary urbanization. *International Journal of Urban and Regional Research*, 37, pp. 909–922.

Moore, S.M., 2013. What's wrong with best practice? Questioning the typification of New Urbanism. *Urban Studies*, 50(11), pp. 2371–2387.

Myers, G., 2014. Unexpected comparisons. *Singapore Journal of Tropical Geography*, 35(1), pp. 104–118.

Nadin, V., 2007. The emergence of the spatial planning approach in England. *Planning Practice & Research*, 22(1), pp. 43–62.

Nation, 2017. I will transform Lilongwe. *The Nation*, [online] 28 January. Available at: http://mwnation.com/i-will-transform-lilongwe-city/ [Accessed 16 April 2018].

Nyasa Times, 2011. Donton Mkandawire died of cancer. *Nyasa Times*, [online] 29 December. Available at: https://www.nyasatimes.com/donton-mkandawire-died-of-cancer-relation [Accessed 15 April 2018].

Parnell, S., 2016. Defining a global urban development agenda. *World Development*, 78, pp. 529–554.

Parnell, S. and Oldfield, S., eds., 2014. *Handbook for cities of the Global South*. London: Routledge.

Parnell, S. and Robinson, J., 2006. Development and urban policy: Johannesburg's city development strategy. *Urban Studies*, 43(2), pp. 337–355.

Peck, J. and Theodore, N., 2012. Follow the policy: a distended case approach. *Environment and Planning A*, 44, pp. 21–30.

Peck, J. and Theodore, N., 2015. *Fast policy: experimental statecraft at the thresholds of neoliberalism*. Minneapolis: Minnesota University Press.

Peck, J., Theodore, N. and Brenner, N., 2009. Neoliberal urbanism: models, moments, mutations. *SAIS Review*, XXIX(1), pp. 49–66.

Peyroux, E., 2016. Circulation des politiques urbaines et internationalization des villes: la stratégie des relations internationales de Johannesburg. *EchoGéo*, 36, [online]. Available at: http://echogeo.revues.org/14642 [Accessed 6 March 2018].

Pieterse, E., 2008. *City futures: confronting the crisis of urban development*. London: Zed Books.

Pinson, G., 2009. *Gouverner la ville par projet: urbanisme et gouvernance des villes européenes*. Paris: Presses de la foundation national des sciences politiques.

Porter, M.E., 1990. The competitive advantage of nations. *Harvard Business Review*, March-April, pp. 73–91.

Raco, M., Parker, G. and Doak, J., 2006. Reshaping spaces of local governance? Community strategies and the modernisation of local government in England. *Environment and Planning C: Government and Policy*, 24, pp. 475–496.

Robinson, J., 2006. *Ordinary cities*. London: Routledge.

Robinson, J., 2011a. Cities in a world of cities: the comparative gesture. *International Journal of Urban and Regional Research*, 35, pp. 1–23.

Robinson, J., 2011b. The spaces of circulating knowledge: city strategies and global urban governmentality. In: E. McCann and K. Ward, eds. *Mobile urbanism: cities and policymaking in the global age*. Minneapolis: University of Minnesota Press. pp. 15–40.

Robinson, J., 2013. 'Arriving at' urban policies/the urban: traces of elsewhere in making city futures. In: O. Söderström et al., eds. *Critical mobilities*. Lausanne and London: EPFL and Routledge.

Robinson, J., 2016a. Thinking cities through elsewhere: comparative tactics for a more global urban studies. *Progress in Human Geography*, 40(1), pp. 3–29.

Robinson, J., 2016b. Comparative urbanism: new geographies and cultures of theorizing the urban. *International Journal of Urban and Regional Research*, 40(1), pp. 219–227.

Robinson, J., forthcoming. Comparative urbanism: tactics for global urban studies.

Roman-Velazquez, P., 2014. Claiming a place in the global city: urban regeneration and Latin American spaces in London. *Political Economy of Technology, Information & Culture Journal (EPTIC)*, 16(1), pp. 68–83.

Roy, A., 2009. The 21st century metropolis: new geographies of theory. *Regional Studies*, 43(6), pp. 819–830.

Roy, A. and Ong, A., 2011. *Worlding cities*. Oxford: Wiley-Blackwell.

Rydin, Y., Thornley, A., Scanlon, K. and West, K., 2004. The Greater London Authority – a case of conflict of cultures? Evidence from the planning and environmental policy domains. *Environment and Planning C: Government and Policy*, 22, pp. 55–76.

Savitch, H. and Kantor, P., 2004. *Cities in the international marketplace: the political economy of urban development in North America and Western Europe*. Princeton, NJ: Princeton University Press.

Schmid, C. et al., 2018. Towards a new vocabulary of urbanization processes: a comparative approach. *Urban Studies*, [online early] Available at: http://journals.sagepub.com/doi/abs/10.1177/0042098017739750?journalCode=usja [Accessed 6 March 2018].

Scott, A.J., ed., 2001. *Global city-regions: trends, theory, policy*. Oxford: Oxford University Press.

Shatkin, M., 2017. *Cities for profit: the real estate turn in Asia's urban politics*. Ithaca, NY: Cornell University Press.

Silverman, M., Zack, T., Charlton, S. and Harrison, P., 2005. *Review of Johannesburg's city strategy: summary of key innovations in existing city strategies and learning from their implementation*. Central Strategy Unit, Johannesburg, July 2005.

Söderström, O., 2014. *Cities in relations: trajectories of urban development in Hanoi and Ouagadougou*. Oxford: Wiley-Blackwell.

South African Cities Network, 2011. SADC city creditworthiness project, capacity building progress report, 11 March 2011. Made available by Johannesburg City Council officials.

Stone, C., 1989. *Regime politics: governing Atlanta, 1946–1988*. Lawrence: Kansas University Press.

Tewdwr-Jones, M., Gallent, N. and Morphet, J., 2010. An anatomy of spatial planning: coming to terms with the spatial element in UK planning. *European Planning Studies*, 18(2), pp. 239–257.

Theodore, N., 2007. Closed borders, open markets: day laborers' struggle for economic rights. In: H. Leitner, J. Peck and E. Sheppard, eds. *Contesting neoliberalism: urban frontiers*. New York: Guilford. pp. 250–265.

The Planning Inspectorate, 2014. *Report on the examination in public into the further alterations to the London Plan. Report to the mayor of London*. [online] Available at: https://www.london.gov.uk/what-we-do/planning/london-plan/past-versions-and-alterations-london-plan/further-alterations-london[Accessed 31 October 2017].

Thornley, A., Rydin, Y., Scanlon, K. and West, K., 2005. Business privilege and the strategic planning agenda of the greater London authority. *Urban Studies*, 42(11), pp. 1947–1968.

Thorpe, W., 2011. A very civil partnership. *United Cities*, March, pp. 44–46.

Travers, T., 2004. *The politics of London: governing an ungovernable city*. Basingstoke: Palgrave Macmillan.

Tuvikene, T., Neves Alves, S. and Hilbrandt, H., 2016. Strategies for relating diverse cities: a multi-sited individualising comparison of informality in Bafatá, Berlin and Tallinn. *Current Sociology*, 65(2), pp. 276–288.

UCLG, 2013. *Peer learning: UCLG mentoring stories, Johannesburg and Lilongwe, 2008–2012*. [online] Available at: https://issuu.com/uclgcglu/docs/jb_lilongwest-ory [Accessed 6 March 2018].

UN-Habitat, 2011. *Malawi: Lilongwe urban profile*. Nairobi: UNHCS.

van Horen, B., 2000. Informal settlement upgrading: bridging the gap between the de facto and the de jure, *Journal of Planning Education and Research*, 19(4), pp. 389–400.

Vicari, S. and Molotch, H., 1990. Building Milan: alternative machines of growth. *International Journal of Urban and Regional Research*, 14, pp. 602–624.

Ward, K., 1996. Rereading urban regime theory: a sympathetic critique. *Geoforum*, 27(4), pp. 427–438.

Ward, K., 2006. 'Policies in motion', urban management and state restructuring: the trans-local expansion of business improvement districts. *International Journal of Urban and Regional Research*, 30(1), pp. 54–75.

Ward, K., 2010. Towards a relational comparative approach to the study of cities. *Progress in Human Geography*, pp. 1–17.

Weber, R., 2015. *From boom to bubble: how finance built the New Chicago*. Chicago, IL: University of Chicago Press.

Wood, A., 2014. Moving policy: global and local characters circulating bus rapid transit through South African cities. *Urban* Geography, 35(8), pp. 1238–1254.

# 7 Whose urban agency is it anyway?

*Michele Acuto*

Have cities really become global political actors? With more than two hundred and counting city networks active today (Acuto, 2016), major efforts like the C40 Climate Leadership Group or Rockefeller 100 Resilient Cities mobilizing millions of dollars of investment and city leaders like Michel Bloomberg and Anne Hildalgo taking a centre stage in major international processes like the Paris Agreement on climate change, the question seems all but rhetorical. Yet in the past decade there have been few critical and scholarly reflections on this claim to fame. International relations (IR) and urban studies remain relatively silent in theorizing what we could call 'global urban governance'. At the same time the general public literature on the 'triumph of the city' (Glaeser, 2011) and the reasons why mayors should 'rule the world' (Barber, 2013) have burgeoned following the expanding variety of global urban initiatives in practice that most academia has partly sidelined. Central to this globalization of urban governance, and rarely discussed in urban studies (Parnell, 2016; Revi, 2017), a number of major treaties, fora and reforms around the United Nations system have been afoot in the last decade, culminating with the various dimensions of the UN Sustainable Development Agenda 2030 (Birch, 2017). This process has now been enshrined in a number of global frameworks like the Paris Agreement on climate, the Shanghai Declaration on health and sustainable development, the UN's new urban agenda and of course the UN's Sustainable Development Goals (SDGs). Equally critical, and rarely discussed in international relations, have been the political transformations shaping urban governance in cities and states, such as devolution shifts, geopolitical moves and transnational pulls, which are reshaping the way we govern urban settlements big and small. In terms of academic timing, then, *The City as Global Political Actor* comes at an apt time for the juncture of urban and international thinking to offer key insights regarding this emergence of the city on a global stage.

When we think of the global pull of cities, we have often automatically considered the 'city' as a *place* rather than confronted its agency – redressing this bias, but also unravelling the inherent contradictions in its possible answers, is a fundamental goal of this volume. The book has opened up critical challenges, beyond the now well-established 'urban age' fascination (Gleeson, 2012; Brenner and Schmid, 2014), with the 'cult of the mayor' and the 'time of the city'. What I would like to stress in this 'postscript' is that questions of (global) agency are fundamentally questions of power and politics. I do this with the intent of opening up a more thorough engagement between international thinking and urban studies, but also to call for a more explicit move to providing scholarly evidence-based argumentations as to the global role of cities beyond the more populist writing that inspired the genesis of this book. In doing so, I hope *The City as a Global Political Actor* becomes another important (academic) step of that growing scholarly community that strives to piece together a picture of *global* urban governance. Over the past two years there have been visible moves in key scholarly outlets like *Science and Nature* to speak across vastly different disciplines and to 'globalize' urban research whilst maintaining clear policy ('real world') relevance (Acuto, Parnell and Seto, 2018; Bai et al., 2018). These interventions have argued for an urban research that steps beyond situated interests and opportunist agendas, whilst remaining 'fit for policy', with efforts towards greater impact-oriented scholarly pluralism that can be found in those very same UN fora highlighted above (e.g. Acuto and Parnell, 2016; McPhearson et al., 2016). Here a collaborative focus on global urban governance jointly developed between urban studies and international relations is critical, I would argue, for the development of an interdisciplinary and applied urban research that is not just about, but also *for* (as sound scientific advice) and *in* (as active participant) urban politics.

## Rising cities?

Major multilateral processes like the SDGs or the Sendai Framework for Disaster Risk Reduction, all offer an affirmative answer to the question of whether cities are recognized as globally relevant. Yet they might also point at worrisome evidence: 'pro-urban' writing, as Pinson recognized in this volume, often campaigns for cities to take the lead 'versus' states. This sees cities confronting the limitations of state action (on the likes of climate, health, equality) and the straightjacketing presented by a state-centric international system (Curtis, 2016). Yet much of the pro-urban discourse presents insufficient scholarly evidence, is generally presented on anecdotal basis, and puts forward proposals whose nature, despite the far-reaching rhetoric, rarely upsets the contemporary neo-liberal status quo. It also tends to step away from the international thinking that is needed to understand how that system works in the first place – with limited analysis, for instance, of questions of international law, sovereignty and citizenship key to the skeleton of the international system (Blank, 2007). Just like political geography, IR has for instance developed far more refined 'global governance' views of international politics (theorizing across scales and including civil society, for example) than the often neo-liberal frameworks that guide sensationalist pro-urban writing (Weiss and Wilkinson, 2014). Paradoxically, much of this writing often also forgoes ample strides already made in urban studies to, for instance, theorize the nature of the globalization of cities (Derudder et al., 2012), deconstruct the 'tango' of city-state relations (Taylor, 2013; Curtis, 2016), analyse the complexity of scalar relations in (urban) politics (Keil and Mahon, 2010), or appreciate the complexity of the growth coalitions driving cities (Hall and Hubbard, 1996) – all of which are well represented throughout this volume. Questions on the status of urban agency in world politics, but also proposals that reach beyond traditional neo-liberal politics and with deep urban understandings, remain scarce. The risk here is to repeat the assertion, as with the negotiations leading to the UN's 2016 'New Urban Agenda', that there is a divergence between states and cities – rather than opening up to truly revolutionary shifts in the ways local governments contribute to global governance, and the ways academia accounts for the global political relevance of cities.

This is well embodied in Barber's *If Mayors Ruled the World* (2013), which sits at the heart of the justification for this volume. Barber, very much in a pro-urban fashion, posits the potential of mayors to 'lead' in world affairs and not surprisingly several of the authors in this volume have taken some issue with this reading. The matter here is that mayors all over the world appear to have emerged in the past few decades as a prominent alternative to international processes and top-down national solutions – a sense well captured by Barber's subtitle 'Dysfunctional Nations, Rising Cities'. Barber asserts that cities are 'democracy's best hope' and that on this principle cities 'should govern globally'. Barber's campaigning statement for mayors to take over global governance is a proposal that, if not entirely new, has in some cases been seen as provocative for international relations research and practice. Yet this sensationalist statement has gathered mixed responses from the authors of *The City as Global Political Actor*.

Barber's core proposition is that of the establishment of a 'global parliament of mayors' whose central role is to 'render coherent an intercity order' that is already in place across boundaries and geopolitics, as much as to 'give the metropolis a megaphone and allow its voice to be heard' by international political audiences. Barber goes to quite some lengths in defining a roadmap for a selected pool of cities to spearhead the congregation of city networks and mayors and form an assembly. Capturing the imagination of scholars and practitioners alike, especially in international relations, we could easily slip into long discussions on the feasibility of a parliament. Yet this would in turn lose sight of critical *urban* questions that the everyday reality of cities presents us with and cannot, as much contemporary urban research testifies to, be disregarded: how to address the mounting polarization and splintering of social structures in cities the world over? How to prevent the 'self-selected' pool of global cities and mayors, like New York or London, to dominate over a subclass of secondary and tertiary settlements both in the North and the South? How to contingently develop network solutions globally and offer spaces for creativity and expression locally, beyond the often suffocating pervasiveness of neo-liberal solutions? It is undeniable that while being effective in its advocacy, Barber's work still needs more thinking through before it can be used to address these well-established concerns from contemporary urban studies and practitioners; at the same time it makes clear that we should not substitute solid urban studies for basic international thinking.

The Parliament has in fact also been tested on the ground, with a preliminary meeting in Amsterdam in 2014 and then in Stavanger in 2017, gathering only 45 cities and struggling to gain traction as a truly 'global' assembly. As with the limits of 'city-state' thinking, it might not be particularly revolutionary to try and kick-start a 'UN of cities'. In this Barber is not, as he professes, 'too radical' in his statement. Rather the idea of a parliament of mayors is after all quite conservative: it offers an extension of existing connections, and a non-deliberative arena for city coalition and exchange that, while potentially extremely valuable, is a fairly common liberal institutionalist solution in international relations, and testifies once again to the unavoidable entrenchment of city leadership and contemporary neo-liberal order. Corijn in this volume suggests that the well-meaning democracy-driven effort of the Parliament can reach out beyond these limits to include civil society but this presents fundamental challenges. For instance, it is this very same direction of action that has marginalized cities (and especially local governments) to be often wrongly packaged together with civil society, rather than recognizing their governmental role and potential. As with the General Assembly of Partners to the UN New Urban Agenda, or the associate status of cities to amongst others UN-Habitat and the Paris Agreement on climate, the status of cities is still uncertain, under-recognized and stuck with merely a ceremonial role ('partners' and 'observers'), and this, in spite of their growing unspoken soft international personality. Equally, when celebrated as in Barber's book, the role of cities is presented in an even more dissociating 'cities vs states' tone, as echoed in Ward's and Robinson's chapters. Ideas like the 'parliament' are then placed, possibly rhetorically, in opposition to dysfunctional states. Yet practically, insofar as they do not develop any significant rupture with current neo-liberal frames or elaborate on the implications of this urban agency for core systemic elements like sovereignty and citizenship, they are well in line with the very same landscape in which states operate. In short, this is a confrontation in words but not always in deeds between the city and the state. As Pinson discusses in his chapter on the 'city actor thesis', the relationship between cities and territorial states is not a zero-sum game, as we are encouraged to think by pro-urban scholars and essayists and research in historical sociology of state building. Rather, context matters here and evidence about said context is fundamentally lacking – something Steve Rayner and I (2016), together with colleagues at the UCL City Leadership Lab, have begun to address by offering evidence of the global architecture of city networking.

Clear across the board, in both this volume as much as in several reviews of Barber's work, an appreciation is to be found of how this argument, and this type of writing, have made a dent in the public and practitioner imaginary, despite its shallow international or urban research bases. As Sam Roberts of the *New York Times* noted already, the book is perhaps one of the most 'messianic' of several 'urban manifestoes' advanced in the past decade to campaign for the role of cities in tackling global challenges. In this volume, Kevin Ward rightly notices how the power of many of these arguments lies in their general public appeal. As much as the carefully crafted informational infrastructure of blogs, co-eds, interviews, presentations, press releases, podcasts and websites broadcasts usually underpin the latest 'pro-urban' arguments (see also Pinson in his chapter in this volume), these are outreach mechanisms that are still somehow rarely sought after by the average urban scholar (in international or urban studies).

## (Global) agency in context

Cities, therefore, are by no means a passive 'venue' for private or multilateral interests, nor a purely subjugated context of 'higher' politics. On the contrary, cities can be effective actors taking part in the dynamics of global governance. Certainly, the rhetoric espoused by most of the major advocacy initiatives involving cities tends to testify to, and bolster, this direct capacity. As I have already argued elsewhere (Acuto, 2013), the extent and persistence over the last two decades of the internationalization of city-led initiatives is a clear sign that cities can partake actively in the architecture of world politics beyond inter-*national* relations. A typical representation of this is the often-cited assertion that, while 'nations talk, cities act' (Curtis, 2016). Variously attributed to a number of city leaders, the statement embodies much of the ethos of the leadership of Bloomberg in his tenure (2010–2013) as chair of the C40 Climate Leadership Group – perhaps the best example of the expanding popularity of cities in global policy agendas. In short, cities are 'out there' in world politics, lobbying, linking, planning and cooperating. As Ward highlights in his chapter, this often implies a loose (non-systematic and anecdotal) understanding of the 'intercity civic infrastructure' underpinning the internationalization of urban governance. This frequently simplistic view, Ward reminds us, is generally lacking in appreciation of the advances of the 'policy mobility' literature that has, in geography for instance, offered a better understanding of policy transfers not always being merely linear but also deeply spatial processes 'in which urban policies are open to morph and mutate as they travel'. A similar argument could be made for work on international relations.

The capacity to link transnationally, negotiate and collaborate with both multilateral and multinational private worlds, and carry 'international' agendas certainly deserves some particular attention in an essay focused on the impact of cities on global order. Today, we could rightfully argue that an important portion of these networking activities can be identified as 'city diplomacy' in that they constitute mediated 'international' relations between rightful representatives of polities (cities in this instance), and that they result in agreements, collaborations, further institution-building and cooperation across boundaries (van der Pluijm and Melissen, 2007). Within the framework of the C40, 'ambassadors' of cities (elected mayors, or their peers) like London or Seoul negotiate common frameworks and partake in collective action on behalf of their 'city-zens' (Holston, 1999). All of this evidence points to the city's capacity to connect, to catalyse action by networking, and to deploy the powers of collaboration. This is of course quite a different take from the more familiar and traditional approach to the sovereign ideal of state power: wielding network power means in practice that the city's influence is never really stabilized and is always shared with other actors, peers and flows. This requires us to devise research approaches which (and communicate to the general public in ways that) understand the city and 'its' agency in the broader context of global governance. This is critical when it comes to the 'city actor thesis'; IR offers a well-developed parlance on world political matters, structures and trends. For instance, dating back to at least the 1950s, the IR theory literature has already dealt extensively with matters of international personhood (Wendt, 2004), discussing the necessary responsibility and representation ('diplomatic') relationships between political communities, the ways in which collective political actors acquire a persona, views and interests, and present these to other actors in the 'international arena'. Little of this is taken into account or presented, tacitly or explicitly, when discussing the role of mayors internationally or the implications of having cities feed directly into global governance processes around, for instance, resilience and global health.

This is, however, not a call for substituting international for urban thinking. Rather, in this context it becomes even more pressing to explore the 'possibilities of engaged pluralism' (Sheppard and Plummer, 2007) – 'engaged' in the sense of an open-ended attempt to learn from each other, and 'pluralist' because it favours avoiding the search for the 'best' approach to 'the one' problem. This can allow us to better understand the globalization of urban governance, or as Kevin Ward put it in his chapter, 'the bringing together of the global and the urban'. Cities are now relating directly to global instruments, treaties and commitments, often bypassing states – as proven by the well-known steps to implement the Paris Agreement by cities or to adhere to agreements such as the Convention on the Elimination of all Forms of Discrimination Against Women (CEDAW) and other human rights principles by cities. The encounter between urban and international political thinking might then yield productive discussions. For instance, from a more comprehensive view of 'global governance' as a complex array of actors beyond just the state, accounting for multinational corporations, international organizations, regions (sub- and supra-state) and

NGOs, cities today might benefit from what Peter Katzenstein (1985) called the 'small economies advantage'. In his IR classic *Small States in World Markets,* Katzenstein examined the successes of relatively small nations of Western Europe, showing how these have managed to stay economically competitive while at the same time preserving their political institutions. In a striking parallel to the growingly globalized context of cities, Katzenstein noted how these states, too dependent on world trade to impose protection and lacking the resources to transform their domestic industries, found a 'third' solution to international prosperity: networking and negotiating the boundaries of their governance with those very forces that might threaten them in the first place. As with many successful 'entrepreneurial' cities, Katzenstein's states offered a rapid and flexible response to market opportunity stemming from what he called 'democratic corporatism' – a mixture of ideological consensus, centralized politics, and complex bargains among politicians, interest groups, and bureaucrats not too far from the idea of the 'entrepreneurial city' (Hall and Hubbard, 1996). Rather than seeking to exert power 'over' global markets and larger political actors, small states, and I would argue also many global(izing) cities, mastered the arts of network power and public-private (but also public-public with other layers of government) partnership. Central to this reasoning is, however, a constructivist view that is almost antithetical to Barber's (and others) 'cities vs states'. Small (and big) states are relational entities whose agency and 'international' development is connected to other entities populating world politics. This is well represented in this volume by Robinson's discussion of the relational nature of city development (and strategies) and the necessary comparative gestures required for urban scholars to grasp 'cities in a world of cities' (Robinson, 2011).

Here Katzenstein's 'small states' lesson for the question of cities and global order is even deeper than a possible comparison between small states and global cities. As he pointed out several years after the publication of his work on this theme, a core issue at play in this reasoning is that there is a great deal of difference in 'understanding-a-thing-on-its-own' (sic) and 'understanding-a-thing-in-context': much like our discussion of cities in global orders, *Small States* was ultimately about placing the object of analysis in its political, economic and historical reality.

## Whose urban agency is it anyway?

As global agendas embrace the 'urban', it is not just intergovernmental organizations, or indeed academia, that are betting on cities and issuing influential 'thinking' on the urban age. Many would, in fact, argue that it is the growing interest, commitment and analysis on the part of the private sector which is shining the spotlight on cities and turning them into central foci of our time. Indeed, the expression 'urban age' itself, brought to fame by the eponymous series of conferences hosted by the London

School of Economics' Cities programme, was largely made possible by the support of Deutsche Bank – certainly a private sector pioneer in this type of research investment. Philanthropies and foundations are also progressively focusing their support on the urban drivers of the 21st century. Front and centre is the spot occupied today by Bloomberg Philanthropies, but similar efforts can be found internationally. This is no trivial consideration: if we can quite easily answer in the affirmative as to the global recognition cities now wield, their effective global powers appear to be validated only if we consider them *in partnership* with these other entities, which we could call 'global urban brokers', represented for instance by consultancy companies, advisors or internationally mobile academics involved in the production, mobility and deployment of a global urban discourse. These are today key drivers of the internationalization of the 'business of cities' (Clark and Moonen, 2016) as the domain of both the 'ideas industry' (Drezner, 2017) but also as a reconfiguration of the built environment and construction industry into a broader 'cities'-focused realm, and of urban governance more generally.

The growing urban emphasis of major corporate and industry interests like Siemens, Honeywell, JP Morgan and many others bears testimony to the growing perception that cities are the key playground of and inevitable partners in 21st century economics. Here we see the powers of coalition that cities hold, as catalysts (and thus leaders, in my view) of global initiatives. However, this also highlights the growing challenge by cities to retain influence in an age of fragmentation and business privilege, and their central dependence on the global brokerage of these actors to 'spring' and scale-jump (Brenner, 2000) their governance beyond the local.

This is illustrated by the two recent reports *Powering Climate Action* and *Climate Action in Megacities 3.0*, issued jointly by C40 Cities and ARUP on the powers that cities have for effective global climate action. On the one hand, the report points out how 'cities have the tools to play their part' in global climate governance and how they are in a 'unique position to catalyse wider climate action'. C40's interpretation of power, here, is the 'degree of control or influence' mayors exert over assets (such as buses) and functions (such as economic development) across all city sectors. Yet, despite the tendency by cities to share similar power 'profiles', the report points to the fact that the way power is wielded is more important than the dimensions of power a city might have: the most 'controlling' cities (as a type of urban governance) are in fact not necessarily the ones delivering the most 'transformative' climate action. On the contrary, it is C40 cities that focus on legislation (as local institutional frameworks) and even more importantly *collaboration* (with other cities, as much as the private sector) that most effectively achieve this (C40 and ARUP, 2016). This is not just in terms of vision setting and networking of policies. Rather it also has very tangible physical manifestations as it prompts cities to take up different forms of procurement practices and gives access to international markets (and vice versa, with obvious impact on local market actors), as well as results often times in the circulation not just of ideas, but of experimentation on the streets of cities in both the Global North and South (Bulkeley and Castán Broto, 2013). The capacity of cities to locate themselves, but also and even more importantly to leverage their positioning as 'thing-in-context' in global governance becomes critical here: as the C40 studies points out, it is not the 'big' cities with a broad scope of power and large ownership of assets, but rather the more networked ones that eventually put in place the greatest number of actions on the ground and the widest territorial reach of these across their conurbations. Mixtures of networked transnational relations are key – with other cities as much as with business, civil society and multilateral organization. As Robinson rightly argues in this volume, then, a thorough understanding of the production of cities' global political agency must start from a better appreciation of the 'diversity of circuits and processes through which a territorialized urban agency is configured'. As Barthold also notes in the case of C40 in this volume, what is fundamental in this globalization of cities' political agency is the role of global urban brokers who facilitate the networked institutionalization of 'knowledge/policy interfaces', allowing for the scale jumps I indicated above.

The influence of these brokers in shaping the city's participation in more-than-local political dynamics, but also the neo-liberal straightjacket that the current skeleton of global governance imposes on local government, all demand more nuanced takes on the global pull of mayors and their peers. In this sense, a fair outlook on city diplomacy should not disregard that these urban politics can be replicated in the seemingly 'flat' domain of city diplomacy and transnational urban networking. Sofie Bouteligier (2013), for example, has voiced concerns on the split existing in transnational municipal initiatives (like Metropolis) between a 'giving end' and a 'receiving end' that largely coincides with a North-South divide between cities and that tends to result in a 'poverty of influence on the side of the Global South'. Echoing the 'policy mobility' literature, Bouteligier warns us that this persisting divide might problematize the transfer of 'best practices' in urban policy, which are nowadays critical in the transnational exchanges between cities and indeed between cities and corporate interests. Models of success in the North are seemingly repurposed for very different conditions in the 'South', or at times even more worryingly 'success stories' from the South travel with consultants and industry standardization between emerging global cities with little critical scrutiny. Bouteligier's concern, certainly echoed by a growing and well-established community of postcolonial urban scholars, is that Asian and African cities are not empowered (enough) through transnational municipal networking, and do not have strong enough ownership of what happens in these organizations. We could argue that to some degree this North-South bias has been changing in the past decade: cities like Rio de Janeiro (chairing C40 until 2016), Istanbul (leading on UCLG), or Mexico City and Santiago (active across many different networks) have progressively surfaced to the forefront of global urban governance. Yet of course this growth on the global scene is not free from challenges. Certainly, Wenz's assessment of Cape Town's ascent in this volume is telling, as she points to the need (very much like in the North) for proactive backing for this effort to 'go global' by local urban governance elites and the centrality of global actors to enrol these cities in their circles.

This, however, brings up a critique that is reiterated in a social network analysis of the C40 Cities by Taedong Lee (2014). Lee pointed out how, in fact, assumptions about horizontal network relations and mutual inter-actions for city networks are incorrect. If the aim of such networks is to facilitate the learning of best practices, learning ties are not necessarily equal and, on the contrary, analysis of networks like C40 might suggest how networking (for learning or for resource pooling) might be unevenly distributed. Lee's analysis illustrates, for instance, a tendency towards separation between those cities that seek information and those that pro-vide information. These 'groupings' tend to be further facilitated by ho-mophily of culture (language and regional proximity) and higher levels of climate change policy performance. In 2012, Lee's suggestion to redress this bias was to form 'advisory committees' with multiple stakeholders within networks like C40 (Lee and Van de Meene, 2012), and indeed the Group has already moved in this direction by developing, since 2013, sub-networks with thematic purposes (like waste management or low-emission vehicles), and by electing to Chairperson a 'Global South' voice like the mayor of Rio de Janeiro Eduardo Paes. Yet, the road ahead is obviously still marked by the need to improve the balance between North and South, 'core' global cities and peripheral metropolises, and this needs to be balanced across that very wide landscape of city networking and di-plomacy highlighted in the first sections of this chapter. The problem, as Lee, Bouteligier (2013) and many others have pointed out, and as I have indicated in my own writing too, is not just that (to paraphrase Boute-ligier) these new global governance arrangements (re)produce problem-atic hierarchical divisions. Rather, a lack of attention and swift action on these divides might, in the long term, be likely to turn into insurmounta-ble path-dependencies when dealing with current urbanization challenges from climate, to development and health. Yet, as highlighted in Wenz's chapter on Cape Town, some possibilities for rebalancing are also ger-minating in Global South cities' engagements with global governance. Where, as Wenz puts it, 'avid inquisitiveness for engaging with "best prac-tice" examples from elsewhere' and tapping into global networks is with an 'equally strong sense of situated agency', encapsulated in her chapter in Cape Town's desire to become a model city in its own right, global action and politics might be better balanced with local needs. Yet, to date, limits in the encounters of urban and international expertise prevent us from de-veloping an effective appreciation of the way global-local urban *political* dynamics function and change through time.

## For a global urban governance (research) agenda

In order to prevent these possibly catastrophic path-dependencies, and to enhance the repertoire of frames for understanding global urban governance, urban and international relations, scholars certainly need to engage critically with the 'pro-urban' literature as has been done in the different contributions to this book. Yet they must also take a step forward in communicating to both the academe and (if not first and foremost) to policymakers and the general public solid evidence-based assessments of the challenges of global urban governance – a pressing research agenda we cannot shy away from at this point in time. As several of us engaged in the implementation of the 'New Urban Agenda' and the SDGs noted at a recent UN General Assembly high-level meeting in 2017, and as evidence from numerous other sources (Birch, 2016; Revi, 2016; Parnell, 2016) makes clear, we need to thoroughly rethink the role and status of cities within global governance in general, and the UN System in particular. However flawed it may be, the United Nations still remains pivotal in convening, spurring, and sanctioning global urban action (e.g. Revi, 2017). Building a better understanding of this context, and of the complex governance arrangements that the multilateral sector implies, is critical. Along with very few other actors, like global philanthropies (e.g. Ford Foundation) and a scarce number of governments (e.g. Germany) and multipurpose city networks (e.g. UCLG and Cities Alliance), the UN agencies are perhaps the best available tool for rebalancing the politics of global urban governance and their inherent 'metrocentric' biases (Bunnell and Maringanti, 2010) in practice and research. This is critical because the growth of city diplomacy, with hundreds of city networks, and its impact, with thousands of transnational initiatives, are unequivocal, and are met with growing recognition in major multilateral fora. The field is rich and primed for multi-stakeholder engagement between the academe and practice (Birch, 2017). Equally, the need here is as much scholarly as it is of practical everyday urban policymaking, as the global-local connections increasingly forged by cities as international actors ultimately shape the very mundane ways we, for instance, take out the trash in the morning and walk across streets the world over.

## References

Acuto, M., 2013. *Global cities, governance and diplomacy: the urban link*. London: Routledge.

Acuto, M., 2016. Give cities a seat at the top table. *Nature*, 537, pp. 611–613.

Acuto, M. and Parnell, S., 2016. Leave no city behind. *Science*, 352(6288), p. 873.

Acuto, M. and Rayner, S., 2016. City networks: breaking gridlocks or forging (new) lock-ins? *International Affairs*, 92(5), pp. 1147–1166.

Acuto, M., Parnell, S. and Seto, K.C., 2018. Building a global urban science. *Nature Sustainability*, 1(1), pp. 2–4.

Bai, X., Dawson, R.J., Ürge-Vorsatz, D., Delgado, G.C., Salisu Barau, A., Dhakal, S., Dodman, D., Leonardsen, L., Masson-Delmotte, V., Roberts, D.C. and Schultz, S., 2018. Six research priorities for cities and climate change. *Nature*, 555(7694), pp. 23–25.

Barber, B.R., 2013. *If mayors ruled the world: dysfunctional nations, rising cities*. New Haven, CT: Yale University Press.

Birch, E.L., 2016. A midterm report: will habitat III make a difference to the world's urban development? *Journal of the American Planning Association*, 82(4), pp. 398–411.

Birch, E.L., 2017. Inclusion and innovation: the many forms of stakeholder engagement in habitat III. *Cityscape*, 19(2), p. 45.

Blank, Y., 2007. Spheres of citizenship. *Theoretical Inquiries in Law*, 8(2), pp. 411–452.

Bouteligier, S., 2013. Inequality in new global governance arrangements: the North–South divide in transnational municipal networks. *Innovation*, 26(3), pp. 251–267.

Brenner, N., 2000. The urban question: reflections on Henri Lefebvre, urban theory and the politics of scale. *International Journal of Urban and Regional Research*, 24(2), pp. 361–378.

Brenner, N. and Schmid, C., 2014. The 'urban age' in question. *International Journal of Urban and Regional Research*, 38(3), pp. 731–755.

Bulkeley, H. and Castán Broto, V., 2013. Government by experiment? Global cities and the governing of climate change. *Transactions of the Institute of British Geographers*, 38(3), pp. 361–375.

Bunnell, T. and Maringanti, A., 2010. Practising urban and regional research beyond metrocentricity. *International Journal of Urban and Regional Research*, 34(2), pp. 415–420.

C40 Climate Leadership Group and ARUP, 2016. *Powering climate action*. London: Arup.

Clark, G. and Moonen, T., 2016. *World cities and nation states*. West Sussex: Wiley & Sons.

Curtis, S., 2016. *Global cities and global order*. Oxford: Oxford University Press.

Derudder, B., Hoyler, M., Taylor, P.J. and Witlox, F. eds., 2012. *International handbook of globalization and world cities*. Cheltenham: Edward Elgar Publishing.

Drezner, D., 2017. *The ideas industry*. Oxford: Oxford University Press.

Glaeser, E., 2011. *Triumph of the city*. New York: Penguin.

Gleeson, B., 2012. Critical commentary. The urban age: paradox and prospect. *Urban Studies*, 49(5), pp. 931–943.

Hall, T. and Hubbard, P., 1996. The entrepreneurial city: new urban politics, new urban geographies? *Progress in Human Geography*, 20(2), pp. 153–174.

Holston, J., 1999. *Cities and citizenship*. Durham, NC: Duke University Press.

Katzenstein, P.J., 1985. *Small states in world markets: industrial policy in Europe*. Ithaca, NY and London: Cornell University Press.

Keil, R. and Mahon, R. eds., 2010. *Leviathan undone? Towards a political economy of scale*. Vancouver: UBC Press.

Lee, T., 2014. *Global cities and climate change*. London: Routledge.

Lee, T. and Van de Meene, S., 2012. Who teaches and who learns? Policy learning through the C40 cities climate network. *Policy Sciences*, 45(3), pp. 199–220.

McPhearson, T., Parnell, S., Simon, D., Gaffney, O., Elmqvist, T., Bai, X., Roberts, D. and Revi, A., 2016. Scientists must have a say in the future of cities. *Nature*, 538(7624), pp. 165–166.

Parnell, S., 2016. Defining a global urban development agenda. *World Development*, 78(C), pp. 529–540.

Revi, A., 2016. Afterwards: habitat III and the sustainable development goals. *Urbanisation*, 1(2), pp. x–xiv.

Revi, A., 2017. Re-imagining the United Nations' response to a twenty-first-century urban world. *Urbanisation*, 2(2), online first, https://doi.org/10.1177/2455747117740438.

Robinson, J., 2011. Cities in a world of cities: the comparative gesture. *International Journal of Urban and Regional Research*, 35(1), pp. 1–23.

Sheppard, E. and Plummer, P., 2007. Toward engaged pluralism in geographical debate. *Environment and Planning A*, 39(11), pp. 2545–2548.

Taylor, P.J., 2013. *Extraordinary cities: millennia of moral syndromes, world-systems and city/state relations*. Cheltenham: Edward Elgar Publishing.

van der Pluijm, R. and Melissen, J., 2007. *City diplomacy: the expanding role of cities in international politics*. The Hague: Netherlands Institute of International Relations' Clingendael'.

Weiss, T.G. and Wilkinson, R., 2014. Global governance to the rescue: saving international relations? *Global Governance*, 20(1), pp. 19–36.

Wendt, A., 2004. The state as person in international theory. *Review of International Studies*, 30(2), pp. 289–316.

# Part II
# Exploring city political agency around the globe

# 8 Greening the global city

## The role of C40 cities as actors in global environmental governance

*Sabine Barthold*

### Introduction[1]

> In international relations, the C40 are representing around 300 million people and the economic power of a medium sized country. It's really almost like a virtual state if you think about it. A country with shared values.
>
> (Interview Mayor's Advisor on Climate Change,
> Toronto, December 2012)

'We are still in' – that statement was issued by more than 1,000 US mayors, governors, universities and businesses in an open letter to the world as immediate reaction to the announcement by President Donald Trump on 1 June 2017 that the US would withdraw their signature from the Paris Agreement. In December 2015, at the COP21 of the UNFCCC in Paris, 195 states, including the US, had committed themselves to climate actions that would limit global warming to a maximum of 2°C in comparison to pre-industrial levels. One day after President Trump's statement, 338 US mayors, calling themselves *Climate Mayors*, announced that they committed themselves to the Paris Treaty regardless of their national government's position (Halper, 2017). This event illustrates a relatively new phenomenon in international politics that scholars and practitioners of global governance have repeatedly observed and reasoned about: the new self-confidence with which city governments appear on the stage of Global Environmental Governance (GEG). As Noah Toly (2010, p. 138) argues, in multiscalar governance, 'cities and the environment may, in fact, be the most powerful mechanisms for affecting and effecting governance outcomes'. But why have urban actors and local governments been so effective in 'scale-jumping' particularly in the field of global environmental governance?

This chapter discusses the agency of urban actors in multilevel climate governance through the lens of the C40 Cities, a global organization of mayors of 80+ megacities that not only facilitates the sharing of knowledge, technologies and policy tools between cities, but also actively bolsters the interests of global cities within the architecture of Global Climate

Governance. 'Global Cities' are nodes in global flows of capital, goods, people and information, and their power positions often go beyond the geographical regions in which they are located (Taylor, 2004). The mapping of hierarchies and networks of cities as an emerging geography of contemporary globalization and uneven development has been documented in a number of major research projects (see Sassen, 1991; Short and Kim, 1999; Taylor, 2000, 2004; Luke, 2003; GaWC, 2012); however, studies that examine the role of cities in political globalization and their political agency in international affairs have for a long time been widely absent from the dominant debates in international relations, which are generally centred around state agency within international institutions, organizations and regimes, as well as the state's responses towards emerging non-state actors under the label of 'global civil society' (Lipschutz, 1992, 2005; Keck and Sikkink, 1998; Salamon et al., 1999; Brown, 2000). As Michele Acuto (2010, p. 427) points out,

> global cities are capable of illustrating political alignments involved in urban redevelopment and city decision-making, through their continuous positioning as strategic sites of globalization and participation in the global networks of world politics. Such a view requires an understanding of the city not solely as place but also as an agent in world affairs.

In this sense, globalization is rather a 'new cartography, where states still have their spaces but are squeezed by important classes of non-state actors' (Segbers, 2010, p. 43). These new classes of actors include those involved in urban politics and policymaking who 'act beyond their own cities to practice or perform urban globalness' (McCann and Ward, 2011, p. xvii). Betsill and Newell (2010) have argued that the institutionalization of inter-urban knowledge production and policy transfer through trans-municipal networks (TMNs) has changed the role of cities and local governments within the architecture of GEG significantly, providing cities with the opportunity to set norms and standards on environmental issues (Bulkeley et al., 2003; Betsill and Bulkeley, 2004, 2006). The question is, what are the interests, strategies and underlying rationalities of urban governance coalitions to promote sustainable or 'green' urbanism? And what motivates global city mayors and other urban actors to establish themselves as 'leaders' in global climate governance?

Methodologically, the study is based on a mixed-methods approach including content and discourse analytical tools. I draw on a large body of quantitative and qualitative information generated by C40 to endorse global cities as 'climate leaders'. The documents I gathered include testimonials published by C40 itself on the history, organization and goals of the network, reports on initiatives and programmes promoted by the network and by member cities, as well as by C40 partners like Arup, the Clinton Climate

Initiative (CCI), the World Bank, the World Resources Institute (WRI) or The °Climate Group. Furthermore, I collected data from grey literature like C40 meeting reports and press releases, local government documents, as well as web and social media content. In the period 2012–2016, I furthermore conducted a number of semi-structured interviews with current and former members of C40 staff, representatives of city administrations and mayor's offices, municipal sustainability or climate and energy departments, city council members, local environmental organizations and community activist groups in three different member cities of the C40.

In the following sections, using the case study of C40 Cities, I will examine the growing ambitions of urban actors to set norms and standards in global climate governance. In the first part, C40 cities – the politics of policy transfer, I will elaborate on the politics of policy transfer among cities through global city networks. I argue that systematic knowledge and technology transfer and the global dispersion of ideas and practices between cities through networks like C40 allow urban actors to have a growing influence on norm setting in global environmental discourses. Part two, C40 and the greening of the global city, will scrutinize the ambitions of C40 to promote 'green' urbanism as a driver for economic growth and a branding strategy for the 'entrepreneurial city'. I will show that urban environmental programmes and climate initiatives are not simply meant to preserve ecological resources in the city, but are increasingly used to generate investment opportunities for global capital and drive urban economic development. Section three, city diplomacy – C40 as actor in global environmental governance, discusses the vertical networking aspirations of C40 engaging in 'city diplomacy' and the authority that the network gained in climate governance by partnering with international organizations and global corporations. C40 cities and the 'worlding' of green urbanism section is focussed on 'bright green' urbanism as a 'worlding' (Roy and Ong, 2011) strategy for cities around the globe and the influence of C40 on global development discourses through the fostering of eco-modernization policies. I will conclude the chapter by discussing the implications of the rising influence of city networks like C40 in multi-scalar governance for the place of the urban in the new cartography of global climate governance.

## C40 cities – the politics of policy transfer

The process of 'planetary urbanization' (Lefèbvre, 1970; Brenner, 2014) traversed a symbolic threshold in December 2008 when – according to the UN Population Division – for the first time in history half of the world's population was living in cities[2]. Estimates say that more than two-thirds of mankind will be urban dwellers by 2050, and most of them will be poor. The highest urbanization rates can be found in non-Western societies, where large numbers of rural populations are migrating to the slums and informal settlements of African and Asian megacities. Already today, urban areas consume about 80% of energy and more than 75% of natural resources globally and they account for about 75% of the world's carbon emissions (see Kamal-Chaoui and Robert, 2009). Long deemed the epitome of ecological degradation and social inequalities (Davis, 2006), urbanization is increasingly being framed as a beacon of hope for a viable way out of global environmental crises like climate change. The UN Population Fund stated: 'If cities create environmental problems, they also contain the solutions. The potential benefits of urbanization far outweigh the disadvantages: the challenge is in learning how to exploit its possibilities' (UNFPA, 2007).

The number of transnational city networks that formed around environmental issues has increased significantly during the past two decades[3]. In global discourses on the environment, cities and city networks act as nodes, or 'knowledge-policy interfaces' (Chilvers and Evans, 2009) in the horizontal and vertical diffusion of ideas and policies. For a long time, environmental knowledge was provided mainly by 'classic' experts from universities and research institutes. But increasingly, 'stakeholders' like non-profit organizations, local citizen initiatives, and community activism groups are integrated into environmental governance regimes by addressing local environmental issues and implementing neighbourhood and community action programmes. However, there are certain boundaries between different bodies of knowledge ranging from urban grassroots activism to the abstract knowledge domains of engineers, urban planners and theorists, which address the often competing demands for the use of urban space, land property values, public and private interests. Stakeholders are struggling over the politics of pollution, environmentalism and landscape design as well as issues like energy, biodiversity and changing regional agriculture. These concerns often revolve around normative notions of life quality, environmental justice and sustainability.

The call for cities and local governments to become 'sustainability leaders' or 'climate champions' has resonated widely among multiple scales of governance and across private and public sectors. Urban governments are developing 'innovative' environmental and climate protection programmes and cities often serve in this process as laboratories for 'pioneering technologies' and policies (Bulkeley, Castán Broto and Edwards, 2015). Anxious not to miss out on the latest trends, municipalities are developing new governance structures and urban planners are, as McCann and Ward (2011) have shown, 'scanning the globe' for the increasingly mobile policy strategies that help them embrace (often competing) economic, social and ecologic demands. These ready-made 'off the shelf programmes' are often developed by technical experts or consultancy firms. The 'fast' knowledge and policy transfers between cities (Peck and Theodore, 2010) are accelerated by international institution building and policy networks that form global regimes of science, regulation and capital investment.

C40 Cities Climate Leadership Group was initially founded as the 'C20' in 2005 as the initiative of London Mayor Ken Livingstone (Mayor of London,

2005; The °Climate Group, 2005). Livingstone's initial aim was to bring together the mayors of 20 of the world's largest megacities of both the developed and the developing world, to address the issue of climate change in a parallel event to the G8 summit in Gleneagles. He had the idea of forming a 'buying club' of major cities that could collectively negotiate lower prices for the procurement of 'green technologies' like LED street lighting or hybrid buses from global manufacturers since 'large cities have sizeable economies that are ideal markets to incubate, develop and commercialize greenhouse gas reducing and adaptation technologies, including those to improve energy efficiency, waste management, water conservation and renewable energy' (C20, 2005, p. 1). Shortly after the first summit in London, Mayor Livingstone and the William J. Clinton Foundation, a non-profit charitable organization founded by the former president of the United States Bill Clinton, negotiated a partnership between C40[4] and the newly created Clinton Climate Initiative (CCI) on the issue of climate change adaptation and mitigation in large cities. In summer 2006 Livingstone, on behalf of C40, and CCI signed a Memorandum of Understanding (Large Cities Climate Leadership Group and CCI, 2006) that initiated a long-term partnership between the C40 Cities and the CCI acting as implementation partner for urban climate programmes (see also William J. Clinton Foundation, 2006).

The Clinton Foundation had used this business-oriented approach before in its campaign on fighting HIV/Aids in developing countries: they brought big pharmaceutical firms to the table with a consortium of national and regional governments to negotiate discounts on bulk purchases of HIV/Aids medication or testing utensils. CCI now plans to transfer this market-based governance model to large cities and multinational companies that have offered 'green' technologies and services for the members of C40. Furthermore, CCI built connections to private banks, international financial institutions and institutional investors to provide sources for capital and develop financing mechanisms. The first programme based on this model was the Energy Efficiency Building Retrofit Program (EEBRP) in 2007. The idea behind the programme was to increase the global market for energy-efficient products and services through the guaranteed demand of a large number of cities. CCI collaborated with many of the world's biggest Energy Service Companies (ESCOs), like Honeywell, Siemens and Johnson Controls Inc., and assisted cities and ESCOs to access sources for financing that would allow them to undertake large-scale retrofit projects. CCI organized standardized contractual models for commercial bank loans with five banks (ABN AMRO, Citi, Deutsche Bank, JP Morgan Chase and UBS) that committed to provide US$1 billion each for the necessary investments. They further negotiated terms and prices from technology producers to which cities could then get access in a purchasing alliance. Similar models were set up for rapid transit bus systems, electric vehicles or IT technologies for smart grid infrastructures. C40 partners, for example, with private corporations like Volvo or Daimler Benz and not-for-profit organizations like EMBARQ to promote the extension of Bus Rapid Transit systems in member cities. Other initiatives include public-private partnerships with Philips for LED city lighting and with Siemens for the provision of Smart Grids.

However, the purchasing alliance strategy turned out to be problematic for local governments because they are often bound to strict legal and ethical guidelines for public procurement and have to be politically responsive to the needs and demands of local businesses in decisions over public spending. The preferential treatment of large multinational companies, which gain the most profit from economies of scale, meant significant disadvantages for local businesses and industries. As a city government executive told me (Interview with former Toronto mayor's climate advisor, Toronto, December 2012), this purchasing alliance has from the beginning been a major source of discomfort within the network:

> One of the reasons that it didn't work was that most cities have very strict rules how they spend their money, how they award contracts. Most cities [...] have competitive bids. [...] It's hard for an organization of cities by cities. It is up to the cities, but it's very hard for a global organization to say 'we are going to partner with Siemens, we are going to partner with Philips' unless you can show there is really no strings attached. [...] All that has to go through the Steering Committee and [they] are usually very uncomfortable with any sort of formal structured relationship with the private sector.

C40 Cities has a set of characteristics that differentiate the organization from other TMNs that engage in GEG such as ICLEI (Local Governments for Sustainability) or Cities for Climate Protection (CCP). First, other than in most TMNs, which often encompass large numbers of smaller municipalities, membership to C40 is limited to big or globally important cities. Membership is based on invitation by the network members. Currently, more than 80 cities are members of the network, differentiated in two status categories: (1) *megacities* and (2) *innovator cities*. Megacities make up the core of C40's membership, cities in this category must either have a population of 3 million or more, and/or a metropolitan area population of 10 million or more, or must be one of the top 25 global cities, ranked by current GDP output, at purchasing-power parity (PPP). Members with megacities status retain sole access to C40 leadership and governance opportunities, such as serving as C40 Chair, as members of the C40 Steering Committee and the C40 Board (see C40, 2012). The second category, innovator cities, was introduced to integrate smaller cities that are of value for the network as significant model cities for their progressive environmental strategies and innovations in local sustainable urban development.

And secondly, C40 is designed as an organization of mayors rather than as a trans-municipal policy network. Within other TMNs local governments are often represented by senior staff of institutions of the city administration, such as planning departments or environmental, transportation or energy offices. C40, however, is principally fashioned to establish networking opportunities for powerful actors that have authority and take a 'leadership' role. Even though de facto mayoral powers might differ from city to city depending on constitutional arrangements of particular localities, C40 is assuming that mayors of big cities have direct powers in local decision-making and long-term development directions. The focus on mayors therefore shows that the purpose of the network goes beyond facilitating policy and knowledge transfer between cities, but also to establishing direct relations between the local authorities of the world's biggest cities and institutions as well as global finance and multinational corporations. The agency of global cities and the powers of mayors to act independently from and sometimes even against the state and its official diplomatic track in international politics, has been crystallized in the C40 slogan 'Nations talk, cities act.'[5] The narrative that global cities' mayors can establish an alternative model for global environmental governance to produce meaningful outcomes in international climate negotiations was from the very beginning a central distinguishing element of C40.

C40 is offering a platform for the exchange of knowledge and best practices through direct city-to-city cooperation. In the case of C40, the main formats of city-to-city cooperation and knowledge transfer are the biennial C40 Mayors Summits and workshops or webinars for city executives on topics like public transportation, waste management or building efficiency. The summits are not only meeting platforms for local officials and urban experts for the mere purpose of policy learning and 'best practice' exchange, they also provide important networking opportunities for businesses and investors to present their 'cutting-edge' products and technology innovations to urban decision-makers. A local climate change advisor for the municipal government (Interview, Toronto 2012) explained:

> Some of these organizations, like Siemens and Philips, have research and almost foundation arms and they try to engage with C40 cities when they come together at the summits and some of the workshops. But there is generally a discomfort, especially if mayors are there [...] they never wanted to feel like they are in some sort of situation where they are being lobbied.

Yet, the city-to-city exchange is a crucial part of the knowledge transfer within the network and highly appreciated by the member cities. And summits and workshops are essential for facilitating this exchange by 'allowing city staff a forum and a place to talk to each other [...] It is very hard to have those city-to-city exchanges without an organization like this – for practical reasons' (Interview with former mayor, Toronto, September 2012). To assist municipal governments with their 'greening' ambitions C40 provides a number of Regional Directors and Programme Experts. These experts offer free consulting on diverse subject matters including waste, transportation, water and climate adaptation, hybrid and electric bus transport, sustainable urban development, sustainable communities, IT infrastructure and information exchange, electric vehicles and renewable energies. These full-time C40 employees are also indispensable for connecting city governments with private sector companies who offer technologies and services in public-private partnerships (PPPs). This business-based governing strategy shows that 'C40's policymaking is hybridized by the involvement of multinationals, consultancy firms and planning technicians, all taking part in recasting the discursive field in which climate governance occurs' (Acuto, 2013, p. 128).

## C40 and the greening of the global city

In public discourse, globalization is often regarded as an external dynamic that cities and urban populations are confronted with. The research supporting this chapter, however, is based on theories of globalization that highlight the meaning of local places in the making of global space(s) and the network character of global urban communities (Keil, 1998; Swyngedouw, 2004; McCann and Ward, 2010, 2011; Massey, 2011; Robinson, 2011). C40 Cities was established as a network of powerful cities with global significance that already played a key role in the global economy. From a critical geography perspective (Wallerstein, 1974; Castells, 1996; Smith, 2008) the relative dominance of global cities and city regions over others derives from their role for the world system of production, in which information forms the substance of contemporary society and networks shape its organizational forms and structures. With the rapid development of ICT, the World Wide Web, the rise of finance capital and mass consumer culture, the globalized economy was suspected to gradually diminish all spatial distances and location was to become irrelevant to the functioning of global firms. Instead, as Saskia Sassen (1991) showed, in a globalized economy the question of location has become ever more crucial for multinational companies, because those service industries that enable the management of global flows of goods, people and capital tend to be spatially concentrated in so-called 'Global Cities'.

Since the 2000s, the pursuit of a particular, market-oriented version of sustainability has been made popular by powerful political and economic elites as an instrument for broader goals of urban economic growth by way of 'eco-modernization'. Ecological modernization schemes, with their emphasis on energy and resource efficiency, are grounded in the assumption that environmental degradation can be decoupled from economic growth through technological innovation and the marketization of 'green' products and technologies. This 'technological fix' is promising a win-win solution for two major obstacles of urban regulation at once: the limited fiscal basis of city halls and the global environmental crisis, which are in fact both 'side effects' of global capitalist development (Smith, 2008). This eco-modernization perspective has been readily adopted by urban sustainability planning, because it implies that economic expansion and increasingly dense urbanization – as long as they are based on 'green' technologies and 'sustainable' designs – will eventually reduce ecological harm without any sacrifices or significant shifts in contemporary lifestyles or existing sociopolitical structures (Isenhour, 2015).

Coalitions of local governments, urban planners and architects, entrepreneurs, technocrats, global consultancy firms like McKinsey, multinationals like Siemens as well as 'Big Green Groups' (Klein, 2013) are promoting 'natural capitalism' and largely market-based strategies for environmental issues like climate change in cities. Concepts like 'urban sustainability' and 'green urbanism' are thereby presented as consensus-based, technocratic, modern and beyond political struggles and deliberations. However, the ubiquity of 'greening' programmes in cities from the top to the bottom of the global economic and political hierarchy does not simply mean that progressive environmental agendas have been co-opted by mainstream political and economic circles. Nor can it be dismissed as simple 'green-washing' of existing unsustainable practices. Rather, environmental and climate change programmes in cities have been integrated into economic development strategies that capitalize on 'green' products and services as new resources for investment and economic growth.

When 'CEO Mayor' (Lowry, 2007) and billionaire Michael Bloomberg took over as Chair of the network in 2010 he hired the global consulting firm McKinsey to refashion the relatively loose network of cities into a fully functioning organization with full-time staff, an executive team as well as a board of investors (or 'funding partners'). During the financial crisis in 2008, the Clinton Foundation had suffered substantial financial losses and had to cut back spending in various fields of engagement, as did the CCI. Large portions of funding now came from Bloomberg himself when Bloomberg Philanthropies bought out the Clinton Climate Initiatives 'Sustainable Cities' section and integrated it into the C40 Cities organization. Bloomberg also invested heavily in media and marketing for the C40 network and created a PR division that would promote the networks and member cities' activities through their website, TV and print media but also through social media like blogs, Twitter and Facebook. In an interview in 2014 the former Head of C40 Communications explained to me that many of the member cities take advantage of the professional marketing opportunities of C40 to help them brand their city image as 'green' and 'sustainable' both to the international media and to their own constituents.

The element of inter-city cooperation and member support is certainly central to the network's functioning. But in the way that ecological improvements are seen as providing cities with a competitive advantage, C40 also creates a competition space in which cities are ranked and measured based on their 'performance' as climate champions. In 2011 the network partnered with Arup, a global architecture and planning consultancy firm to systematically collect data on member cities' climate actions and establish a baseline for the cities to measure their progress in the reduction of greenhouse gas emissions. The performance of member cities according to these measuring and accounting standards also forms a basis for awards and prizes. Since 2013, C40 hosts the 'Siemens City Climate Leadership Awards' that grants six out of the 10 prizes to C40 member cities who are being recognized by a wide audience as global 'climate leaders'. For Siemens, the partnership with C40 is a strategy to develop their image as a market leader in green and sustainable technologies. But since partnering with individual firms directly would be damaging to the credibility and non-profit image of C40, awards are a way to access funding without being viewed as compromised by market benefits.

In the face of global financial crises and ecological disasters, the promotion of sustainability and of clean, green lifestyles adds symbolic value to places and helps to 'brand' cities as modern, future-oriented and attractive destinations for flows of capital and people. In the urban context, the 'bright green' version of sustainability (Steffen, 2009) – much like the 'creative city' (Florida, 2002) discourse before – can be aligned with pre-existing economic and marketing goals. The growing number of 'green city' rankings show that local environmental and sustainability programmes are increasingly perceived as a key factor for global inter-urban competitiveness (see for example Tamanini, J./Dual Citizen LLC, 2016; Economist Intelligence Unit (EIU)/Siemens, 2012; Kamal-Chaoui and Robert, 2009). Green or sustainable urbanism is here more than just an environmental programme, it is a branding strategy for the 'entrepreneurial city' (Harvey, 1989, 1990; Hall and Hubbard, 1996) that advertises places with emphasis on high standards of life quality and public services and promotes cities as desirable places for business and investment. These 'soft' urban qualities, in which cultural and environmental features of urban spaces and lifestyles play a major role, are key for cities to become more appealing for global businesses and investors, the highly skilled workforce that is the basis for the 'new economy' (Gibbs and Krueger, 2007), and not least, the business and leisure tourists who spend large amounts of the surplus value generated elsewhere in the cities they visit.

## City diplomacy – C40 as actor in global environmental governance

'In the international policy arena', Castán Broto (2017, p. 1) argues, 'climate change has most often been presented as a global problem requiring global solutions', but this understanding of global and local as separate realms of action has changed significantly since the 2009 COP15 in Copenhagen. Seen by many as the ultimate failure of multilateralism in climate change negotiations, COP15 marked the shift away from a regulatory approach to governing climate change that was based largely on the authority of national states towards a broader understanding of multilevel governance based on the coordination of multiple forms of state and non-state action. On the one hand, this shift acknowledged the role of local governments alongside other forms of state control, and on the other hand, recognized the multiple networks of actors like businesses, public-private partnerships, community activist groups and civic organizations that directly or indirectly shape urban trajectories (Castán Broto, 2017, p. 2). And even though the urban is at the centre of these new governance arrangements, urban coalitions that organize around governing climate change transverse the city and its administrative boundaries and create new, networked forms of transnational governance (Bulkeley and Betsill, 2005; Toly, 2008; Betsill and Newell, 2010; Bulkeley, 2010; Bulkeley, Castán Broto and Edwards, 2015).

City networks like the C40 Cities increasingly also engage in vertical networking and thereby transverse the scale of the nation - state in international politics. In December 2015, parallel to the UNFCCC COP21 Climate Conference in Paris, more than 1,000 mayors, local representatives and community leaders took part in the Climate Summit for Local Leaders convened by Paris Mayor Anne Hidalgo and the newly appointed UN Secretary-General's Special Envoy for Cities and Climate Change Michael Bloomberg. While the leaders of the world struggled to adopt a climate change agreement that would succeed the Kyoto Protocol with its limited participation, world cities announced a number of initiatives that sought to define their future role in the architecture of Global Climate Governance. For instance, a global Compact of Mayors brings together well over 2,000 cities with voluntary commitments to reduce emissions by 454 megatons by 2020. Key partners of this covenant include global city networks such as the C40 Cities Climate Leadership Group, ICLEI (Local Governments for Sustainability), and the United Cities and Local Governments (UCLG). One year later, in October 2016, the United Nations Conference on Housing and Sustainable Urban Development (Habitat III) was held in Quito. The New Urban Agenda adopted by the UN General Assembly acknowledged the vital relationship between global urbanization and human development and emphasized the central role that the world's cities will play in the fulfilment or failure of the future transition towards sustainable development (UN General Assembly, 2016).

The partnership with international organizations adds a new dimension to the agenda of C40 Cities called 'city diplomacy'. As collective actors organized in global city networks, cities can gain agency and institutional validation by partnering with national governments and international organizations in GEG. C40 is directly working with global institutions like the World Bank and UN-Habitat. With strong ties to international organizations the network is a facilitator of networking opportunities between cities, global corporations and the 'green technologies' they offer, and international sources for funding. The network multiplies local efforts and connections that cities often have already established and provides them with access to privileged international policy ties, e.g. to international organizations and institutions, which are generally restricted to states and their diplomatic representatives. The global linkage between major cities with a global geographical diversity that is represented in the network provides cities with a legitimacy to speak on global environmental issues and increases their independence from regional and national governments, thereby enhancing their local policymaking independence (Acuto 2013, p. 110). In an interview (Toronto, November 2012), a mayor and former C40 Chair explained this characteristic function of the network as global-local linkage and the norm setting power that derives from it:

> It's why I found C40 really exciting, because we made real change and we were changing the generally understood possibilities at the same time, changing how people thought, how the national governments thought, how the international community thought – and acting at the same time. [...] That's the great gift of mayors' organizations. Once you get a critical mass you can change people's minds and you can do what is needed to be done and that will spread. The knowledge spreads, the actions spread, that energy spreads.

As Acuto (2013, p. 111) argues, this representation of global cities as legitimate actors in global environmental governance is achieved precisely because cities are 'being identified as active components of an effective effort against global warming while also not acting on behalf of their national governments, but rather in the name of their "duty" as key governance scales on environmental issues'. The C40 Networks authority in global governance derives from the twofold characteristic of the city as scale of governance: the legal authority and political representativeness of local governments and the authority of technical knowledge in the realms of planning and urban development. In this sense C40 is functioning both (1) as a network of cities that is pooling and multiplying the particular authority, fiscal competency and technical expertise that distinguishes local governments from other scales of climate governance; and (2) as a collective actor in global environmental governance, providing municipal governments with access and influence on global decision-making procedures and international policy circles that local actors usually lack. In 2014, another important milestone for C40 in setting global norms for urban climate policies was accomplished at the COP 20 in Lima when C40, together with the World Resources Institute (WRI) and ICLEI, launched the Global Protocol for Community-Scale Greenhouse Gas Emissions Inventories (GPC) as a common global standard for GHG inventories in cities. This standard is setting norms for the ways cities measure and report their GHG emissions and account for their reductions in carbon emissions[6]. Adopting the GPC standard on reporting GHG emissions also became the requirement for cities to join the Compact of Mayors[7] that was launched at the UN Climate Change Summit in

New York in September 2014 by Secretary General Ban Ki Moon and his newly appointed Special Envoy for Cities and Climate Change Michael Bloomberg. The Compact is a global agreement on the measuring, documentation and reporting standards for urban climate programmes between cities, major city networks and a number of partners, which include international organizations like UN-Habitat and the World Bank Group, standardization organizations like Car*bonn* and the Carbon Disclosure Project (CDP), environmental think tanks, and consultancy firms like The °Climate Group and the WRI, multinationals like Veolia as well as global 'Big Green' organizations like the World Wide Fund for Nature (WWF). The distinctive features of C40 place the organization between global governances as a classical decision-making process amongst states in global regimes and organizations and the globalization-from-below model that has been discussed under the label of 'global civil society'.

## C40 cities and the 'worlding' of green urbanism

Climate initiatives can be found in cities all over the world, irrespective of their geographical location and development level, but there are differences in the local conditions that shape the way climate governance is being implemented. Generally, urban climate governance discourse distinguishes between mitigation and adaptation, whereby the former is often framed as a global, the latter as a local governance issue (Castán Broto, 2017, pp. 2–3). And even though this division has led to conflicts and trade-offs between adaptation and mitigation efforts and has in some cases hindered integrated action, the debate is shaped by the separation of the two aspects. In affluent cities, with high levels of consumption, climate governance often prioritizes mitigation and the reduction of urban carbon emissions, while cities where large parts of the population live in informal settlements with structural deficiencies in the provision of basic public services focus more often on climate adaptation and the vulnerabilities that are associated with climate change.

Networks like C40 often imply a levelled, non-hierarchical power geometry between member cities of the organization. But since C40 represents a cross-boundary engagement with explicit political and economic functions, 'the Group is prone to replicate the dynamics of world affairs in microcosm' (Acuto, 2013, p. 116). Urban scholars have been critical about the growing influence of global elite networks and the impacts they have on place-making in cities around the world (Hodson and Marvin, 2010; Roy and Ong, 2011). Robinson (2006) has argued that the world of travelling ideas and policies is very distinct from (and often collides with) the specificities of concrete places, their local cultures and the actual life-worlds of local communities and residents. Environmental discourses are territorialized – and often contested – in particular urban settings but are still connected to a dominant global growth paradigm that privileges certain solutions over others in the global circulation of knowledge and policies. As Hodson and Marvin (2010, p. 14) suggest, studying global-urban connections and 'understanding the practices and politics of intervention in the city can therefore help analyse how the global agenda develops by and through the urban, and how the urban at the same time is transformed into a global project'.

Ecological modernization and 'green growth' are heavily promoted within the network as development strategies particularly for rapidly growing megacities of the global South. Urban climate programmes are thereby not simply an end in themselves – 'green urbanism' has become an indicator of a city's modernness, an exercise in 'worlding' – the 'art of being global' – and an attempt towards 'world recognition in the midst of intercity rivalry and globalized contingency' (Roy and Ong 2011, p. 3). In fact, C40s business-oriented strategy with a technocratic focus on energy efficiency by technology generally benefits affluent cities far more than cities in developing countries, because a larger part of GHG emissions derive from the former's disproportionately high energy use. Cities in the global South do not profit from this policy strategy, because they deal with other environmental problems like water and air pollution or poor waste management. Often Southern cities are more concerned with issues of climate change adaption and vulnerabilities as well as social and infrastructural resilience to deal with extreme weather events like flooding, storms or draughts and therefore have other technology needs (see Roman, 2010, p. 78). Resiliency discourses highlight the close relationship between people's ability to cope with disasters and systemic drivers that intensify both poverty and vulnerability. Castán Broto (2017, p. 3) argues,

> The vulnerability of the urban poor is not only dependent on their exposure to climate-change related hazards, but also on the structural conditions that reproduce poverty, such as economic inequality, lack of political representation, deficient access to services, and diminished life opportunities.

Recent discourses on resiliency turn attention away from technological fixes and planning measures toward governance processes that facilitate structural and institutional adaptation to changing ecological environments. As Castán Broto (2017, p. 3) puts it, 'effective adaptation planning in urban areas is akin to a revolution in urban governance that addresses the political, economic and social determinants of poverty and climate change vulnerability'.

The structural dependencies between uneven development and climate change governance are equally relevant for mitigation efforts. In cities, most climate change mitigation initiatives concentrate on production-based carbon emissions. Even though in industrialized countries production-related GHG emissions have significantly decreased over the last 25 years, mainly through domestic deindustrialization and growing investments in clean energy and green technologies and services, overall emissions from international trade have simultaneously amplified. Mitigation policies promoted by C40 concentrate almost exclusively on direct emissions through energy production and transport – policy areas where the co-benefits of climate governance through e.g. cost-reductions and public health increases are highest. This approach neglects a major part of climate change mitigation particularly in affluent cities: $CO_2$ emissions related to (over)consumption (Isenhour, 2015). Today, GHG emissions embedded in consumer goods in OECD countries exceed by far those directly related to production and transportation (Ahmad and Wyckoff, 2003). These emissions are often simply allocated to countries with lesser labour costs and lower environmental standards (Klein, 2014, p. 79). The problem of 'carbon leakage' demonstrates that the decrease in production-related GHG emissions in affluent countries can lead directly to an overall emissions increase related to the manufacturing of export goods, particularly in developing nations like China and India (Babiker, 2005). 'Bright green' sustainability concepts, however, often see policies that are explicitly targeting individual consumption as an affront to capitalist 'free-market logics' (Klein, 2014).

## Concluding remarks

In a complex system of multilevel governance, new players have entered the stage of world politics and new coalitions are forming around complex global issues like climate change. Urban climate governance today includes a broad spectrum of actors that transcend the classical system of nested hierarchies – ranging from local through national to global – in which politics was organized and studied for a long time. The prominence that C40 gained as an organization of cities within international climate governance institutions provides urban actors with a platform to influence global climate and development discourses. The systematic production, transfer and circulation of knowledge and policy models is thereby a means by which C40 has gained the authority to set technical and political norms to push the global environmental discourse in particular directions. C40 has acquired its 'leadership' position by bringing together a variety of economic, cultural and political elites around a globally circulating concept of 'sustainability' that idealizes technological innovation, economic growth, and modernity. This market-based version of urban sustainability that the network promotes by investing considerable amounts of private and public resources into marketing and PR, networking and standard-setting, resonates widely with cities around the world that are anxious to 'fall behind' in the global hierarchy of places and therefore provides C40 with enormous symbolic power in the discourse on sustainable urbanism and global climate governance.

The vertical networking strategies of the network in the form of 'city diplomacy' have provided urban actors with direct influence on global decision-making processes and international policy circles that have traditionally been restricted to national governments and international organizations. Urban governments are thereby not acting on behalf of their national governments but rather in the name of their 'duty' as key sites for the implementation of environmental policies. The political motives of urban actors and municipal governments are often perceived as un-ideological and pragmatic in contrast to the attitudes of their respective national governments' in global governance. But in fact, the 'we are all in this together' rhetoric endorsed by the C40 network tends to obscure the power relations that emerged from a history of colonialism and that continue to be manifest in the unequal development between North and South today. Here, a core-periphery pattern is emerging that is rooted in the very same global system of inequality and Western global hegemony that determines the current state of world politics. For many less affluent cities, environmental programmes and climate initiatives have become a 'worlding' strategy and an indicator for their 'modernness' – in a climate of global inter-urban competition, these features are prerequisites for attracting the highly volatile global capital streams to invest in urban economies.

For C40, with its focus on the sharing of technology and access to financing mechanisms, climate action is a means to advance economic growth and the 'bright green' economy. This policy paradigm ignores (and sometimes worsens) socio-economic inequalities within and between global cities and the particular vulnerability of the urban poor to climate hazards. Policies promoting energy-efficiency and the proliferation of green technologies are not per se a bad thing. But compressing urban sustainability discourses to technical fixes and economic rationalities weakens those initiatives that regard sustainability as a structural problem with multiple social, economic, ecological and cultural dimensions. In this 'bright green' version of 'sustainable urbanism', democratic politics is reduced to stakeholder participation and self-management of individual actors, which is consequentially disciplining, if not entirely excluding, voices that articulate political dissent with the overall problematizing of the objective at stake. This discourse tends to depoliticize the politics of nature by excluding questions of social justice, global inequality and struggles for alternative socio-economic models. There is reason to doubt that capitalist market mechanisms alone can resolve the environmental issues they have helped create in the first place. According to the Global Carbon Project (2017), 2017 was the year with the highest amount of $CO_2$ ever measured in the atmosphere. With carbon emissions rising at this pace, time to keep global temperatures at a safe level is running out fast. Frederic Jameson's famous quote, that today it is easier to imagine the end of the world than the end of capitalism, cuts to the core of the difficulties we face in tackling global climate change and creating a liveable, just, and healthy future for all mankind – perhaps the biggest obstacle to overcome in this struggle is the dominant ideology that is persistently mistaking the end of capitalism for the end of the world.

## Notes

1 The author would like to thank the organizers of the workshop 'The City as Global Political Actor' at UCSIA and the Urban Studies Institute of the University of Antwerp as well as the participants for their lively discussions and helpful comments that contributed to the development of this text. I would also like to thank the editors of this volume for their support and critical remarks, without which this chapter would not have been possible. Last but not least, I would like to thank my doctoral supervisors Dorothee Brantz, Roger Keil and Carola Hein for their tireless encouragement and support, without which I would not be the scholar I am today.

2 However, a recent study released by a team of researchers based at the European Commission suggests that the actual state of world urbanization is in fact much higher than most official estimates suggest. The team around lead researcher Lewis Dijkstra used high-resolution satellite images to determine the number of people living in urban areas and found that already 84 percent of the world's population, or almost 6.4 billion people, live in urban areas. This difference in the figures is mainly due to the fact that urban populations particularly in Africa and Asia appear to be strongly underestimated (see Scruggs, 2018).

3 Already in 2007 Keiner and Kim (2007) counted 49 international city networks working on sustainability topics in the major research project *'Networking Cities and Regions for Sustainability' (NetCiReS)* at ETH Zürich (see Schmid, Keiner and Kim, 2007).

4 By that time the network was named *Large Cities Climate Leadership Group* (2006) and Ken Livingstone was appointed Chair of the new organization. Later in 2006, when the membership had grown to 40 world cities, the name was changed to *C40 Cities – Climate Leadership Group*.

5 The slogan was gaining prominence during the UN Conference on Climate Change/COP 15 in Copenhagen 30 November through 13 December 2009, when C40 Cities organized a side event hosted by the Mayor of Copenhagen that was widely recognized by state delegations and drew a lot of media attention (Interview Toronto, September 2012). Michael Bloomberg also kept using this expression (see Bernstein, 2010).

6 http://www.ghgprotocol.org/city-accounting.

7 https://www.compactofmayors.org.

## References

Acuto, M., 2010. Global cities: gorillas in our midst. *Alternatives: Global, Local, Political*, 35, pp. 425–448.

Acuto, M., 2013. *Global cities, governance and diplomacy. The urban link*. London/ New York: Routledge.

Ahmad, N. and Wyckoff, A., 2003. Carbon dioxide emissions embodied in international trade of goods. *OECD Science, Technology and Industry Working Papers*, 2003/15. Paris: OECD Publishing.

Babiker, M.H., 2005. Climate change policy, market structure, and carbon leakage. *Journal of International Economics*, 65(2), March 2005, pp. 421–445.

Bernstein, A., 2010. Bloomberg takes city sustainability program global. *WNYC*, [online] 14 November. Available at: https://www.wnyc.org/story/286768-bloomberg-takes-city-sustainability-program-global/ [Accessed 31 January 2018].

Betsill, M. and Bulkeley, H., 2004. Transnational networks and global environmental governance. The cities for climate protection program. *International Studies Quarterly*, 2004(48), pp. 471–493.

Betsill, M. and Bulkeley, H., 2006. Cities and the multilevel governance of global climate change. *Global Governance*, 12(2) (Apr.–June 2006), pp. 141–159.

Betsill, M. and Newell. P., 2010. *Governing climate change.* London/New York: Routledge.

Brenner, N., 2014. *Implosions/explosions: towards a study of planetary urbanization.* Berlin: JOVIS.

Brown, C., 2000. Cosmopolitanism, world citizenship, and global civil society. *Critical Review of International Social and Political Philosophy*, 3, pp. 7–26.

Bulkeley, H., 2010. Cities and the governing of climate change. *The Annual Review of Environment and Resources*, 35, pp. 229–253.

Bulkeley, H. and Betsill, M., 2005. Rethinking sustainable cities: multilevel governance and the 'urban' politics of climate change. *Environmental Politics*, 14(1), pp. 42–63.

Bulkeley, H., Castán Broto, V. and Edwards, G.A.S., 2015. *An urban politics of climate change. Experimentation and the governing of socio-technical transitions.* New York: Routledge.

Bulkeley, H., Davies, A., Evans, B. Gibbs, D., Kern, K. and Theobald, K., 2003. Environmental governance and transnational municipal networks in Europe. *Journal of Environmental Policy & Planning*, 5(3), pp. 235–254.

C20, 2005. *Communiqué from C20: World cities climate change summit,* [press release], 5 October 2005. Available at: http://openpolitics.ca/tiki-index.php?page=C20+Climate+Change+Summit+Communique%2C+2005-10-05 [Accessed 21 December 2013].

C40 Cities Climate Leadership Group, 2012. *C40 announces new guidelines for membership categories,* [press release], 3 October 2012. Available at: www.c40.org [Accessed 21 December 2013].

Castán Broto, V., 2017. Urban governance and the politics of climate change. *World Development*, 93, pp. 1–15.

Castells, M., 1996. *The rise of the network society, vol. I. The information age: economy, society, culture.* Oxford: Blackwell.

Chilvers, J. and Evans, J., 2009. Understanding networks at the science-policy interface. *Geoforum*, 40, pp. 355–362.

Davis, M., 2006. *Planet of slums.* New York, London: Verso.

Economist Intelligence Unit (EIU)/Siemens, 2012. *The green city index,* [online] Available at: https://www.siemens.com/entry/cc/features/greencityindex_international/all/de/pdf/gci_report_summary.pdf [Accessed 1 February 2018].

Florida, R., 2002. The rise of the creative class. *Washington Monthly*, (May), pp. 15–25.

GaWC, 2012. *Research bulletin,* [online] Available at: http://www.lboro.ac.uk/gawc/index.html [Accessed 29 September 2012].

Gibbs, D. and Krueger, R., 2007. Containing the contradictions of rapid development? New economy spaces and sustainable urban development. In: D. Gibbs and R. Krueger, eds. *The sustainable development paradox: urban political economy in the United States and Europe.* New York and London: Guilford Publications. pp. 95–122.

Global Carbon Project, 2017. *Carbon budget and trends 2017,* [online] 13 November. Available at: www.globalcarbonproject.org/carbonbudget [Accessed 21 November 2017].

Hall, T. and Hubbard, P., 1996. The entrepreneurial city: new urban politics, new urban geographies? *Progress in Human Geography*, 20(2), pp. 153–174.

Halper, E., 2017. A California-led alliance of cities and states vows to keep the Paris climate accord intact. *Los Angeles Times*, [online] 2 June. Available at: http://www.latimes.com/politics/la-na-pol-paris-states-20170602-story.html [Accessed 22 June 2017].

Harvey, D., 1989. From managerialism to entrepreneurialism: the transformation of urban governance in late capitalism. *Geografiska Annaler*, 71B, pp. 3–17.

Harvey, D., 1990. *The condition of postmodernity*. Cambridge/Oxford: Blackwell.

Hodson, M. and Marvin, S., 2010. *World cities and climate change: producing urban ecological security*. Milton Keynes: Open University Press.

Isenhour, C., 2015. Green capitals reconsidered. In: C. Isenhour, G. McDonogh and M. Checker, eds. *Sustainability in the global city. Myth and practice*. Cambridge: Cambridge University Press. pp. 54–74.

Kamal-Chaoui, L. and Robert, A. eds., 2009. *Competitive cities and climate change*. OECD Regional Development Working Papers, N° 2. OECD Publishing, OECD.

Keck, M.E. and Sikkink, K., 1998. *Activists beyond borders: advocacy networks in international politics*. Ithaca, NY: Cornell University Press.

Keil, R., 1998. Globalization makes states: perspectives of local governance in the age of the world city. *Review of International Political Economy*, 5(4), Winter 1998, pp. 616–646.

Keiner, M. and Kim, A., 2007. Transnational city networks for sustainability. *European Planning Studies*, 15(10), pp. 1369–1395.

Klein, N., 2013. Naomi Klein: 'Big green groups are more damaging than climate deniers', interview with Jason Mark for Earth Island Journal. *The Guardian*, [online] 10 September. Available at: https://www.theguardian.com/environment/2013/sep/10/naomi-klein-green-groups-climate-deniers [Accessed 31 January 2018].

Klein, N., 2014. *This changes everything. Capitalism vs. the climate*. London, New York, Toronto: Penguin Books.

Large Cities Climate Leadership Group & CCI, 2006. Draft memorandum of understanding between the large cities climate leadership group and the William J Clinton Foundation. In: City Council of Melbourne, *Report of the Environment Committee C20: LARGE CITIES CLIMATE LEADERSHIP GROUP*, Agenda Item 5.3, 04/07/2006: 6–8. Available at: http://www.melbourne.vic.gov.au/AboutCouncil/Meetings/Lists/CouncilMeetingAgendaItems/Attachments/2006/EC_53_20060704.pdf [Accessed 11 October 2013].

Lefèbvre, H., 2003[1970]. *The urban revolution / Henri Lefebvre; translated by Robert Bononno; foreword by Neil Smith*. Minneapolis: University of Minnesota Press.

Lipschutz, R., 1992. Reconstructing world politics: the emergence of 'global civil society'. *Millennium*, 21(3), pp. 389–420.

Lipschutz, R., 2005. *Globalization, governmentality, and global politics. Regulation for the rest of us*. New York: Routledge.

Lowry, T., 2007. The CEO Mayor. How New York's Mike Bloomberg is creating a new model for public service that places pragmatism before politics. *Business Week*, [online] 14 June. Available at: https://www.bloomberg.com/news/articles/2007-06-24/the-ceo-mayor [Accessed 31 January 2018].

Luke, T., 2003. 'Global Cities' vs. 'global cities'. Rethinking contemporary urbanism as public ecology. *Studies in Political Economy*, 70, pp. 11–33.

Massey, D., 2011. A counterhegemonic relationality of place. In: E. McCann and K. Ward, eds. *Mobile urbanism. Cities and policymaking in the global world*. Minneapolis: University of Minnesota Press. pp. 1–14.

Mayor of London, 2005. *Mayor brings together major cities to take lead on climate change*, [press release], 10 April 2005. Available at: http://www.london.gov.uk/media/mayor-press-releases/2005/10/mayor-brings-together-major-cities-to-take-lead-on-climate-change [Accessed 16 September 2013].

McCann, E. and Ward, K., 2010. Relationality/territoriality: toward a conceptualization of cities in the world. *Geoforum*, 41, pp. 175–184.

McCann, E. and Ward, K. eds., 2011. *Mobile urbanism. Cities and policymaking in the global world.* Minneapolis: University of Minnesota Press.

Peck, J. and Theodore, N., 2010. Mobilizing policy: models, methods, and mutations. *Geoforum*, 41(2), pp. 169–174.

Robinson, J., 2006. *Ordinary cities. Between modernity and development.* London and New York: Routledge.

Robinson, J., 2011. The spaces of circulating knowledge. City strategies and global urban governmentality. In: E. McCann and K. Ward, eds. *Mobile urbanism. Cities and policymaking in the global world.* Minneapolis: University of Minnesota Press. pp. 15–40.

Roman, M., 2010. Governing from the middle: the C40 cities leadership group. *Corporate Governance*, 10(1), pp. 73–84.

Roy, A. and Ong, A. eds., 2011. *Worlding cities : Asian experiments and the art of being global.* Oxford: Wiley-Blackwell.

Salamon, L.M., Anheier, H.K., List, R., Toepler, S., Wojciech Sokolowski, S. and Associates, 1999. *Global civil society: dimensions of the non-profit sector.* Baltimore, MA: Johns Hopkins University, Institute for Policy Studies' Center for Civil Society Studies.

Sassen, S., 1991. *The global city: New York, London, Tokyo.* Princeton: Princeton University Press.

Schmid, W.A., Keiner, M. and Kim, A., 2007. *Networking cities and regions for sustainability. Final report.* Institute for Landscape and Spatial Planning, ETH Zurich.

Scruggs, G., 2018. Everything we've heard about global urbanization turns out to be wrong: researchers. Reuters, [online] 12 July. Available at: https://www.reuters.com/article/us-global-cities/everything-weve-heard-about-global-urbanization-turns-out-to-be-wrong-researchers-idUSKBN1K21UU [Accessed 17 September 2018].

Segbers, K., 2010. The emerging global landscape and the new role of globalizing city regions. In: M. Amen, N.J. Toly, P.L. McCarney and K. Segbers, eds. *Cities and global governance. New sites for international relations.* Farnham: Ashgate. pp. 33–44.

Short, J.R. and Kim, Y.H., 1999. *Globalization & the city.* Harlow: Pearson/Prentice Hall.

Smith, N., 2008. *Uneven development: nature, capital and the production of space.* Athens: University of Georgia Press.

Steffen, A., 2009. Bright green, light green, dark green, gray: the new environmental spectrum. *WorldChanging.com.* 27 February 2009.

Swyngedouw, E., 2004. Globalisation or 'glocalisation'? Networks, territories and rescaling. *Cambridge Review of International Affairs*, 17(1), pp. 25–48.

Tamanini, J./Dual Citizen LLC, 2016. *The Global Green Economy Index$^{TM}$. Measuring national performance in the green economy*, [online] Available at: http://dualcitizeninc.com/GGEI-2016.pdf [Accessed 1 February 2018].

Taylor, P.J., 2000. World cities and territorial states under conditions of contemporary globalization. *Political Geography*, 19, pp. 5–32.

Taylor, P.J., 2004. *World city network: a global urban analysis*. New York/London: Routledge.

The Climate Group, 2005. *Low carbon leader: cities*, [online] 1 October. Available at: http://www.theclimategroup.org/_assets/files/low_carbon_leader_cities.pdf [Accessed 6 June 2013].

Toly, N.J., 2008. Transnational municipal networks in climate politics: from global governance to global politics. *Globalizations*, 5(3), pp. 341–356.

Toly, N., 2010. Cities, the environment, and global governance. A political ecological perspective. In: M. Amen, N.J. Toly, P.L. McCarney and K. Segbers, eds. *Cities and global governance. New sites for international relations*. Farnham: Ashgate. pp. 137–149.

UNFPA, 2007. *State of the world population 2007. Unleashing the potential of urban growth*. New York: UNFPA.

UN General Assembly, 2016, September 29. *New urban agenda*. Draft outcome document of the United Nations Conference on Housing and Sustainable Urban Development (Habitat III) A/CONF.226/4*.

Wallerstein, I., 1974. *The modern world-system I: capitalist agriculture and the origins of the European world-economy in the sixteenth century*. New York: Academic Press.

William J. Clinton Foundation, 2006. *President Clinton launches Clinton climate initiative*. Press release, 1 August 2006.

# 9 Metropolitan regions as new scales and evolving policy concepts in the European Union's policy context

*Carola Fricke*

## Introduction: metropolitan regions in the European Union policy arena

What role could metropolitan regions play in the European context, given Barber's provoking thesis that mayors should rule the world through a co-operative network of cosmopolitan cities? Could European metropolises become supranational actors? The current status of metropolitan regions in the European Union (EU) indicates that metropolitan regions are rather far from supplanting national or supranational institutions. In fact, metropolitan regions have only recently emerged as viable scales and issues in the EU multi-scalar system.

An indicator of these processes is the increasing visibility of the term 'metropolitan region' – and its synonyms – in official European spatial planning documents, in regional policy regulations and in the Urban Agenda. For instance, metropolitan issues arise in policy documents of European institutions, including the European Commission, the European Economic and Social Committee (EESC), and the European Parliament – as well as other affiliated units such as ESPON and Eurostat. However, most of the current policy discourse addresses metropolitan regions only indirectly, either as functional areas that arise from the interaction between a city centre and its surrounding areas, as nodes of spatial and economic development or as an intermediate level to facilitate city-regional cooperation. Moreover, not only have metropolitan regions become a point of discussion in policy discourse, but certain regions have also begun to adopt an active role in the EU arena by lobbying for recognition in legislative documents and funding programmes through representative offices and networks.

This chapter builds on a long-established academic debate on metropolitan regions and policies in Western European states. This debate gathered momentum in the 1990s, with several researchers focussing on the importance of 'world cities' (Friedmann, 1986), 'global cities' (Sassen, 1991) or 'global city-regions' (Scott, 2001) in the global urban system (see Derudder, De Vos and Witlox, 2015 for a differentiation and comparison between these concepts). In addition, debates on 'city-regions' (see Neuman and Hull, 2009 for an overview) and comparative studies on metropolitan governance in Europe (see Herrschel and Newman, 2002; Salet, Thornley and Kreukels, 2003; Heinelt and Kübler, 2005) added a political perspective on metropolitan regions.

Despite the depth and breadth of the abovementioned debates, only some authors have paid attention to the relevance of metropolitan regions in the supranational context of the EU. Policy-induced studies generally adopt normative or prescriptive perspectives on EU metropolitan policies (Popescu, 2005) and give examples of various types of metropolitan governance (Tosics, 2011). Other authors reflect on the role of metropolitan regions in the EU from a more analytical perspective: Wilks-Heeg, Perry and Harding (2003) discuss the implications of European policies on metropolitan governance from a bird's eye perspective. Wiechmann (2009) captures the development of European policies on metropolitan regions in a descriptive, historical-chronological manner. Others investigate the external positioning and supranational activities of city-regions (Kübler and Piliutyte, 2007; Heiden, 2010).

The aim of this chapter is to add to this literature by exploring the increasing visibility of metropolitan regions, both as issues in EU policy and as scales in the supranational arena. Accordingly, the chapter investigates how metropolitan regions are addressed in various EU policies. To better understand the relationship between metropolitan regions and the larger context of the EU multi-scalar system, the chapter also incorporates a discussion of the main actors' contributions to the construction of a metropolitan dimension.

This chapter argues that the piecemeal incorporation of metropolitan regions into EU discourse can be attributed to processes of rescaling and reframing. On the one hand, the rise of the metropolitan region as a new entity in the EU multi-scalar polity can be conceived as the result of evolving social constructions of spatial scales. On the other hand, the rising relevance of metropolitan issues can be understood as a product of shifting policy frames.

The following analysis will illustrate these arguments by tracing the evolution of conceptions and terminology used to refer to metropolitan regions in EU discourse. European policy concepts will be analysed using three approaches, each of which builds on an existing strand of the academic debate, namely, an internal functional, external economic and governance approach to metropolitan regions. These approaches parallel to some extent the three dimensions of debate in research on city-regions (Harrison and Growe, 2014, p. 28).

In order to empirically analyse the metropolitan dimension in EU policies, this chapter identifies the main documents, statements and funding schemes from urban, regional and spatial policies that explicitly or implicitly refer to metropolitan regions. It traces the development of EU policies addressing metropolitan regions from their origins in the late 1990s, focussing on documents produced during the programming period of 2007–2013. The analysis avoids a chronological order, but rather explores the varying ways in which policies have defined metropolitan regions and related concepts such as city-regions and functional urban areas. Methodologically, this chapter takes an iterative, qualitative-interpretive approach to analysing the terminology in policy documents with regard to academic concepts.

The remainder of this chapter is divided into five sections. The following section discusses the notion of rescaling and reframing as analytical concepts for understanding the development of European metropolitan policies. The section on the parameters for reframing process introduces the relevant political and institutional contexts, summarizing the positions of the main actors involved in reframing metropolitan policies. A comparative section traces the development of three understandings of metropolitan regions in official EU policy, while the next section discusses the analysis's empirical findings with regard to the theoretical background.

## The construction of metropolitan policies as rescaling and reframing

The theoretical approaches of rescaling and reframing both inform the analysis of metropolitan issues in supranational policies. The increasing importance of metropolitan regions in EU policy could be interpreted as the product of political rescaling or as a reframing of policy concepts. The following section discusses the two approaches and how they can be combined.

### *Metropolitan policies as rescaling processes or 'production of scale'*

Understanding geographical scales 'simultaneously as sites and stakes of sociopolitical struggle' (Brenner, 1999, pp. 442–443) allows for an interpretation of metropolitan regions both as new components in the EU multi-scalar system and as reference-points for specific policies. The concept of 'scale' thereby goes beyond territorial understandings of governmental 'levels' and indicates the geographical relations between different institutional fixes. Brenner specifies the concept of scale not

> as a generic label for levels of state spatial organization – for example, the global, national, regional and local scales of the modern inter-state system [...] [but as an approach] in which state scalar selectivity is understood as an expression, a medium and an outcome of political strategies.
>
> (Brenner, 2009, p. 127)

Thus, the concept 'production of scale' should be restricted 'to its most generic function as a catchphrase for summarizing the proposition that geographical scales and scalar configurations are socially produced and politically contested through human social struggle rather than being pregiven or fixed' (Brenner, 2001, p. 600). In the context of metropolitan policies, the construction of scale entails 'the processes through which particular institutional fixes are formed at the metropolitan level' (Cox, 2010, p. 219).

Indeed, several studies argue that the establishment of new forms of governance in subnational city-regions, both in Europe and globally, can be interpreted as examples of state rescaling in the context of neo-liberalization processes (see for instance Brenner, 2003; Brenner, 2004; Cox, 2010). According to Brenner (2003), metropolitan institutional reforms since the 1990s were 'focused upon economic priorities such as promoting territorial competitiveness and attracting external capital investment in the context of supraregional trends such as globalization, European integration and intensified interspatial competition' (p. 15). In other words, the rescaling of metropolitan regions has to be understood in the context of intensified competition between cities and regions beyond the national context. The rise of metropolitan regions can also be interpreted as a product of rescaling processes within the EU. Heiden (2010) adopts the notion of rescaling of statehood to understand international activities of city-regions in Europe. From this perspective, the European orientation of metropolitan regions represents the outcome of rescaling in the context of neo-liberalization as it expresses a shift towards competitiveness and an attempt of positioning in the supranational context.

The establishment of new metropolitan levels alongside more established entities such as cities, therefore, represents a tentative institutional fix in addition to the existing territorial structures of Europe. In one sense metropolitan regions are 'unbounded' policy spaces; yet they are nevertheless constrained by institutional resilience – and at times by resistance from existing governmental levels (Gualini, 2006, p. 894). Policies regarding metropolitan governance thus exist in a state of constant tension with established political-administrative bodies (Cox, 2010, p. 216), a conflict Brenner (2001, p. 607) attributes to the institutional path dependency of established governmental levels. Indeed, the increasing relevance of (new) administrative levels often takes place in the context of a reframing of spatial development or territorial policies (Gualini, 2006, p. 893).

The approach of 'rescaling' and the related notions of scales and institutional fixes contribute to understanding the spatial dimension as well as institutional outcomes of socio-spatial processes. Analysing the production of geographical scale thus requires an understanding of policies and actors beyond established territorial levels. Such a perspective on the outcomes of political-strategic processes rather than structural aspects of traditional governmental systems has its challenges, because a coherent empirical method adapted to the processual character of rescaling has not been developed (Brenner, 2009). Therefore, this chapter argues that examining metropolitan regions through the lens of rescaling can benefit from the incorporation of interpretive perspectives on policymaking in multilevel systems.

### Metropolitan policies as political reframing

The following section intends to complement the analytical perspective proposed by rescaling scholars by combining it with the approach of political reframing, incorporating discourse and practices (Delaney and Leitner, 1997, p. 94). In order to compensate for the methodological and analytical limitations of rescaling, this study thus employs theories of shifting policy frames as social mechanisms. To what extent does the reframing approach contribute to understanding the underlying mechanisms of scalar shifts in metropolitan policies? For understanding the rescaling processes, it seems worthwhile to analyse the arguments and concepts mobilized in the policy discourse which motivate the introduction of metropolitan regions in the EU context.

The development of a metropolitan dimension in EU policies can be conceived as the outcome of a reframing process, as resulting from shifting frames of actors who shape EU policy. This builds on the approach of frame reflection by Rein and Schön (1993), who argue that policy problems are created in part through the act of naming: 'The complementary process of naming and framing socially constructs the situation, defines what is problematic about it, and suggests what courses of action are appropriate to it' (Rein and Schön, 1993, p. 153). Viewed in this context, direct or indirect references to metropolitan regions can be interpreted as an indication of the gradual establishment of a metropolitan dimension in EU policy. For analysing shifting frames of metropolitan policies, the following section examines the nested context of each policy – the sponsoring organization, the proximate policy environment and the role of adjacent institutions (Rein and Schön, 1993, p. 154; Rein and Laws, 2000).

The two approaches of reframing and rescaling seemingly stand in contrast to each other, because they build on different analytical assumptions concerning the driving forces behind political change. The notion of rescaling entails concentrating on the materiality and institutional fixes as outcomes of political struggles related to neo-liberalization processes. In contrast, the notion of reframing suggests focussing on conceptual shifts and symbolic aspects of policies. Despite this seeming tension and their different disciplinary origins – critical geography and interpretive policy analysis, respectively – the two approaches have much in common: both emphasize the character of politics as outcomes of social-strategic processes. The chapter thus combines the two approaches in a two-step analysis. The approach of rescaling builds the larger explanatory context, while the parameters of the reframing approach form the analytical frame for understanding the shifts in European metropolitan policies.

## Parameters for understanding European metropolitan policies as reframing

In a broad sense, metropolitan policies can be defined as public programmes, documents or other practices that refer to the spatial, economic or infrastructural development and governance of metropolitan regions (Zimmermann, 2012, pp. 152ff; Zimmermann, 2016). The EU lacks an explicit legal competence in metropolitan issues and addresses them only indirectly in other policy fields. Metropolitan policies are thus not understood as an independent policy field but rather as a subtopic within the fields of spatial, urban and regional policy.

In the EU, regional policies primarily aim to economically integrate member states through redistributive policies that build on structural funds allocated at the regional level. Such funds allow the European Commission to influence the policymaking and governance in the member states, especially at the subnational level (see among many others Bache, 2006; Bachtler and Mendez, 2007).

While the EU holds no genuine competence in the field of spatial planning, policy statements such as the European Spatial Development Perspective (ESDP, Council of Ministers Responsible for Spatial Planning & European Commission, 1999) or the Territorial Agenda (Council of Ministers Responsible for Spatial Planning and Urban Development, 2007b) represent tentative steps towards the coordination of spatial development in its member states. These statements incorporate various concepts of European space (Dabinett and Richardson, 2005, p. 203).

The EU also holds no legal competence in the areas of urban policies. Nevertheless, an 'Urban Agenda' has recently emerged, building on the recognition of an *acquis urbaine* (Atkinson, 2001). In its earlier developments, EU urban policies articulated a broad understanding of cities and primarily implemented funding programmes. Initially, these URBAN-initiatives focussed specifically on deprived urban areas (see for instance Tofarides, 2003; Dukes, 2008) and did not mention metropolitan regions or city-regional cooperation. In 2016, a relaunch of the Urban Agenda was undertaken by various EU institutions and stakeholders.

Institutionally, European policies for metropolitan regions take place in both formal and informal contexts. The formal legislative procedure involves the European Commission, the European Parliament and the Council of Ministers. Figure 9.1 depicts the hierarchical relationship between the main collective actors. One important collective actor is the Council of the EU, which represents the executive branch of each member state. The Urban Development Group (UDG) facilitates more informal coordination between the relevant ministries of the member states. The European Commission, a supranational executive body, is organized into Directorates-General (DGs); the DG Regional and Urban Policy and Eurostat are most relevant for metropolitan policies. Together with the Urban Intergroup of the

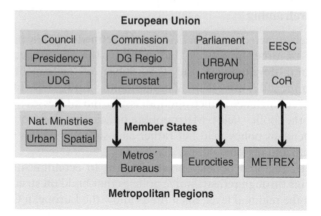

*Figure 9.1* Hierarchical relationships between collective actors in the EU multi-scalar
system.
*Source*: Author's own compilation.

European Parliament, these entities are the core actors in the EU's ordinary
legislative procedure. In addition to these key groups, a number of advisory
bodies also take part in policy formulation. The Committee of the Regions
(CoR), for instance, advises on subnational policy and consists of local and
regional representatives appointed by each member state. The EESC is another advisory body that represents the interests of civil society, employers
and employees.

Moreover, metropolitan regions have established themselves as independent subnational actors embedded in the EU multi-scalar system. Some
individual regions have lobbying offices in Brussels, others have formed networks, including Eurocities, an organization founded in 1986 to represent
the interests of Europe's 'largest cities' (EUROCITIES, 2013), and Metrex,
which represents the interests of 'metropolitan regions and areas' with more
than 500,000 inhabitants (METREX, 2014).

## Comparing concepts of metropolitan regions in EU policies

Analysing the use of the term metropolitan region – as well as related
terms such as metropolitan area, city-region, or functional urban area – in
policy documents can help reveal principal actors' positions and underlying frames. Because there is no agreement on how to define metropolitan
regions, the following overview aims to elucidate the variations in terminology and conceptions of metropolitan regions in policy discourse and
practice.

The corpus for the comparative analysis consists of official documents
published by institutional EU actors, including policy statements, general

communications, programs, research reports and other publicly accessible texts. The documents were selected for their relevance to metropolitan issues, considering the wide potential variation in definitions of the term. The analysis is limited to the most relevant documents published during the funding period 2007–2013, when policies for the new period starting in 2014 were prepared.[1]

Overall, three main concepts of what constitutes 'the metropolitan' can be identified in the policy discourse. They correspond to approaches proposed in recent academic debate on metropolitan regions (see also Harrison and Growe, 2014, p. 27 differentiating an agglomeration, scale and hub-and-spoke perspective on city-regions). To some extent, the academic conceptualizations have inspired the policy discourse via policy-induced studies or research. Nevertheless, the academic analytical concepts reflect different perspectives on existing 'metropolitan realities'. Accordingly, the methodology for document analysis is neither purely deductive nor inductive, but rather iterative-interpretive in carving out variation between empirical policy concepts in correspondence to academic notions.

Table 9.1 gives an overview of the main criteria informing definitions of metropolitan regions and related terms in the academic debate. The first approach focusses on functional interdependence and the internal relationships between a city centre and its surroundings. This line of thought is responsible for the coining of the terms 'metropolitan area' and 'functional urban area'. The second one is an external economic approach that

*Table 9.1* Comparing approaches in academic literature related to metropolitan regions.

| *Approach* | *Defining criteria* | *Related concepts* | *Exemplary authors* |
|---|---|---|---|
| **Internal functional** | Interdependence between city centre and surroundings in terms of flows of commuters or exchanges of services | Metropolitan area; functional urban area | Davoudi, 2008; Parr, 2005; Rodríguez-Pose, 2008 |
| **External economic** | Functions of larger cities in a global or European urban hierarchy | Global city-regions; world cities; global cities | Friedmann, 1986; Knox and Taylor, 1995; Sassen, 1991; Scott, 2001 |
| **Governance** | Coordination between territorial entities, Institutions, organizations | Metropolitan governance; city-regional governance or cooperation | Blatter, 2006; Herrschel and Newman, 2002; Nelles, 2013 |

*Source*: Author's own compilation.

emphasizes the relative importance of metropolitan regions in the global urban system. This approach employs terms such as 'global city-regions', 'world cities' and 'global cities'. The third stream of debate focusses on inter-local coordination between political entities, stressing the concept of governance and political cooperation.

The following interpretive sections build on the three identified academic understandings of metropolitan regions, which reveal the key actors' positions in the European context. This analytical differentiation helps to carve out the main frames mobilized in the policy process.

### *Functional urban areas as empirical concepts in statistics and policy induced research*

The first approach defines metropolitan areas using mostly statistical methods. In the context of EU policies on metropolitan regions, this functional perspective corresponds to statistical methods for identifying metropolitan areas using data that is available at a subnational spatial resolution (for instance, at the NUTS[2] 3 level) such as population density. The evidence-based techniques generally use the terms 'metropolitan area' or 'urban area' instead of metropolitan region. The ESPON 1.4.3 project, for instance, defines metropolitan areas as functionally interdependent urban units: morphological urban areas (MUAs) are identified by high population density, while functional urban areas (FUAs) are labour basins and a surrounding commuting area (ESPON, 2007). Fact sheets published by the Commission refer to 'metro regions' as 'larger urban zones', defined as having more than 250,000 inhabitants (European Commission, Directorate-General Regional and Urban Policy, and Eurostat, 2014). Other Commission publications argue that the EU is becoming more 'metropolitan', a statement they support by citing increasing population and GDP in certain metro regions (European Commission, 2009, p. 1). Finally, the publication on spatial typologies in the EU introduces a statistical definition of metropolitan regions that is similar to the concept of 'larger urban zones' (European Commission, 2011b).

This functional approach to metropolitan areas is often regarded as the most 'neutral' one because it incorporates referencing statistics and scientific research. Though the data collection process can vary significantly between member states, many studies nevertheless attempt to move beyond national boundaries and governmental levels. The Commission, for instance, whose political influence partly relies on policy-induced research and formalized knowledge, argues that the method can provide an 'evidence-based approach' to policymaking. Other actors, including interest groups and networks of metropolitan regions also build their argumentation on the idea that metropolitan areas are empirically existing units.

## Metropolitan regions as nodes in networks or as functional entities in urban hierarchies

The second approach takes an external economic perspective, focussing on the relative importance of metropolitan regions in spatial and economic development within the global or European urban system. In conceptual-isations reminiscent of the hub-and-spokes metaphor, many of these pol-icies frame metropolitan regions as nodes in spatial networks or urban hierarchies.

Examples of this approach mainly consist of declarations by the Council of Ministers in the field of spatial planning. These declarations reflect both the interests of the member state holding the presidency at the time and the results of negotiations between various member states. The ESDP, for in-stance, mentions the role of metropolitan regions in economic development when discussing polycentric spatial development of the European territory (Council of Ministers Responsible for Spatial Planning & European Com-mission, 1999, p. 20). Similarly, the Territorial Agenda (Council of Ministers Responsible for Spatial Planning and Urban Development, 2007b) refers to metropolitan issues by discussing the way that 'city-regions' can contribute to regional development and urban-rural partnerships. The Territorial Agenda 2020 also frames 'metropolitan regions' in a passage on globalization and economic potentials as 'assets for the development of the whole European territory, provided that other regions benefit from their dynamism and are connected through networks' (Council of Ministers Responsible for Spatial Planning and Territorial Development, 2011, p. 4). Finally, the Barca Report (2009) for the Commission argues that metropolitan regions contribute to economic development because of their positive agglomeration effects.

## Metropolitan governance as a concept for inter-municipal cooperation and regional funding

The third approach views metropolitan governance primarily as a concept that could increase inter-municipal cooperation. This perspective defines metropolitan regions as entities with interdependencies that extend across political-administrative borders. It emphasizes the ways that cooperation between core cities and surrounding municipalities can address policy is-sues more effectively than they can be dealt with at other administrative levels. In the context of the EU, this understanding of metropolitan regions as political entities co-evolved with the debate on governance, and also with more practical research on forms of city-regional cooperation. For in-stance, the Commission-sponsored study 'Cities of Tomorrow' (European Commission, 2011a) introduces metropolitan regions as a new level for pol-icy implementation, arguing that there is a need for governance (European Commission, 2011a, pp. 35ff).

Further examples of this approach can be found in statements related to the Urban Agenda. The Leipzig Charta (Council of Ministers Responsible for Spatial Planning and Urban Development, 2007a), for instance, suggests integrated strategies at the city-regional level (pp. 2–3) that aim to strengthen metropolitan and city-regional cooperation (p. 3). The Toledo Declaration (Council of Ministers Responsible for Spatial Planning and Urban Development, 2010) lists metropolitan partnerships between neighbouring communities as a crucial strategy to maintain territorial cohesion (p. VII). The European Parliament's report on Territorial Agenda and Leipzig Charta (European Parliament, Policy Department Structural and Cohesion Policy, 2007) also emphasizes the need for balanced, mutual partnerships between urban and rural areas instead of the development of solely metropolitan regions (European Parliament, Committee on Regional Development, 2008). The CoR moreover frames metropolitan regions in terms of governance in a 2011 document, referencing the role of cooperation between the metropolis and its hinterland (Committee of the Regions, 2011, p. 126) However, the CoR's positioning vis-à-vis metropolitan issues is somewhat reconciliatory. This reserved attitude could be related to the interests of the CoR's members – established local and regional authorities – who prefer maintaining their capacities instead of strengthening the metropolitan scale.

Conceiving metropolitan regions as political entities that can build cooperation between territorial units lays the groundwork for incorporating them into regional funding regulations. In the funding period 2007–2013, the European Regional Development Fund (ERDF) offered few opportunities for metropolitan regions; a summary of its urban dimension (European Commission, 2008) emphasizes the role of cities and metropolitan areas for economic development but does not explicitly reference city-regional cooperation or metropolitan regions. In contrast, the 2014–2020 ERDF funding regulation introduces new opportunities for city-regions (Scholze, 2014) – and thus for metropolitan regions – under the strategic development of inter-municipal cooperation, referred to as 'functional areas'. For instance, article 7 of the ERDF regulation (1299/2013) dedicates 5% of the funding to 'sustainable urban development projects' which include integrated territorial investments (ITI) in functional regions (and also CCLD and Urban Innovative Action).

## Discussion: the piecemeal emergence of metropolitan regions in the EU

The above overview of empirical understandings of metropolitan regions proposed by various actors underlines the growing visibility of metropolitan issues in the EU context. The following section discusses the extent to which these conceptual positions on metropolitan issues can be interpreted as outcomes of rescaling and reframing processes.

### Metropolitan regions as varying concepts in EU policies

A comparison of concepts of metropolitan regions in EU documents, funding and research reveals that the issue has become more important in recent years. However, metropolitan issues have not yet been established fully in the agenda – and metropolitan policies are a long way from forming an independent policy field. Official EU documents contain only implicit or indirect references to metropolitan regions, and use more neutral expressions such as 'functional region' or 'urban authorities'. The EU holds no regulatory power in the area of metropolitan policies, thus both redistributive and symbolic instruments are tentatively contributing to shaping an implicit metropolitan dimension.

For instance, the use of the term 'metropolitan area' signals a cautious approach to the issue, as this expression impinges less on the sovereignty of the EU member states to define their own territorial structure (referring to the principle of subsidiarity). The term 'metropolitan region', in turn, expresses positive or normative expectations. Positions on how metropolitan areas should be defined in relation to existing administrative boundaries, moreover, vary significantly between national contexts. Also, the mismatch between the political-administrative and the functional-economic region fuels demands for more comprehensive concepts.

The variety of concepts of 'the metropolitan' moreover shows that the EU is a long way from settling on one core definition. This can be interpreted as both a shortcoming and an advantage. The vagueness of the term could be interpreted as a sign of lacking conceptual clarity. However, its flexible character also leaves room for multiple meanings. While definitions of the term vary between actors and national contexts, the concept remains, like the terms 'polycentricity' and 'city-region', a 'plastic term' (Rossignolo and Toldo, 2008) – a word that is open to various interpretations. The vague quality of the term can thus allow for innovation, and perhaps greater success in policymaking (Rodríguez-Pose, 2008, p. 1036).

### Metropolitan regions as issues of reframing

The overview of metropolitan concepts contributes to better understand the role of EU and non-EU actors in framing metropolitan issues. The above discussed understandings of metropolitan regions are not mutually exclusive, as some actors build on a combination of understandings. Nevertheless, we can carve out the tendencies of actors' positioning towards metropolitan regions. Some, such as the EESC and interest groups such as Metrex and Eurocities actively contribute to introduce the issue into the agenda. However, supranational actors such as the European Parliament avoid overemphasizing the role of metropolitan areas, and frame them instead as one territorial level embedded in a multi-scalar system that also includes municipalities, cities and regions. In order to deal with opposition to the concept of metropolitan regions, a coalition of actors attempted to bridge the gap between urban and rural interests by introducing the concept of rural-urban partnerships.

Other influential actors outside formal EU institutions include specific member states' governments, researchers and other political organizations such as the OECD. These groups have helped develop the conception of 'the metropolitan' through the publication of studies and declarations, and also influenced decision-making through the formation of partnerships and coalitions. Thereby, more detailed analysis of the reframing processes of metropolitan issues in the EU context (Fricke, forthcoming) shows that particular understandings of metropolitan regions and the support of metropolitan regions as a relevant policy issue depends on individuals and research organizations as frame sponsors (see Rein and Schön, 1993, pp. 151 and 158).

### Metropolitan regions as outcomes of rescaling

The analysis also allows to interpret the increasing relevance of metropolitan regions as the outcome of rescaling processes. Metropolitan regions can be understood as 'new institutional fixes' in a multi-scalar system or as an issue of sociopolitical struggles. However, metropolitan regions present in the EU context are often those whose administrative boundaries overlap with other governmental levels or which are institutionalized in governance structures. The contingency of metropolitan rescaling processes becomes visible when the positions of involved actors are examined in-depth, uncovering tensions between competing understandings and in the variety of proposed attributes to define 'the metropolitan'. Besides, the concepts currently evolving in the EU context are likely to have some impact on the distribution of competences and resources in future policy negotiations and are thus linked to more material processes and policy struggles captured by the notion of rescaling. Definitions of metropolitan regions that assume they have a positive role in economic and spatial development can have an even larger impact on power exerted by existing governance levels. The at times conflicting definitions of the metropolitan dimension in EU policies thus expose the way that metropolitan policies are involving politics of scale and social-constructivist processes.

While comparing positions in the policy process can contribute to an understanding of the complexity of rescaling metropolitan regions, more concrete evidence of the evolving nature of metropolitan policies can be found in the statements of the EESC. In its earliest opinion on metropolitan regions, published in 2004 (European Economic and Social Committee, 2004), the EESC emphasizes the importance of metropolitan areas for economic development, but also points out their vulnerability to social inequality and environmental problems. By justifying the relevance of metropolitan regions with their importance for economic development, the EESC opinion mirrors the political-economic perspective of the world city approach. The opinion acknowledges the variety of possible definitions of metropolitan areas, referring to the mismatch between socio-economic interdependencies and administrative boundaries, which can be interpreted as acknowledging contingent frames. The subsequent EESC's opinion on metropolitan areas

(European Economic and Social Committee, 2007) gives suggestions on how to address metropolitan regions in EU policies, by calling upon various institutions such as the Commission, Eurostat and ESPON, and proposing the establishment of a structured debate or a Green Paper. The document also refers to examples of metropolitan regions and policies within EU member states. The EESC's most recent opinion goes even further by recommending 'the establishment of a High Level Group (HLG) or Task Force on metropolitan developments alongside the Commission's existing Interservice Group on Urban Development' (European Economic and Social Committee, 2011, p. 1). The opinion also suggests further steps to institutionalize the issue in the European agenda through research and knowledge exchange.

This example demonstrates that the growing relevance of metropolitan issues in the EU is more than a question of reformulating a territorial system through the introduction of a new level for policy implementation. Instead, the above interpretation of the EESC's opinions points to reframing processes involved in the development of a metropolitan dimension in the EU context. For justifying the need of forming a metropolitan dimension in EU policies, the EESC's opinions thereby draw on a combination of analytical notions of metropolitan regions and 'best practice' examples from policies in member states. Analytical or even critical scholarly perspectives on metropolitan regions thereby enter the political discourse by being translated into aspirational political concepts. Moreover, the emergence of metropolitan regions in the EU can be interpreted as a 'politics of scale' in which specific actors, including the Commission and other interest groups, 'engage with a specific scalar division of labour of the state' (Cox, 2010, p. 219).

## Conclusions

The aim of this chapter was to explore the emergence of metropolitan regions in EU policies, as well as the role of various institutional actors in shaping conceptions of 'the metropolitan'. Broadly, the document analysis revealed the extent to which metropolitan regions are still primarily alluded to in implicit references made in policy statements and funding regulations. This demonstrated the extent of variation in conceptions and definitions of metropolitan regions in the political realm. Comparing conceptions showed that understandings of metropolitan regions corresponded to three approaches proposed in academia: as functional urban areas delineated by statistical indicators; as nodes in networks or functional entities in urban hierarchies; and as political regions for inter-municipal cooperation. Thus, European policies for metropolitan regions build on diverse ideas of what constitutes a metropolitan region, and the understanding of the term 'metropolitan' varies between actors. The term's multifaceted character has contributed to a rather fragmented policy discourse, in which several characteristics and concepts for metropolitan regions exist in parallel. For example, an evidence-based and statistical approach to functional urban areas is regarded by some actors as neutral, while others infuse the term with more normative expectations by understanding metropolitan regions as political entities of inter-municipal cooperation or an emerging scale in a multilevel system.

Besides, the chapter set out to explore the rescaling and reframing of metropolitan regions in the context of EU policies. Due to their rather recent appearance, metropolitan regions as political actors do not have a formalized role in the EU arena and lack a fixed position in formal institutions, in contrast to other subnational levels of government. Thus, they have limited influence on the policy process; their impact is visible only implicitly. However, via networking and lobbying, metropolitan regions have succeeded in establishing themselves as an issue and a scale for policy implementation in the context of EU policies. Yet, metropolitan regions remain only one subnational level among other entities such as cities and regions. However, the engagement of metropolitan actors in the supranational context also represents a window of opportunity for overcoming local or national constraints by using the EU's symbolic and material recognition to enhance their own governance capacities.

In this sense, the analysis confirms the thesis that metropolitan regions are currently becoming new scales in the EU. The empirical analysis reveals that the piecemeal rise of metropolitan regions can be interpreted as an outcome not only of political rescaling, but also of political reframing. The growing relevance of metropolitan regions as institutional fixes in the EU polity can be conceived as the social construction of spatial scales; a process characterized by power struggles between various actors' interests over terminology. This increasing visibility of 'the metropolitan' as a problem or opportunity that asks for common European action can thus be interpreted as a process of shifting policy frames.

Further research is needed on the role that knowledge about 'the metropolitan', policy-induced research and statistical analysis plays in justifying the relevance of the issue. In fact, several studies by ESPON, Eurostat and the Commission have been integral in developing a specific understanding of what constitutes a metropolitan region, such reports are occasionally referenced in policy documents. Additional studies on metropolitan regions produced by the OECD, Eurocities and Metrex appear to have also influenced the debate on metropolitan regions. The prominent role of case studies, research and statistics in policy development indicates that shifts in the conception and framing of 'the metropolitan' are caused in part by learning between EU policymakers and external actors. This thus raises questions concerning the performativity of particular academic concepts and their translation into the political realm. More detailed studies on the emergence of a metropolitan dimension in Europe indicate that particular individuals connecting the academic and political sphere can be seen as 'passeurs' of ideas between contexts and as 'travelling agents' for the mobility of political concepts (Fricke, forthcoming). Further analysis of this topic should therefore explore the role of exchange between communities of practice in shaping a common body of knowledge around conceptions of 'the metropolitan'.

# Notes

1 Additional information was collected from interviews with representatives from EU institutions, but did not directly enter the present analysis.
2 The NUTS classification (nomenclature des unités territoriales statistiques; nomenclature of territorial units for statistics) is a hierarchical system for categorizing the administrative entities of the EU for comparison between member states.

# References

Atkinson, R., 2001. The emerging 'urban agenda' and the European spatial development perspective: towards an EU urban policy? *European Planning Studies*, 9(3), pp. 385–406.

Bache, I., 2006. The politics of redistribution. In: K.E. Jørgensen, M.A. Pollack and B. Rosamond, eds. *Handbook of European Union politics*. London and Thousand Oaks: Sage. pp. 395–411.

Bachtler, J. and Mendez, C., 2007. Who governs EU cohesion policy? Deconstructing the reforms of the structural funds. *Journal of Common Market Studies*, 45(3), pp. 535–564. doi:10.1111/j.1468-5965.2007.00724.x.

Barca, F., 2009. *An agenda for a reformed cohesion policy: a place-based approach to meeting European Union challenges and expectations.* Independent report prepared at the request of Danuta Hübner, Commissioner for Regional Policy. Available at: http://ec.europa.eu/regional_policy/archive/policy/future/barca_en.htm [Accessed 2 February 2018].

Blatter, J.K., 2006. Geographic scale and functional scope in metropolitan governance reform: theory and evidence from Germany. *Journal of Urban Affairs*, 28(2), pp. 121–150. doi:10.1111/j.0735-2166.2006.00264.x.

Brenner, N., 1999. Globalisation as reterritorialisation: the re-scaling of urban governance in the European Union. *Urban Studies*, 36(3), pp. 431–451.

Brenner, N., 2001. The limits to scale? Methodological reflections on scalar structuration. *Progress in Human Geography*, 25(4), pp. 591–614. doi:10.1191/030913201682688959.

Brenner, N., 2003. Standortpolitik, state rescaling and the new metropolitan governance in Western Europe. *disP - The Planning Review*, 152, pp. 15–25.

Brenner, N., 2004. *New state spaces: urban governance and the rescaling of statehood.* Oxford and New York: Oxford University Press.

Brenner, N., 2009. Open questions on state rescaling. *Cambridge Journal of Regions, Economy and Society*, 2(1), pp. 123–139. doi:10.1093/cjres/rsp002.

Committee of the Regions, 2011. *Urban governance in the EU: current challenges and future prospects.* Available at: http://urban-intergroup.eu/wp-content/files_mf/corurbangoverancefinal.pdf [Accessed 2 February 2018].

Council of Ministers Responsible for Spatial Planning & European Commission, 1999. *European Spatial Development Perspective (ESDP): towards balanced and sustainable development of the territory of the European Union.* Luxembourg: Office for Official Publications of the European Communities.

Council of Ministers Responsible for Spatial Planning and Territorial Development, 2011. *Territorial agenda of the European Union 2020: towards an inclusive, smart and sustainable Europe of diverse regions.* Agreed at the informal ministerial meeting of ministers responsible for spatial and territorial development. Gödöllő,

Hungary. Available at: http://ec.europa.eu/regional_policy/en/information/publications/communications/2011/territorial-agenda-of-the-european-union-2020 [Accessed 2 February 2018].

Council of Ministers Responsible for Spatial Planning and Urban Development, 2007a. *Leipzig charta on sustainable European cities: final draft.* Available at: http://ec.europa.eu/regional_policy/archive/themes/urban/leipzig_charter.pdf [Accessed 2 February 2018].

Council of Ministers Responsible for Spatial Planning and Urban Development, 2007b. *Territorial agenda of the European Union: towards a more competitive and sustainable Europe of diverse regions.* Agreed on the occasion of the informal ministerial meeting on urban development and territorial cohesion. Leipzig. Available at: http://ec.europa.eu/regional_policy/en/information/publications/communications/2007/territorial-agenda-of-the-european-union-towards-a-more-competitive-and-sustainable-europe-of-diverse-regions [Accessed 2 February 2018].

Council of Ministers Responsible for Spatial Planning and Urban Development, 2010. *Toledo declaration: informal ministerial meeting on urban development.* Available at: http://urban-intergroup.eu/2011/09/spanish-presidency-2010/ [Accessed 2 February 2018].

Cox, K.R., 2010. The problem of metropolitan governance and the politics of scale. *Regional Studies*, 44(2), pp. 215–227. doi:10.1080/00343400903365128.

Dabinett, G. and Richardson, T., 2005. The Europeanization of spatial strategy: shaping regions and spatial justice through governmental ideas. *International Planning Studies*, 10(3–4), pp. 201–218. doi:10.1080/13563470500378549.

Davoudi, S., 2008. Conceptions of the city-region: a critical review. *Proceedings of the ICE - urban design and planning*, 161(2), pp. 51–60. doi:10.1680/udap.2008.161.2.51.

Delaney, D. and Leitner, H., 1997. The political construction of scale. *Political Geography*, 16(2), pp. 93–97. doi:10.1016/S0962-6298(96)00045-5.

Derudder, B., De Vos, A. and Witlox, F., 2015. Global city/world city. In: B. Derudder, ed. *International handbook of globalization and world cities*. Cheltenham: Elgar. pp. 73–82.

Dukes, T., 2008. The URBAN programme and the European urban policy discourse: successful instruments to europeanize the urban level? *GeoJournal*, 72(1–2), pp. 105–119. doi:10.1007/s10708-008-9168-2.

ESPON, 2007. *Project 1.4.3: study on urban functions. Final report.* [pdf] Luxembourg. Available at: www.espon.eu/sites/default/files/attachments/fr-1.4.3_April2007-final.pdf [Accessed 2 February 2018].

EUROCITIES, 2013. Eurocities: about us. [online] Available at: www.eurocities.eu/eurocities/about_us [Accessed 2 February 2018].

European Commission, Directorate-General for Regional Policy, 2008. *Fostering the urban dimension: analysis of the operational programmes co-financed by the European Regional Development Fund (2007–2013).* Working Document. Belgium: Brussels.

European Commission, Directorate-General for Regional Policy, 2009. *Metropolitan regions in the EU* (Regional Focus No. 01/2009).

European Commission, Directorate-General for Regional Policy, 2011a. *Cities of tomorrow: challenges, visions, ways forward.* Luxembourg: Brussels.

European Commission, Directorate-General for Regional Policy, 2011b. *Regional typologies: a compilation* (Regional Focus No. 01/2011).

European Commission, Directorate-General Regional and Urban Policy, & Eurostat, 2014. *Defining urban areas in Europe. Three levels of urban areas based on population distribution and commuting.* Available at: http://ec.europa.eu/

eurostat/documents/4313761/4311719/Metro_reg_Defining_urban_areas [Accessed 2 February 2018].

European Economic and Social Committee, 2004. *Opinion on European metropolitan areas: socio-economic implications for Europe's future.* (2004/C 302/20).

European Economic and Social Committee, 2007. *Opinion on European metropolitan areas: socio-economic implications for Europe's future.* (2007/C 168/02); ECO/188. Available at: www.eesc.europa.eu/?i=portal.en.eco-opinions.14236 [Accessed 2 February 2018].

European Economic and Social Committee, 2011. *Opinion on metropolitan areas and city regions in Europe 2020: own-initiative opinion.* (2011/C 376/02).

European Parliament, Committee on Regional Development, 2008. *Report on the follow-up of the territorial agenda and the Leipzig charter: towards a European action programme for spatial development and territorial cohesion (2007/2190(INI)): session document.* Available at: www.europarl.europa.eu/sides/getDoc.do?pubRef=-//EP//TEXT+REPORT+A6-2008-0028+0+DOC+XML+V0// EN [Accessed 2 February 2018].

European Parliament, Policy Department Structural and Cohesion Policy, 2007. *Follow-up of the territorial agenda and the Leipzig charter: towards a European action programme for spatial development and territorial cohesion: ad-hoc note.* Brussels. Available at: www.europarl.europa.eu/thinktank/en/document. html?reference=IPOL-REGI_ET%282007%29397237 [Accessed 2 February 2018].

Fricke, C., forthcoming. *The European dimension of metropolitan policies. Policy learning and reframing of metropolitan regions in the context of Europeanisation.* PhD. Berlin: Technische Universität Berlin.

Friedmann, J., 1986. The world city hypothesis. *Development and change*, 17(1), pp. 69–83. doi:10.1111/j.1467-7660.1986.tb00231.x.

Gualini, E., 2006. The rescaling of governance in Europe: new spatial and institutional rationales. *European Planning Studies*, 14(7), pp. 881–904. doi:10.1080/09654310500496255.

Harrison, J. and Growe, A., 2014. From places to flows? Planning for the new 'regional world' in Germany. *European Urban and Regional Studies*, 21(1), pp. 21–41. doi:10.1177/0969776412441191.

Heinelt, H. and Kübler, D. eds., 2005. *Metropolitan governance: capacity, democracy and the dynamics of place.* London: Routledge/ECPR Studies in European Political Science.

Herrschel, T. and Newman, P., 2002. *Governance of Europe's city regions: planning, policy and politics.* London: Routledge.

Knox, P.L. and Taylor, P.J. eds., 1995. *World cities in a world-system.* Cambridge and New York: Cambridge University Press.

Kübler, D. and Piliutyte, J., 2007. Intergovernmental relations and international urban strategies: constraints and opportunities in multilevel polities. *Environment and Planning C: Government and Policy*, 25(3), pp. 357–373.

METREX, 2014. About METREX. [online] Available at: www.eurometrex.org/ ENT1/EN/About/about_METREX.php [Accessed 2 February 2018].

Nelles, J., 2013. Cooperation and capacity? Exploring the sources and limits of city-region governance partnerships. *International Journal of Urban and Regional Research*, 37(4), pp. 1349–1367. doi:10.1111/j.1468-2427.2012.01112.x.

Neuman, M. and Hull, A., 2009. The futures of the city region. *Regional Studies*, 43(6), pp. 777–787. doi:10.1080/00343400903037511.

Parr, J., 2005. Perspectives on the city-region. *Regional Studies*, 39(5), pp. 555–566. doi:10.1080/00343400500151798.

Popescu, I., 2005. The metropolitan dimension of Europe: from cities to urban regions. *Administratie Si Management Public*, (5), pp. 69–73. http://www.ramp.ase.ro/en/index.php?pageid=3&archiveid=5

Rein, M. and Laws, D., 2000. Controversy, reframing, and reflection. In: W.G.M. Salet and A. Faludi, eds. *The revival of strategic spatial planning*. Amsterdam: Royal Netherlands Academy of Arts and Sciences. pp. 93–108.

Rein, M. and Schön, D., 1993. Reframing policy discourse. In: F. Fischer and J. Forester, eds. *The argumentative turn in policy analysis and planning*. Durham, NC: Duke University Press. pp. 145–166.

Rodríguez-Pose, A., 2008. The rise of the 'city-region' concept and its development policy implications. *European Planning Studies*, 16(8), pp. 1025–1046. doi:10.1080/09654310802315567.

Rossignolo, C. and Toldo, A., 2008. The polycentric 'vocation' of European territories: towards the construction of the Italian north west and other networking stories. In: R. Atkinson and C. Rossignolo, eds. *The re-creation of the European city. Governance, territory and polycentricity*. Amsterdam: Techne Press. pp. 65–89.

Salet, W., Thornley, A. and Kreukels, A. eds., 2003. *Metropolitan governance and spatial planning: comparative case studies of European city-regions*. London and New York: Spon Press.

Sassen, S., 1991. *The global city: New York, London, Tokyo*. Princeton, NJ: Princeton University Press.

Scholze, J., 2014. *City regions: the possibility of using EU structural funds for the development of urban-rural partnership in the 2014–2020 funding period*. Comparative Policy Analysis. Brussels.

Scott, A.J. ed., 2001. *Global city-regions: trends, theory, policy*. Oxford: Oxford University Press.

Tofarides, M., 2003. *Urban policy in the European Union: a multi-level gatekeeper system*. Aldershot, Hants and Burlington, VT: Ashgate.

Tosics, I., 2011. *Governance challenges and models for the cities of tomorrow*. Issue paper commissioned by the European Commission (Directorate General for Regional Policy).

van der Heiden, N., 2010. *Urban foreign policy and domestic dilemmas: insights from Swiss and EU city-regions. ECPR Press monographs*. Colchester: ECPR Press.

Wiechmann, T., 2009. Raumpolitische Diskurse um Metropolregionen in Europa - Eine Spurensuche. In: J. Knieling, ed. *Metropolregionen und Raumentwicklung: 3 // 231. Metropolregionen. Innovation, Wettbewerb, Handlungsfähigkeit*. Hannover: Verlag der ARL. pp. 101–132.

Wilks-Heeg, S., Perry, B. and Harding, A., 2003. Metropolitan regions in the face of the European dimension. In: W. Salet, A. Thornley and A. Kreukels, eds. *Metropolitan governance and spatial planning. Comparative case studies of European city-regions*. London and New York: Spon Press. pp. 20–38.

Zimmermann, K., 2012. *Institutionalisierung Regionaler Kooperation als Kollektiver Lernprozess? Das Beispiel Metropolregion Rhein-Neckar. Modernes Regieren: Band 9*. Baden-Baden: Nomos.

Zimmermann, K., 2016. Metropolenpolitik in Polyzentrischen Regionen. In: R. Danielzyk, A. Münter and T. Wiechmann, eds. *Planungswissenschaftliche Studien zu Raumordnung und Regionalentwicklung: Vol. 5. Polyzentrale Metropolregionen*. Detmold: Rohn. pp. 100–121.

# 10 Building city political agency across scales

## The Johannesburg International Relations Strategy

*Elisabeth Peyroux*

### Introduction

There is a small but growing body of literature that explores the political agency of the city at the international and global scale, more particularly its increased role and impacts on international affairs, global governance and diplomacy. While cities' international activities are not new (see the history of city-to-city cooperation, trans-municipal movements and networks in Europe, paradiplomatic activities by subnational entities in the USA) scholars are examining the key drivers and the mechanisms by which cities have been empowered in the last decades, mainly in the fields of international relations, political science and comparative politics, with insights from urban studies (Acuto, 2013). They question the significance of their new functions, capabilities and influences within the context of urban transformation, state rescaling, the reconfiguration of the international system and the trend towards networked forms of transnational urban governance (Pinson and Vion, 2000; Viltard, 2008, 2010; Nijman, 2009; Acuto, 2013; Bouteligier, 2014; Curtis, 2014; Ljungkvist, 2014). This research complements other work from geographers, who explore the role of cities in global environmental governance, more particularly how municipal networks give power and influence to cities for addressing key global challenges in a context of multi-level governance (Betsill and Bulkeley, 2006; Bulkeley 2010).

This chapter aims at enriching the discussion on the internationalization of cities that takes place within the fields of geography and urban studies by critically assessing the hypothesis of the growing empowerment of cities at the international scale, as formulated by scholars from international relations and political science. It draws on the analysis of the Johannesburg's International Relations Strategy, whose objective is to strengthen the international status, role and influence of the city through a wide range of activities: city-to-city cooperation, participation in transnational urban alliances, associations and networks, as well as in the global system of governance through direct engagement with international organizations and forums (City of Johannesburg/CoJ, 2012, 2016b). The chapter also aims at enriching the discussions by bringing attention to the geopolitical dimension of international city strategies from a Global South perspective, as an important factor that drives and shapes the political agency of a city (Peyroux, 2016). This underlines the importance of considering the positionality of (urban) space from which city stakeholders act (Sheppard, 2002).

The first section of the chapter reviews current debates on the political agency of a city at the international and global scale, mostly dealt with in political science and international relations. It underlines both the relevant insights offered by such fields as well as the limitations and gaps. The second section analyses the Johannesburg International Relations Strategy released by the metropolitan government in 2012.[1] It shows how political history, inter-governmental relations, in particular city/state relations, economic processes, as well as the strategic positioning of the city within different geopolitical spheres of influences shape Johannesburg's international activities. It emphasizes how the strategy is embedded in the reconfiguration of power between Northern and Southern countries and the changing origin and nature of urban expertise. The geopolitical dimension has to be understood in the context of the growing economic, but also diplomatic, influence and power of Southern countries and cities, particularly within the BRICS alliance (Brazil, Russia, India, China, South Africa) (Véron, 2013). Whereas selective city-to-city cooperation is an important component of the strategy, the chapter underlines how the political agency of Johannesburg at the international and global scale is best understood as the capacity to influence the production of urban norms and values based on South African urban experiences and practices. This highlights the relevance of a networked reading of political agency. While it is important to take a 'Southern standpoint' in order to consider policy innovations, relations and flows circulating within the Global South (Robinson, 2011), it is also of importance to reflect upon the new positionalities of Southern cities on the international scene and the way this produces a new categorization and hierarchy of cities in the South. The conclusion highlights the need for developing an interdisciplinary research agenda that can increase our understanding of the links between urban development processes, transnational dynamics and global policymaking.

## Framing and theorizing the political agency of cities at the international and global scale

The internationalization of cities has been an ongoing topic of investigation for scholars from many disciplines (urban geography, political and economic geography, urban studies, sociology, anthropology, political science and recently international relations) over the past decades and has largely contributed to renewing our understanding of and conceptualization of cities as sites, places and actors, and as nodes embedded in multi-scalar networks. The various dimensions of the internationalization processes (the internationalization of urban economies, the internationalization of urban policies, the internationalization of local governments) have been addressed by different, though overlapping, sets of literature: the world/global city literature, the policy transfer and 'urban policy mobility' literature, the literature on the 'international action of cities' and city diplomacy. While coming from different disciplinary and theoretical perspectives, these bodies of literature all examine the complex interconnections and interdependencies between urban spaces, the rescaling of statehood and urban governance, and the relevance of territorial processes in a world increasingly conceptualized in terms of flows and (city) networks.

More recently, an emerging literature from political science and international relations has started exploring 'city political agency' at the international and global scale, more particularly in the environmental field (Acuto, 2013; Barber, 2013; Bouteligier, 2014; Curtis, 2014; Aust, 2015). By looking at global practices of cities and questioning their increased influence in international affairs, they have challenged traditional notions and approaches and brought new insights into the fields of international relations.

### *The increased political activities of cities at the international and global scale*

The conceptualization of the *political agency* of cities rests on the empirical observation that cities have an 'active *presence*' in world politics, international affairs and global governance (Ljungkvist, 2014, p. 2). While cities' international *activities* are not new, they have gained prominence over the past decades. City-to-city cooperation has become a worldwide policy instrument for promoting local development through exchanges between cities (North/South, South/South), in particular through the influence of international organizations and networks, such as United Cities and Local Governments of Africa (UCLGA), Africities and Metropolis. The goals have changed over time with the evolving geopolitical context: from promoting friendship and solidarity to contributing to urban development and economic growth. There is a *growing participation* of cities in transnational city networks and associations, international forums that cities use as platforms for developing further cooperation with cities or as a place to represent themselves, defend their interests, and a site for political lobbying in order to promote their views and policy orientations. As underlined by van der Pluijm (2007, p. 20):

> The aim of UCLG for instance, the global association of municipalities, is to gain an official status at the UN in order to promote and protect the interests of cities worldwide in all of the issues in which the UN deals. Cities seeks to have a greater influence in decision-making in international organisations.

Cities are increasingly participating in *global governance* (see in particular the debates around cities and global environment governance) (Betsill and Bulkeley, 2006; Bulkeley, 2010; Curtis, 2014). ICLEI (International Council for Local Environmental Initiatives) and UCLG are examples of city-based coalitions and engagements (Acuto, 2013, p. 183) that establish an *urban presence* within the realm of global governance. C40 is considered the 'most significant case of global city agency in global environmental politics' (Acuto, 2013, p. 99). Other aspects of the active political presence of cities relate to the high visibility of their mayors in world affairs (Ljungkvist, 2014). Cities also develop direct interaction with regional and international organizations (UNICEF, UNESCO, UN-Habitat) and collaborate on international laws and global politics (Nijman, 2009). The New Urban Agenda that was adopted at the United Nations Conference on Housing and Sustainable Development in 2016 confirmed the key role that cities are expected to play in implementing the Sustainable Development Goals (SDGs) (Barnett and Parnell, 2016).

### From international activities and presence to city political agency, power and influence

Despite the focus on cities' international activities there is still little theorization of the international city political agency. The literature on 'urban policy mobility' in geography and urban studies has been mostly concerned with the internationalization of urban policies, exploring how and why policy ideas move and travel, how elements from elsewhere are embedded in, and reshape local policymaking. They have focused on the nodes, circuits and networks that connect people and ideas, on the agents of mobility and transfer (individuals, epistemic communities, city officials, local governments). They have looked at the adoption and re-appropriation of urban policies 'in motion', often located in the context of the neo-liberalization of urban policies (Ward, 2006; Peck and Theodore, 2010; McCann and Ward, 2011; Didier, Peyroux and Morange, 2012; Peyroux, Pütz and Glasze, 2012; Wood, 2014; Michel and Stein, 2015). Work on city-to-city cooperation has focused on the historical construction of trans-municipal partnerships and networks, particularly within the European context (Saunier, 2006; Clarke, 2012), on city policies and decentralized cooperation policies and their recent evolutions under changing economic and geopolitical contexts (Husson, 2000; Pasquier, 2012; Söderström, Dupuis and Leu, 2013; Söderström, 2014). They have explored the efficiency of intercity partnerships in terms of local capacity building and urban governance strengthening in the South (Bontenbal, 2009; Bontenbal and van Lindert, 2009), their role as vectors of European integration (Pinson and Vion, 2000) or as a learning place (Campbell, 2009; van Ewijk and Baud, 2009). The literature exploring city's international activities in a broader perspective (including a wider repertoire of actions) has paid attention to the contextual factors determining the internal features of the 'international agency' of cities (Lefevre and d'Albergo, 2007), with the purpose of comparing cities across Europe and assessing their autonomy vis-à-vis global forces. The conceptualization, however, remains linked to the political status of the city within the national and local institutional system, as both enabling and constraining city actions.

There has been little theorization of the political agency of the city at the international and global scale within the field of international relations (Ljungkvist, 2014). First, the integration of non-state actors in the field is recent. Second, acknowledging the power of cities in international relations destabilizes the assumptions that international relations make about the international system (Curtis, 2014): it challenges the state-centred (Viltard, 2010) and territorial perspective (Acuto, 2013, p. 25), the fact that subnational governments are under the (sole) influence and direction of national government (Betsill and Bulkeley, 2006). However, it has been acknowledged that the state is no longer the primary, single actor in international relations, foreign policies and diplomacy, that there are a plurality of non-state actors, particularly in the field of global environmental governance – transnational networks, epistemic communities, networks of firms and NGO, but also subnational actors developing their own transnational environmental policies (Compagnon, 2013).

In this context three analytical challenges have captured research interests: assessing the shift of power between the state and the city (in terms of autonomy and independency); conceptualizing the agency of cities with in transnational networks; and assessing the outcomes and impacts of current practices of cities at international and global level. This underlines two inter-related issues:

- The rising importance of cities in world affairs, international relations, global governance and diplomacy, linked to the changing international system, evolving modes of global governance and decision-making processes (the political and institutional processes that shape city agency).
- The 'paradoxical importance of the local and urban in the contemporary global and globalizing age' (Ljungkvist, 2014, p. 53) (the framing of global challenges): 'Urban issues and cities are becoming understood as increasingly central and relevant in global politics' (Ljungkvist, 2014, p. 52); 'Global challenges are made into urban issues' (Ljungkvist, 2014, p. 54); 'Contemporary global "risks" are described as having increasingly urban facets' (Ljungkvist, 2014, p. 49).

The leading questions from the perspective of international studies are: what is the source of the growing power of cities? What are the key drivers of the empowerment of cities? (Curtis, 2014). The discussion on the mechanisms that empower the city at the international level (resources, capacities, means) assumes that (global/izing) cities are already powerful. This assumption builds on the literature on the global city and its networks and extends it by bringing in the political role of cities (Bouteligier, 2014). The explanation is rooted in global urbanization and urban transformation processes, in the 'political implication of new urban forms' (Curtis, 2014, p. 4), which are regarded as factors for understanding city power. Scholars acknowledge the 'horizontal and vertical stretching of urban space', the 'transnational reach in a context of global urbanization'. The 'extension of city's capabilities and influence allows them a new functionality within the international system, including the power to engage in novel forms of global governance' in a context where there are 'growing convictions that states have failed to tackle some of the most pressing governance issues' (Curtis, 2014, p. 4).

The definition and characterization of the 'political agency of city' (Acuto, 2013, p. 5) is dealt with in two separate, though overlapping sets of literature. A first set of literature in political science and international relations deals with the political agency of cities, indeed local governments vis-à-vis the state, mostly around the notions of 'city diplomacy', defined as a form of 'management of decentralized international relations' (van der Pluijm, 2007, p. 11; Viltard, 2008, 2010), 'diplomacy from below' (Krippendorf, 2000) or 'municipal foreign policy'. This work focuses on *inter-governmental relations* and the redistribution of competencies between the different layers of government. The research explores the causes, motivations, forms and nature of city engagements (bilateral or multilateral) and the nature of the relationships between central governments and subnational governments, trying to assess whether there are competition, cooperation, complementary or replacement processes (van der Pluijm, 2007; Viltard, 2008, 2010). Scholars formulate hypotheses about the fragmentation of the state, the loss of state sovereignty vis-à-vis local governments and other new transnational actors, the 'decomposition of the Westphalian system' (Viltard, 2018, p. 511). Networks are here defined as 'diplomatic scenes' (Viltard, 2010, p. 599).

A second set of literature discusses the role and capacities of cities, mayors and transnational networks, as part of a *new geography of global governance* (Acuto, 2013). Cities and their relations to states, UN agencies and civil society are regarded as part of the *'development of a global architecture'* (Acuto, 2013, p. 27) of 'multi-stakeholders arrangements that organize world politics' (Acuto, 2013, p. 19). The political agency of the city is questioned in relation to the networked forms of urban governance, in particular how the organizational forms of trans-municipal networks attribute power to cities, even those that are not global (Bouteligier, 2014). C40 is used as an example to show how such a network engages, empowers and resources cities, giving them the capacity to tackle environmental challenges, to position themselves as leaders, to become 'vital actors' in global governance (Bouteligier, 2014, p. 62). This is accomplished through providing access to information, knowledge and partners (cities, multinational companies, foundations), the exchange of best practices, development of partnerships and implementation of concrete projects, and, most importantly, through allowing cities to shape the activities and direction of the network via the establishment of 'subtle power relations' and the creation of specific understandings about social practices that can lead to the emergence of hegemonic, dominant patterns of thought and action (Bouteligier, 2014, p. 63). Trans-municipal networks, conceptualized as 'collective agents of global projects', can also allow a city to pool global influences and shape norms in global environmental governance. The participation alone in such networks is however not sufficient to grant such powerful or influential positions: skills are needed to take advantage of the networks. Finally, studies on the political agency of cities underline the political power of individuals, more particularly, the catalytic effects mayors can have on global governance in key areas (water management, climate change, gender…), and how their influence is framed by the discursive production of new 'international identities' for the cities they represent (Acuto, 2014). The issue of city identities and representations is all the more important to consider, as it shows how Southern cities gain central political positions that may challenge 'West-centricity in city diplomacy' (Acuto, 2014, p. 72).

The literature on cities and climate change governance brings relevant complementary insights by exploring the relations of power and influence between subnational and national state and non-state actors, questioning the emergence of non-hierarchical models of governance and examining how transnational networks of municipal governments (such as the Cities for Climate Change Protection programme) give power and influence to cities and contribute to the creation of new spheres of authority (Betsill and Bulkeley, 2006).

There are contrasting conclusions about the empowerment of cities and their capacity to act at the international and global scale. Such conclusions are seldom rooted in empirical work and based on evidence. Furthermore, a sound methodological framework to address these issues is still missing. Scholars working on city diplomacy acknowledge the resilience of state and international relations (Viltard, 2008). Acuto (2013, p. 148) underlines that despite its extension, the geography of global governance still remains 'international': 'Rather than *re*placing central governments, city-centric linkages *dis*place the predominance of nation states with more fluid, city-oriented, and cross-cutting political connections (…)'. Betsill and Bulkeley (2006) argue that we are witnessing the emergence of a plural mode of governing that does not necessarily mean the weakening of the state, but rather a redefinition of its scope and scale of activity. Other authors, on the contrary, argue that major cities along with global institutions bypass the state (Nijman, 2009).

These studies lack conceptual clarity: the notion of 'actors' embraces the whole cities or local governments indiscriminately without considering the complex and heterogeneous nature of the city, the diversity of its interests and representations, its precarious limits and the contingency of its politics (Acuto, 2013). In fact, most of the works deal with the internationalization of local governments, their representatives or elected officials, including the mayor (Barber, 2013; Beal and Pinson, 2014) or with the strategy of networks or alliances that represent them (e.g. UCLG). They do not acknowledge the different, even diverging positions and interests that cities may have within such networks and alliances and how this may affect their capacity to act. The analysis of networks should take into consideration the wide variety of public and private actors that compose them (such as Cities Alliance, which includes local government representatives, national governments, NGOs and multilateral organizations such as the World Bank and UN-Habitat).

Similarly, it is important to make the distinction between town twinning, city-to-city cooperation, activities within transnational networks and participation in global governance, as these activities have different genealogies, contexts and time frames. They are part of various institutional and political dynamics and, as we have seen, they relate to different debates regarding the political agency of cities.

Finally, there are other driving forces and other motivations to consider in the exploration of the city political agency. Most of the literature does not take into consideration the role of the changing geopolitical contexts, how regional integration and regional alliances influence the role and positioning of cities, or how this shapes the production of norms and values. The understanding of geopolitics in the context of city strategies does not relate to conflicts as such or to claims over national or local territories and sovereignty issues. We are considering the competition and rivalries between cities in terms of their international positioning within a specific geopolitical context, as shown by the case study of Johannesburg. Political lobbying within transnational networks and organizations has to be understood in the context of emerging countries claiming different forms, relationships and norms of development cooperation that build on Southern experiences and expertise (Mawdsley, 2012a, b). We agree with the need to consider 'how shifts in contemporary geopolitics are complicit in any reframing of the geographies of urban theory' (McFarlane and Robinson, 2012, p. 767).

## Building city political agency across scales: the Johannesburg International Relations Strategy

The political agency of Johannesburg in the international and global scale first involves exploring the relationship between the city and the state: here, the position of the city on the international scene should not be understood in terms of rivalry, but of alignment and mutual economic, political and geopolitical reinforcement between the national and local authorities.

### *Local government and international relations in the context of Johannesburg*

The end of the diplomatic isolation of South Africa in 1994, the lifting of international sanctions and boycotts, which had been imposed since the

1960s, and the new legislation on local governments have allowed South African cities to develop and formalize their international cooperation activities. In South Africa the decentralized three-tier system of government is composed of the national, provincial and local 'spheres'. Established by the 1996 Constitution as 'separate', 'interdependent' and 'interconnected', they are governed by an imperative of inter-governmental cooperation. Local governments can develop international relations under two conditions: they must help achieve the objectives fixed to them by the constitution; and they must be implemented in collaboration with the other two spheres of government. They should therefore meet an internal objective – to contribute to local economic development and the strengthening of urban management capacity in line with the Growth and Development Strategy (the GDS 2040) – and external objectives – to strengthen the links between South Africa and other countries and to enhance its international position in line with the national geopolitical agenda.

In Johannesburg, international activities were initially conducted as part of a 'policy of municipal international relations' adopted by the central government in 1999 (The Municipal International Relations Policy Framework – MIR) (de Villiers, 2005; Ruffin, 2013). The South African government and the African National Congress (ANC) had recognized the importance of twinning for developing local government capacity and promoting a positive image of the new South Africa (de Villiers, 2005). MIR was established as an incentive and a non-binding framework for municipalities. It was defined as 'a relationship between two or more communities from two different nation - states, in which one of the key players is a municipality. These links may include non-governmental organizations, community based organizations and private associations' (DPLG, 1999, p. 3, quoted in de Villiers, 2005, p. 248). Confusion about the nature of key players, however, dominates (DPLG, 1999, p. 3, quoted in de Villiers, 2005, p. 248). The foreword of the policy document emphasizes the role of municipalities: the MIR 'promotes partnerships between South African municipalities and municipalities around the world to ensure maximum learning, synergy and promotion of our national interests, including investment promotion' (DPLG, 1999, p. 3, quoted in de Villiers, 2005, p. 248). In Johannesburg MIR primarily concerned relationships between municipalities. In the late 2000s, the Johannesburg metropolitan municipality was involved in several town twinning agreements with local governments in the North and South and was participating in two partnerships, whose objectives were to transfer management skills and expertise to the municipalities of Addis Ababa (Ethiopia) and Lilongwe (Malawi) (for more details on city-to-city partnerships, see Harrison, 2015).

The MIR was criticized for its limited implementation, its ceremonial dimension and the lack of coordination between actors (Interview 1, official of the CSU: Policies and Strategies, Johannesburg, 20/09/2011). In 2012, the Central Strategy Unit (CSU) developed its own 'Strategy of International

Relations' (SRI) (CoJ, 2012). Placed under the authority of the Office of the Executive Mayor, this unit is responsible for strategic planning, performance management, and the implementation of the long-term Johannesburg Growth and Development Strategy (GDS 2040). The metropolitan authority, established in 2001 following the territorial and administrative reforms inherited from the apartheid structures, wanted to consolidate the strategy in order to make it more efficient than the previous one (Interview 1, 20/09/2011). Based on an assessment of existing partnerships, the strategy establishes a method for selecting the partner cities based on economic and geopolitical criteria, a clear programme of action, structured around well-defined objectives, with concrete and measurable outcomes. Four types of activities are identified: 'cities to cities engagement' (twinning, collaboration on projects, 'mentoring' and 'peer programmes'); 'networking'; 'intergovernmental relations'; and 'knowledge management and learning', demonstrating the command of information and communications technology.

### The preservation of municipal autonomy

The analysis of the 2012 strategy shows that rather than there being a formal collaboration between the different spheres of government, the metropolitan municipality's objectives, priorities and values are simply aligned with the central and provincial government. In fact, the metropolitan municipality operates in an 'institutional vacuum' (Interview 1, 20/09/2011), which enables it to achieve a fair degree of autonomy. The Department of International Relations and Cooperation (DIRCO) (2010–2013 Strategic Plan) and the National Development Plan (NDP) Vision for 2030, which set the policy and strategy for the national government, provide no guidelines regarding the role that metropolitan municipalities should play in the international positioning of South Africa (CoJ, 2012, p. 57): the vertical integration of international activities of the three spheres is not described. It is up to the City of Johannesburg, through the nature and content of its international commitments, to reflect the objectives, priorities and values of the national and provincial government. The 2016 assessment of the Johannesburg International Relations Strategy, however, highlighted the need for formalizing the relationships between the DIRCO, which is the constitutionally mandated authority to lead and oversee South Africa's foreign policy, and the CoJ, regarding its engagement in international travel (adherence to operational requirements, sharing programmes and informing about the hosting of foreign delegations, submitting post-engagement reports) (CoJ, 2016a). Convergence of approaches has been facilitated by the leading position of the ANC holding the majority both in the national government and in the City of Johannesburg (at the time of the drafting of the 2012 strategy).[2] The 'inter-governmental relations' component developed as part of the SRI aims at filling this institutional gap by ensuring a better alignment of each of the institutional players' activities. The City of Johannesburg intends 'to integrate the role of cities (...) on the national agenda as an essential component for achieving the objectives of growth and development of the region' (CoJ, 2012, p. 62).

Up to 2016, the concrete implementation of the strategy remained the responsibility of the metropolitan municipality under the leadership of the Central Strategy Unit (CSU). The strategy has been housed at the Group Strategy, Policy Coordination and Relations (GSPCR) since 2016. It involves a small number of officials from various departments or elected officials, including the mayor, who initiate and participate in field trips (study tours), sign cooperation agreements and implement or support projects in the partner cities. The 2016 assessment noted the predominant role of the mayor in the implementation of the strategy and the limited knowledge of city officials with regard to the content of the strategy outside the GSPCR and the heads of city departments (CoJ, 2016a). First the CSU, then the GSPCR, developed cooperation between municipalities – not directly between communities. Besides such activities with partner cities that fall under the strategy, municipal departments are also developing direct involvement with international development agencies (such as UN-Habitat) or foreign cities as part of content-driven projects or programmes. Over time, the partnerships have shifted; initially conceived as instruments of friendship and exchange in the context of the past struggle against apartheid and the democratic transition, they are now regarded as a tool for local capacity building (Interview 3, an official of the CSU: Management, Johannesburg, 29/09/2011). There is a plan to open the partnerships up to private actors in order to draw economic benefits from the relationships between cities, notably with the BRICS cities (Interview 2 official of the CSU: Policies and Strategies, Johannesburg, 10/21/2013). The 2012 strategy underlines the interest of developing business opportunities in the context of city-to-city cooperation, and this is reinforced in the new phase of the strategy as part of supporting business-to-business, economically-led activities that in turn lead to investment in South Africa (Interview 6 with official of the GSPCR, 04/05/2017).

The strategy relies on a powerful marketing tool, the JIKE (Jo'burg Innovation & Knowledge Exchange), which receives requests for study tours and also hosts foreign visiting delegations. The JIKE promotes and disseminates the Johannesburg expertise nationally and internationally through newsletters (Insight Knowledge) and brochures showcasing its 'good practices'. Such practices are based on the narrative of the 'successful transition' that Johannesburg achieved (1997–2001) when the city shifted from racially segregated municipalities to a single unified metropolitan authority (the so-called 'Johannesburg Transformation story') (Interview 4, official from JIKE: Knowledge Exchange Program, Johannesburg, 13/09/2011). This transition was, however, hotly contested at the time because of its interventionist and binding character, and its entrepreneurial and neo-liberal orientation

(Didier, Peyroux and Morange, 2012). The International Relations Strategy is supported by the activities of the Visitor and Resource Centre (RC & V), opened in 2002 by the mayor of Johannesburg.

The strategy has limited financial resources (it doesn't have its own budget; the funding of activities are supported by relevant departments involved in city exchanges) but it relies on the financial, technical and mediation support of a national network (the South African Local Government Association – SALGA), international organizations of local governments (UCLG), international organizations (World Bank) and the European cooperation agencies (German Agency for International Cooperation – GIZ), as is the case with Addis Ababa (Interview 5, independent consultant, Johannesburg, 21/09/2011).

### Categorization and prioritization of cities in bilateral cooperation

The Johannesburg SRI is designed as an extension of the City Development Strategy (GDS 2040). The objective of sharing experiences between cities is to improve governance, local administration and the provision of services for Johannesburg and the partner cities in order to meet the development goals set by the GDS 2040 (Peyroux, 2015), and more broadly, to contribute to the growth and development of the African continent. The strategy is based on a different categorization of African and BRICS cities.

Acknowledging the competition with emerging cities of West Africa, the strategy aims at positioning Johannesburg as a 'World Class African City of the Future' (CoJ, 2012, p. 6). As the regional capital of Southern African Development Community (SADC) and a 'strategic economic hub', Johannesburg intends to affirm its position as a 'leader city' and maintain its 'competitive advantage' across Gauteng, SADC and the African continent (CoJ, 2012, p. 10). As a newcomer among the BRICS cities (2010), Johannesburg also aims at occupying a strategic position within the emerging economies and at 'branding' the city and the whole country 'as the leading city of the country' (CoJ, 2012, p. 8).

Partner cities are identified and selected based on their strategic value to Johannesburg and South Africa. This selection is based on a detailed analysis of international trends in urbanization, economic development and growth, and of the challenges they pose to cities in a globalized world. Partner cities are selected according to geographical criteria (location), economic (weight, size, demographic growth potential, economic assets) and geopolitical interests (priority given to South/South relations).

### Consolidating a leadership position within the African continent

Exchanges with African cities aim at supporting urban growth and development in the context of regional integration and the New Partnership for Africa's Development – NEPAD. It is based on a relational understanding of urban and regional economies, considered as interrelated in the context of economic globalization, particularly within the African continent (the example of African migrants coming to South Africa in order to escape from unstable political regimes and high poverty is cited). The cities are selected in order to ensure a strategic position of Johannesburg within the regional and continental economy: these cities are either 'corridors' or 'economic clusters' structuring regional development process, or cities with strong demographic potential and economic development prospects (Luanda, Lagos and Maputo). Cities prioritized in the next phase of the strategy include Dakar, Brazzaville and Kigali (CoJ, 2016b).

According to Johannesburg, sharing experiences of economic and urban policy can lead to the reduction of 'the vulnerability of urban areas vis-à-vis global change' (CoJ, 2016b, p. 6). Building capacity in the local government sector, particularly in Africa, is regarded as an imperative to 'reap the benefits of economic growth and urbanization' (CoJ, 2016b, p. 6). Johannesburg considers itself in a good position to provide assistance to African cities, which are considered less able to respond to the challenges they are facing (the consequences of rapid urbanization, urban sprawl, the growth of informal settlements and the persistence of poverty and inequality) because of their poor resources and management capacity. The attention paid to Africa is in line with the priorities of the central government, which intends to 'meet the developmental needs of developing countries'. The explicit commitment to the African agenda and to Southern Africa was re-affirmed in the DIRCO 2015–2020 policy. Past cooperation agreements include a mentoring programme with the cities of Addis Ababa and Lilongwe in order to help them prepare their City Development Strategy. Because of the importance of the assistance provided by the City of Johannesburg to the municipality of Lilongwe (Interview 1, 20/09/2011), these partnerships represent more than just a transfer of skills; they can actually serve to co-construct local policies.

These partnerships between cities offer the means to disseminate the norms and values promoted by Johannesburg and South Africa. The City of Johannesburg promotes a 'pro-poor' policy, which, in its view, means addressing the persistent inequality that the rise of the middle class cannot hide (CoJ, 2016b, p. 7). It encourages public participation in policymaking (as was the case in the partnership with Lilongwe) (Interview 1, 20/09/2011). Johannesburg wants to remain independent from international donors' funding in order to keep a critical mind (CoJ, 2016b). The action of 'development agencies' that drives African cities' development agendas are criticized for 'failing to take into account the complex needs of [our] cities' (CoJ, 2012, p. 7).

## *A strategic positioning within the BRICS alliance*

BRICS cities are recognized as 'global cities of the future', 'hubs of competitiveness' that will play an increasingly influential role on the international stage (CoJ, 2012, p. 8). These cities are associated with resilience capacities with respect to the 2008 crisis, with major changes in their economic structure (including the rise of the middle class) and efforts in technology development and infrastructure (CoJ, 2012, p. 8). Johannesburg acknowledges that the city is not up to the standards of the BRICS cities (in terms of population size and the Human Development Index) and aims at developing relationships with the most dynamic and innovative cities (São Paulo, New Delhi, Shanghai) through 'peer to peer relationships' that provide mutual benefits, particularly in Information and Communication Technology (China), housing (Brazil) and textiles (India) (Interview 1, 20/09/2011). The strategy of Johannesburg reinforces the government's political agenda vis-à-vis the BRICS, which is considered too weak at the present time, by positioning South Africa as a 'catalyst' for the South/South relations (CoJ, 2012, p. 54). Johannesburg is in line with the Discussion Paper on International Relations (2012), which considers that the alliance with the BRICS represents 'a platform for alternative ideas to the Washington consensus and neoliberal policies' (CoJ, 2012, p. 54). In that regard, Johannesburg promotes a universal and public service provision. The strategy document argues that the city financial strategies should rely on elected municipalities and not solely on public-private partnerships (CoJ, 2012, p. 45). This 'alternative' to neo-liberalism remains hotly debated (Didier, Peyroux and Morange, 2012; Parnell and Robinson, 2012). The next phase of the strategy intends to extend the scope of South/South cooperation beyond the BRICS focal area, where there is 'evidence of thriving economy, investment in infrastructure and improvement in quality of life' (CoJ, 2016b, p. 35).

### A reassessment of partnerships with Northern cities

The partnerships with Northern cities have been reassessed in the current context of declining financial resources allocated to town twinning as a consequence of the 2008 crisis. Johannesburg also takes into consideration its changing needs. The North/South relations are no longer a priority (CoJ, 2016b, p. 9). Since the restructuring of the metropolitan authorities (1997–2001), Johannesburg no longer considers itself in a learning phase in the traditional areas of cooperation (financial management, urban regeneration and safety) developed in the past with cities, such as Birmingham, London and New York (Interview 3, 29/09/2011). Johannesburg now states that partnerships must be 'smart'. Tangible benefits are expected in the fields of health, low-carbon urban environment, the green economy and 'smart cities' (CoJ, 2012, p. 67). Because of their status as regional capitals and their innovative practices as 'smart cities', a privileged partnership is being built with both Bilbao and New York (CoJ, 2012, p. 80).

Developing privileged bilateral relations with BRICS cities highlights a shift in the perception of urban hierarchies from the City of Johannesburg: 10 years ago the international strategy was meant to position the city as a 'sustainable global city' with regards to the 'big cities' or 'alpha cities' of the 'first world', such as London or New York. The emerging economies of BRICS cities have changed the situation (Interview 3, 29/09/2011).

### *Active participation in transnational networks and alliances*

The 'networking' activities play an important role in this strategy and continue to remain the focus in the next phase because it allows the gaining of traction despite low resource levels (CoJ, 2016a). Because of the international exposure and the opportunities that they provide to Johannesburg, they are even considered the most important activities (Interview 2, 10/21/2013). Participation in flagship events, such as conferences and forums (C40, Metropolis) offers a platform for presenting and therefore publicizing and disseminating Johannesburg 'best practices' in planning, urban management and governance. Such best practices include financial tools, such as green bond and pooled financing, for which Johannesburg is used as a reference (CoJ, 2016a). These participations have been successful: the participation in the COP 21 in Paris in December 2015 was crowned with a C40 Cities Award for city leadership in the fight against climate change (Green Bond initiative). Johannesburg was directly solicited by Cities Alliance, UCLG and SALGA to play the role of 'mentor' for the city of Lilongwe, following a presentation of its City Development Strategy in an international conference (this partnership was awarded a prize in 2011 by Metropolis, UCLG and a Chinese municipal government) (Interview 1, 20/09/2011). Local government networks do create partnership opportunities between cities with UCLG, facilitating the linking of cities and the replication of 'best practices' (CoJ, 2012, p. 62). The mentoring programme has also been extended to other cities in South Africa, Malawi, Mozambique and Namibia (Interview 1, 20/09/2011). Johannesburg has built up a leadership position in the field of metropolitan management: Bamako recently asked Johannesburg's assistance in the fields of waste management, rural electrification and water (Interview 6, official, former External Relations Unit, Johannesburg, 11/09/2011).

The political agency of Johannesburg is best understood as the capacity to influence the production of urban norms and standards. Johannesburg promotes the participation in events likely to 'forge public opinion' and to have an impact on perceptions ('opinions formers', 'events' and 'content-based' events) over 'mega-events' (Interview 2, 10/21/2013). The city hosted the annual meeting of Metropolis 'Caring cities' (which brought together 500 participants) in 2013, and in 2014 hosted the 5th biennial C40 Mayors Summit gathering mayors from across the world. The city actively participated in the COP 17 in Durban in 2011, when the mayor of Johannesburg, then president of SALGA, used this forum to put forward the position of the citizens of Gauteng and feed the national policy (Interview 6, 11/09/2011).

The City gives priority to 'high level' networks (Metropolis, UCLG), to networks or organizations that promote regional integration and South/South relations (The African Union, Africities, UCLGA, BRICS Friendship Cities and Local Government Forum) and forums that promote smart cities, innovation and the green economy (C40, ICLEI). The city lobbied for the election of the mayor and the appointment of members of the Mayoral and City Manager Committees to key positions. Johannesburg, a member of ICLEI, hosted its secretariat and one of its mayors was a member of the International Office. Its previous mayor (2011–2016) is a member of the C40 Steering Committee, co-chair of Metropolis and was elected president of UCLG in 2016. Johannesburg also intends to play a strong advocacy role with organizations based in the North in order to facilitate the integration of the city in networks supporting smart cities practices.

Political lobbying within such networks and organizations has to be understood in the context of emerging countries claiming different forms, relationships and norms of development cooperation that build on 'Southern' experience and expertise (Mawdsley, 2012a, b). While this can provide the vehicle for promoting alternative conceptions of urban development, the Johannesburg case study shows that such discursive claims can also hide a more subtle form of reproduction of hierarchies between cities in emerging economies and cities in developing countries. By showing how city political agency may be related to geopolitical issues, Johannesburg shows how categories and hierarchies of cities are being recomposed through the political and economic leadership that some Southern cities may have over others. The capacity of Johannesburg to actively engage in international networks should be compared and assessed in order to establish whether we are witnessing the growing importance of Southern cities or just a new leadership by a handful of well-resourced Southern metropolitan authorities.

## Conclusion

Recent work in geography, urban studies, and political science have expanded our understanding and theorization of the internationalization of cities by looking into urban economies, networks and flows, the interconnection and interdependencies between urban spaces and cities, scales and inter-city relations, urban governance and transnational urban policymaking. The political role and agency of cities and local governments in international politics and global governance remain under-researched, calling for a wider engagement in disciplinary fields and theoretical frameworks – within and outside the field of geography and urban studies. While current theoretical debates in the field of international relations can enrich our conceptualization of the political agency of cities, its theorization remains insufficiently rooted in empirical studies. Geography and urban studies must tackle this challenge in order to fully grasp the link between urban change and transnational political dynamics. This engagement with the international relations and geopolitics literature opens up relevant avenues for further research in terms of analytical focus, theory and methodology:

First, we need to place the city into a wider network of diplomatic, political and geopolitical interests that are too often only ascribed to nation states. This calls for confronting city international strategies with national geopolitical agendas in order to assess potential alignment or divergence of position between the different levels of government. This, however, should not be reduced to the sole analysis of inter-governmental relations as part of a conventional multilevel governance perspective, but rather reframed as part of trans-scalar governance (Betsill and Bulkeley, 2006; Compagnon, 2013) that takes into account the specificities of cities as non-state actors vis-à-vis current conceptualizations of other non-state actors, such as NGOs and multinational firms (Aust, 2015). Cities are places where local experiences and expertise can be leveraged, shared and enhanced as part of bilateral or multilateral cooperation. Cities are actors that can act beyond their territorial boundaries: as 'networking agents' (Acuto, 2013, p. 151) they can contribute to solving global issues in crucial areas (climate change among others) by shaping the international norms and values that are produced and diffused through transnational networks (van der Pluijm, 2007). Cities are also instruments of foreign policies and national diplomacy that can help contribute to building privileged relationships with strategic countries according to national geopolitical and economic interests.

Second, we should investigate the uneven involvement of cities in policy networks and explore the reasons for that: is it linked to their economic position (as global or megacities)? To their capacity for technological and/or social innovation, irrespective of their size? To their capacity to build up alliances with other cities around shared norms and values? To their alignment to national state positions?

Third, we need to develop methodologies for assessing (1) how cities contribute to or influence decision-making processes at the international and global level; (2) what networks are contributing via cooperation; and (3) their efficiency in translating global concerns into localized action, as well as their capacity to develop and implement global norms and practices (Aust, 2015). This chapter calls for engaging in other relevant work, such as the 'new geography of development' (Mawdsley, 2012a), in order to deepen our analysis of the international and global positionalities of cities beyond the traditional North/South divide.

## Notes

1 The analysis is based on interviews conducted in Johannesburg in 2011, 2013 and 2017 with officials of the Johannesburg metropolitan municipality in charge of the preparation and the implementation of the strategy, and on a content analysis of the political document adopted in 2012 (City of Johannesburg – CoJ, 2012).
2 The election of a mayor from the opposition party (the Democratic Alliance) in August 2016 changed the power relations between the city and the national government.

## References

Acuto, M., 2013. *Global cities, governance and diplomacy. The urban link.* London: Routledge.
Acuto, M., 2014. An urban affair. How mayors shape cities for world politics. In: S. Curtis, ed. *The power of cities in international relations.* London and New York: Routledge. pp. 69–88.

Aust, P.H., 2015. Shining cities on the hill? The global city, climate change, and international law. *The European Journal of International Law*, 26(1), pp. 255–278.

Barber, B.J., 2013. *If mayors ruled the world: dysfunctional nations, rising cities*. New Haven, CT and London: Yale University Press.

Barnett, C. and Parnell, S., 2016. Ideas, implementation and indicators: epistemologies of the post-2015 urban agenda. *Environment and Urbanization*, 28(1), pp. 87–98.

Beal, V. and Pinson, G., 2014. When mayors go global: international strategies, urban governance and leadership. *The International Journal of Urban and Regional Research*, 28(1), pp. 302–317.

Betsill, M. and Bulkeley, H., 2006. Cities and the multilevel governance of global climate change. *Global Governance*, 12(2), pp. 141–159.

Bontenbal, M., 2009. Strengthening urban governance in the South through city-to-city cooperation: towards an analytical framework. *Habitat International*, 33(2), pp. 181–189.

Bontenbal, M. and van Lindert, P., 2009. Transnational city-to-city cooperation: issues arising from theory and practice. *Habitat International*, 33(2), pp. 131–133.

Bouteligier, S., 2014. A networked urban world. Empowering cities to tackle environmental challenges. In: S. Curtis, ed. *The power of cities in international relations*. London and New York: Routledge. pp. 57–68.

Bulkeley, H., 2010. Cities and the governing of climate change. *Annual Review of Environment and Resources*, 35, pp. 229–253.

Campbell, T., 2009. Learning cities: knowledge, capacity and competitiveness. *Habitat International*, 33, pp. 195–201.

City of Johannesburg, 2012. An integrated international relations agenda for the city of Johannesburg. Group Strategy, Policy Coordination and Relations. City of Johannesburg.

City of Johannesburg, 2016a. An assessment of the City of Johannesburg(s (Coj's) International Relations approach – and emerging recommendations. Prepared for the CoJ, Group Strategy, Policy Coordination and Relations (GSPCR) by Indlela Growth Strategies. Johannesburg.

City of Johannesburg, 2016b. The City of Johannesburg's 2016 International Relations Strategy. Common purpose through internationalisation of local government. International Relations Unit. Group Strategy, Policy Coordination and Relations (GSPCR), City of Johannesburg.

Clarke, N., 2012. Urban policy mobility, anti-politics, and histories of the transnational municipal movement. *Progress in Human Geography*, 31(1), pp. 25–43.

Compagnon, D., 2013. Chapitre 38. L'environnement dans les RI. In: T. Balzacq et F. Ramel, eds. *Traité de relations internationales*. Paris: Presses de Sciences Po (P.F.N.S.P.). pp. 1019–1052.

Curtis, S. ed., 2014. *The power of cities in international relations*. London and New York: Routledge.

de Villiers, J.C., 2005. *Strategic alliances between communities, with special reference to the twinning of South African provinces, cities and towns with international partners*. Dissertation presented for the degree of Doctor of Philosophy (Business Management and Administration). University of Stellenbosch.

Didier, S., Peyroux, E. and Morange, M., 2012. The spreading of the city improvement district model in Johannesburg and Cape Town: urban regeneration and the neoliberal agenda in South Africa. *The International Journal of Urban and Regional Research*, 36(5), pp. 915–935.

Harrison, P., 2015. South–south relationships and the transfer of 'best practice': the case of Johannesburg, South Africa. *International Development Planning Review*, 37(2), pp. 205–223.

Husson, B., 2000. La coopération décentralisée, légitimer un espace public local au Sud et à l'Est. *Transverses n° 7.* Centre International d'Etudes pour le Développement Local (Ciedel).

Lefevre, C. and d'Albergo, E., 2007. Why cities are looking abroad and how they go about it. *Environment and Planning C: Government and Policy*, 25, pp. 317–326.

Ljungkvist, K., 2014. The global city: from strategic site to global actor. In: S. Curtis, ed. *The power of cities in international relations.* London and New York: Routledge. pp. 32–55.

Mawdsley, E., 2012a. The changing geographies of foreign aid and development co-operation: contributions from gift theory. *Transactions of the Institute of British Geographers*, 37, pp. 256–272.

Mawdsley, E., 2012b. *From recipients to donors: the emerging powers and the changing development landscape.* London: Zed.

McCann, E. and Ward, K., 2011. *Mobile urbanism: city policymaking in the global age.* Minnesota: University of Minnesota Press.

McFarlane, C. and Robinson, J., 2012. Introduction – experiments in comparative urbanism. *Urban Geography*, 33(6), pp. 765–773.

Michel, B. and Stein C., 2015. Reclaiming the European city. Business improvement districts in Germany. *Urban Affairs Review*, 51(1), pp. 74–98.

Nijman, J.E., 2009. The rising influence of urban actors. Amsterdam Center for International Law (ACIL). *The Broker*, 17, pp. 13–17.

Parnell, S. and Robinson, J., 2012. (Re)theorizing cities from the Global South: looking beyond neoliberalism. *Urban Geography*, 33(4), pp. 593–617.

Pasquier, R., 2012. Quand le local rencontre le global : contours et enjeux de l'action internationale des collectivités territoriales. *Revue française d'administration publique*, 1(141), pp. 167–182.

Peck, J. and Theodore, N., 2010. Mobilizing policy: models, methods, and mutations. *Geoforum*, 41, pp. 169–174.

Peyroux, E., 2015. Discourse of urban resilience and 'inclusive development' in the Johannesburg Growth and Development Strategy 2040. *The European Journal of Development and Research*, Special issue on « Inclusive Development », 27(4), pp. 560–573.

Peyroux, E., 2016. Circulation des politiques urbaines et internationalisation des villes: la stratégie des relations internationales de Johannesburg. *EchoGéo*, [online] 36, online since 30 June 2016. Available at: http://echogeo.revues.org/14623 [Accessed 23 October 2017]. doi:10.4000/echogeo.14623.

Peyroux, E., Pütz, R. and Glaze, G., 2012. Business Improvement Districts (BIDs): Internationalisation and contextualisation of a 'travelling concept'. (Guest editorial.) *European Urban and Regional Studies*, 19(2), pp. 111–120.

Pinson, G. and Vion, A., 2000. L'internationalisation des villes comme objet d'expertise. *Pôle Sud*, 13, pp. 85–102.

Robinson, J., 2011. Cities in a world of cities: the comparative gesture. *International Journal of Urban and Regional Research*, 35(1), pp. 1–23.

Ruffin, F.A., 2013. Municipal international relations: the South African case of metropolitan eThekwini. *Loyala Journal of Social Sciences*, XXVII(1), Jan-June, pp. 119–141.

Saunier, P.Y., 2006. La toile municipale aux 19–20 siècles : un panorama transnational vu d'Europe. *Urban History Review. Revue d'histoire urbaine*, XXIV(2), pp. 163–176.

Sheppard, E., 2002. The spaces and times of globalization: place, scale, networks, and positionality. *Economic Geography*, 78(3), pp. 307–330.

Söderström, O., 2014. *Cities in relation. Trajectories of urban development in Hanoi and Ouagadougou*. Oxford: Wiley-Blackwell.

Söderström, O., Dupuis, B. and Leu, P., 2013. Translocal urbanism: how Ouagadougou strategically uses decentralised cooperation. In: B. Obrist et al., ed. *Living the African city*. Basel: SGAS & Lit Verlag. pp. 99–117.

van der Pluijm, R. and Melissen, J., 2007. *City diplomacy: the expanding role of cities in international politics*. [pdf] The Hague: Netherlands Institute of International Relations, Clingendael Diplomatie Paper n°10. Available at: www.clingendael.org/sites/default/files/pdfs/20070400_cdsp_paper_pluijm.pdf [Accessed 23 October 2017].

van Ewijk, E. and Baud, I., 2009. Partnerships between Dutch municipalities and municipalities in countries of migration to the Netherlands; knowledge exchange and mutuality. *Habitat International*, 33(2), pp. 218–226.

Véron, J.B., 2013. Les BRICS en Afrique: ambitions et réalités d'un groupe d'influence. *Afrique contemporaine*, 248(4), pp. 7–9.

Viltard, Y., 2008. Conceptualiser la 'diplomatie des villes' ou l'obligation faite aux relations internationales de penser l'action extérieure des gouvernements locaux. *Revue française de science politique*, 3(58), pp. 511–533.

Viltard, Y., 2010. Diplomatie des villes: collectivités territoriales et relations internationales. *Politique étrangère*, (3), pp. 593–604.

Ward, K., 2006. Policies in motion, urban management and state restructuring: the trans-local expansion of business improvement districts. *International Journal of Urban and Regional Research*, 30, pp. 54–75.

Wood, A., 2014. Moving policy: global and local characters circulating bus rapid transit through South African cities. *Urban Geography*, 35(8), pp. 1–17.

# 11 A closer look at the role of international accolades in *worlding* Cape Town's urban politics

*Laura Nkula-Wenz*

## Introduction

This chapter takes its point of departure from the intricate ways in which Cape Town's urban politics have been shaped by a recent throng of international *accolades* – a term frequently used by the City of Cape Town when tallying the prizes and awards it has received over the years, and a term defined by the Oxford Dictionary as 'an award or privilege granted as a special honour or as an acknowledgement of merit'. Examples include serving as a key host city during the 2010 FIFA World Cup, having Table Mountain listed as one of the world's New Seven Wonders of Nature, and being named the World Design Capital for 2014. In using Cape Town's World Design Capital designation as the primary empirical lens, this chapter seeks to make sense of and ultimately contribute to a better understanding of the increasingly inter-scalar nature of urban politics in African cities. In relation to the book's aim of viewing the city as an emergent *global* political actor, I argue that it is crucial to provide a nuanced understanding of the diverse political practices and governmental logics folded into the notion of 'going global' – including but also exceeding the drive for economic growth (Bunnell, 2013; Lauermann, 2016).

Conceptually, the argument builds upon recent theoretical advances in postcolonial urban scholarship that have suggested the notion of *worlding* to investigate so-called 'Southern'[1] cities through their emerging inter-urban connections and increasing engagements with global forms in circulation (Roy and Ong, 2011). In turn, international urban accolades are seen as 'world-conjuring projects' (Ong, 2011, p. 1) or – as I have come to understand them – distinctive *worlding devices*. As several authors have already pointed out, the bidding for international accolades and the hosting of associated events presents a crucial intersection for both introspective and 'extrospective' (Peck and Tickell, 2002; McCann, 2013) political practices and governmental logics, which – I concur – are best captured through a relational framework of urban analysis.

The argument is substantiated through findings from my recently concluded PhD project, in which I closely followed and engaged with the complex aspirational politics that evolved over the course of Cape Town's four-year-long journey towards becoming the World Design Capital 2014. The research comprised 64 semi-structured interviews with informants from local and provincial government, public-private partnerships, the private sector, and civil society organizations, as well as participant observation at over 50 World Design Capital-related meetings and events between October 2010 and February 2014.

As I will show, being designated as the first African World Design Capital provided local policy elites with a powerful device for mobilizing ideas, knowledge, objects, and people, presenting itself as a fulcrum of inter-scalar politics and shifting governmental discourses. This analysis also falls in line with recent arguments around the geographic expansion of urban politics – one 'that is globally ambitious, but with ambitions rooted in local development concerns' (Lauermann, 2016, p. 9). Consequently, this chapter in particular grapples with the role that the World Design Capital designation has played as a hinge for harnessing local political legitimacy through ostensibly global recognition.

This question is especially pertinent in light of the fact that Cape Town is South Africa's largest opposition-run metro, located in the country's only opposition-run province, the Western Cape. As such, the city is repeatedly invoked by the Democratic Alliance as its ultimate showpiece for governmental competencies, thus functioning as an indispensable tool for legitimizing the party's primary claim that it is better equipped to run the country than the African National Congress. Thus, as McCann (2013, p. 14) has aptly argued, 'extrospective' logics of policy boosterism do not exist in a vacuum but are always accompanied by 'an introspective politics of persuasion', geared towards gaining buy-in from the local population through a carefully curated self-image. Based on this, I propose that an analysis of Cape Town's World Design Capital process provides important insights into

> how urban elites' extrospective stance toward policy [i.e. attracting international investment] is balanced and bolstered by an introspective politics that seeks to generate pride in and support for local policies by continually informing the local population of their global influence and praise.
>
> (McCann, 2013, p. 22)

However, as I will further discuss below, Cape Town's ambitions to cast itself as the first *African* World Design Capital was not without its local contentions, as it once more exposed the city's historically ambiguous relationship with its continental location.

In summary, this chapter contends that unpacking the diverse range of rationales that were driving Cape Town's ambitions towards becoming Africa's first World Design Capital enables us to add further nuance and critical perspective to how cities 'arrive at' certain political practices and governmental logics as proposed by Jennifer Robinson (2013, and this volume). Overall, it addresses the following questions in three parts: firstly, how can the notion of *worlding* be used as an epistemological lens to understand the contentious interplay between global ambitions and local developmental agendas in so-called Southern cities? Secondly, how do international accolades function as burgeoning *worlding devices*? And thirdly, what kind of political practices and new imaginaries derived from the collective push towards *worlding* Cape Town 'by design'?

## The basic tenets of *worlding*

In order to understand international urban accolades as distinctive *worlding devices* for inserting one's city in the international sphere, as well as locally asserting its ostensibly global status, it is first of all important to grapple with the meaning of this emergent notion. As an intellectual approach, *worlding* goes back to the postcolonial critique of Gayatri Chakravorty Spivak (1999), and has been subsequently unlocked for urban theory by Abdoumaliq Simone (2001), and more recently by Ananya Roy and Aihwa Ong (2011) in their seminal book *Worlding Cities: Asian Experiments and the Art of Being Global*. It is generally used to describe the production of cities through sometimes ephemeral and frequently trans-local connections and global forms in circulation (including those of capitalism) (Roy, 2009, p. 823). Thus, it is also closely related to the contemporary study of mobile urbanism and global urban policy mobilities (McCann and Ward, 2011; Temenos and McCann, 2013).

McCann, Roy, and Ward (2013, p. 584) define *worlding* as a 'heterodox project' that seeks to 'disrupt the established maps of global urbanism'. In other words, the urge to focus on 'worlding' rather than 'world' cities derives from an epistemological frustration with conventional world city analysis and the way it has tended to reify the hegemonic status of a selected few Euro-American global cities as seemingly universal benchmarks of urban theory-making (Sassen, 1991; Taylor, 2006). At the same time, it is important to note that this critique is not seeking to dismiss the contribution of world and global city approaches to urban studies *per se*. Instead, it rather problematizes the ways in which popularized city hierarchies together with Western notions of modernity have frequently led to circumscribed readings of cities in the global South as passive, inert recipients of both urban theory and policy innovations. Comaroff and Comaroff (2012, p. 1) note

> the non-West, now the global south, is presented primarily as a place of parochial wisdom [...] of unprocessed data [...] as reservoirs of raw fact: of the historical, natural, and ethnographic minutiae from which Euro-modernity might fashion its testable theories and transcendent truths.

Consequently, *worlding* challenges this 'additive or predictive assimilation of the Southern experience' into existing urban theory frameworks (Roy, 2009, p. 821), arguing that cities everywhere have to be taken seriously as potential locations of urban theory production (Simone, 2010; Roy, 2014; Chattopadhyay, 2012).

As McCann, Roy, and Ward (2013, p. 584f) argue: 'In urban studies, worlding must be understood as an *intervention* in this structured divide between theory and ethnography' [emphasis added]. In rebuking a singular and linear logic of urban transformation, *worlding* thus stresses the processual character of city-making, i.e. the fact that cities everywhere are able to 'create global connections and global regimes of value' (McCann, Roy and Ward, 2013, p. 585; see also Söderström, 2014). Furthermore, Ong (2011, p. 12) explains, 'worlding [...] is linked to emergence, to the claims that global situations are always in formation'. Yet, *worlding* does not follow any grand historical logic of world-making or world city-ness for that matter, but is rather concerned with the multiple instances of 'being in the world' (Mbembe and Nuttall, 2004, p. 347) through a variety of constantly unstable relationships and precarious connections (Simone, 2001, p. 23). Most importantly, *worlding* processes do not occur on a single scale. Rather, they are inherently multi-scalar – a 'mix and match of different components' (Ong, 2011, p. 12) – in a global re-composition that is taking place in the vast space between local 'idiosyncratic compromises' (Simone, 2001, p. 18) and global narratives of urban transformation. After all, 'worlding cities are mass dreams rather than imposed visions' (McCann, Roy and Ward, 2013, p. 585), made up of 'a mix of speculative fiction and speculative fact [...] as practitioners aim to build something they believe is for the better' (Ong, 2011, p. 12). As such, *worlding* also always encompasses both 'extrospective' political practices tied to ideas of global competition and economic growth, as well as introspective politics that focus on a more diverse array of context-specific agendas (McCann and Ward, 2012; Lauermann, 2016). Taking this as a cue for my own research, I have come to understand Cape Town's World Design Capital process not as a kind of Africanized replica of the globally pervasive creative city script, but instead as an experimental and at times contested renegotiation of global, national, and local references, agendas, and desires by a diverse range of urban actors.

Thus, the main aim of my empirical work was to unravel the intricate mesh of inter-scalar governmental technologies, relational social practices, and competing political logics that formed around the powerful collective ambition of positioning Cape Town as the continent's first World Design Capital. However, before sharing some of my findings on how Cape Town negotiated and leveraged this powerful imaginary, it is important to spell out the role of international urban accolades as increasingly important drivers of local urban development agendas, particularly in cities of the so-called 'global South'.

## International urban accolades as *worlding devices*

Given the long history of research on (mega-)events in urban studies (Getz, 1991; Gold and Gold, 2008), it is rather curious that studies investigating their specific role as networked knowledge relays and vehicles of inter-urban policy mobility have only gained prominence fairly recently (Surborg, Vanwynsberghe and Wyly, 2008; Boyle, 2011; Cook and Ward, 2011; Lauermann, 2014). In consequence, more attention still needs to be paid to the way in which such international urban accolades propel inter-scalar processes of learning and knowledge creation, the geographical expansion of policy networks, as well as their *ad hoc* materializations that cannot be neatly captured within the top-down/bottom-up dichotomies defined by conventional urban governance analysis.

A more refined approach is particularly necessary given the expanding geographies of international (mega-)events: China's growing international-ization symbolized by its Olympic Summer Games in 2008, South Africa's stint as first African host country for the 2010 FIFA Soccer World Cup, as well as India's Commonwealth Games in the same year, followed by Brazil's double coup (or some would say 'double trouble') of organizing both the Soccer World Cup 2014 and the Olympic Games in Rio de Janeiro 2016, all point towards the heightened international attention for emerging markets, particularly those of the BRICS group. While this new thrust towards lo-cations in the Southern hemisphere remains fuelled by a stubborn belief in mega-events as catalysts of economic growth, recent studies have also pointed towards ways in which local development coalitions strategically leverage these bids to shape local political agendas that include but also go beyond economic growth imperatives (Pillay and Bass, 2008; Lauermann, 2014).

Alongside these well-established and globally mediatized sport-centred events, large-scale cultural happenings and urban cultural award schemes have been equally on the rise, fuelled by the international impetus towards the commodification of arts and culture on the back of the peripatetic cre-ative city paradigm (Peck, 2012). These events can be roughly divided into three categories: the first includes educational, professional, and artistic events such as large-scale conventions, world expos, biennales, and global summits (Hiller, 1995); the second comprises traditional heritage events such as popular festivals and carnivals (Getz, 1991; Quinn, 2005; Waitt, 2008); while the third category features institutional cultural accolades and awards, bestowed upon a city by an international organization in recog-nition of its efforts to preserve, display, and develop its distinct cultural and creative pedigree. The third category has been growing steadily in numbers, not least due to the popularity of the European Union's Capital of Culture – a vanguard initiative within the realm of culture-led urban awards-cum-events. Spurred by its rise as one of the most 'strategic weap-ons in the cultural arms race' (Richards, 2000), other organizations also started to create similar designations with even broader international reach, e.g. UNESCO's Creative Cities Network, inaugurated in 2004, or the World Design Capital, a designation conferred by the World Design Organization (formerly the International Council of Societies of Industrial Design, Icsid), further discussed below.

While most traditional festivals such as Pamplona's Bull Run or established art shows like the Venice Biennale are inherently place-bound and difficult to recreate elsewhere, the designations in the third category form part of a 'growing international "awards and credentials industry"' (McCann, 2013, p. 22) and are thus explicitly designed to move from city to city in more or less fixed time cycles. In turn, they are globally mobile and highly inter-referential by default, fuelling both city-to-city competition and collaboration under the banner of their increasingly popular and thus evermore desirable brand identity (Franke, 2002; Evans, 2003).

With regards to embodied practices of inter-urban learning and 'city diplomacy' (Acuto, 2013) international accolades of both the cultural and sporting kind usually combine event-led policy tourism, where so-called 'policy entrepreneurs' (Hoyt, 2006) from previous host cities are invited to share their first-hand experience *in situ*, with 'visit-led' policy tourism, where delegations from the current host city inspect the sites of previous eventing success (Jacobs, 2012, p. 415). Particularly in terms of cultural accolades as increasingly prominent vehicles of 'global cultural brokerage' (Yeoh, 2005, p. 955), the spectrum of actors involved in these cosmopolitan relationships extends even further beyond technocrats and business elites to also include 'low-skilled, low-income migrant workers; specialists in expressive activities; and world tourists' (Yeoh, 2005, p. 955).

It can be argued that these types of coalitions are not solely bound together by rationalist calculations of economic benefit, but equally by normative ideas and shared visions around how a 'better quality city' (McCann, 2004) could be catalysed through the respective event or award scheme. Furthermore, and in spite of the fact that the most prestigious urban accolades are associated with rigidly normed protocols and standardized organizational routine, several authors have already argued that local political processes cannot simply be regarded as reproductive blueprint scenarios, impressed onto the local context by some elusive outside force (Surborg, Vanwynsberghe and Wyly, 2008; Vanwynsberghe, Surborg and Wyly, 2013; Lauermann, 2016).

Instead, the insidious 'festivalization of urban governance' in Southern cities (Steinbrink, Haferburg and Ley, 2011) presents itself as a continuously circulating governmental regime of inter-urban expertise, multinational business interests, as well as context-specific experiences that come together in time and place-sensitive situated networks. As Myers (2011, p. 185) has rightly noted '[...] taken as a whole, the festivalization of African cities [...] marks the profoundly cosmopolitan, globalized, imaginative, generative, and dynamic character of the continent's "always moving spaces"'. At the same time, this relational reading of strategic linkages between local and global, internal and external objectives does not imply that coercive practices have ceased to exist. After all, large-scale international urban events evidently remain a preferential docking-station for accelerating neo-liberalization and advancing ideals of urban entrepreneurialism with the help of powerful, post-political, consensus-driven visions of urban renewal (Swyngedouw, 2011). Hence, in paraphrasing Horne (2007, p. 85), we can justly ask: given all the unpredictability and uncertainty surrounding major international events, why do cities still compete for the right to host them?

In recourse to my previous argument, I suggest that if we start rethinking international accolades as powerful *worlding devices* and thus as transversal 'mooring posts' (Hannam, Sheller and Urry, 2006) of globally mobile urban development paradigms, we might find that one answer lies – somewhat counter-intuitively – in exactly this unpredictable offering of 'speculative experiments that have different possibilities of both success and failure' (Roy and Ong, 2011, p. xv). Moreover, as Lauermann (2016, p. 1897) has recently pointed out,

> by strategically referencing the contingent nature of the event – the likelihood that a temporary project might fail to produce its stated objectives – temporary projects may be used to create a 'state of exception' in urban politics. This mitigates risk associated with speculative governance tactics, and opens a space to define development agendas informally through a series of temporary projects.

In light of the fact that particularly Cape Town's municipal government has been repeatedly critiqued for its technocratic managerialism, risk-aversion, and 'silo-mentality' from both inside and outside the institution (van Donk et al., 2008; Kersting, 2012), processes like the World Design Capital bid can effectively function as a safety net that allows municipal officials to engage in more open-ended political experiments outside of rigid public performance protocols. This also opens our view towards the 'social life' of international accolades beyond but certainly not divorced from actually existing processes of urban neo-liberalization (Peck and Theodore, 2012).

Lastly and with particular reference to 'Southern' urban contexts, it remains important to reiterate that aggregating global recognition through international accolades has been framed as a significant experience of modernity, particularly for those cities commonly imagined as 'less modern' (Roche, 2002). However, as we see conventional criteria of Western urban modernity caving and crumbling against our experiences with multiple 'glocal' urban modernities (Swyngedouw and Kaïka, 2003), we have also come to realize that their precarious reinvention in the 21st century metropolis is fundamentally based upon an open-ended loop of speculative trial and error (Roy, 2009, 2014). After all, as Ong (2011, p. 12) shrewdly remarks, 'tinkering with a spectrum of urban ideas and forms is an art of being global'. Indeed, the spectacular allure of international accolades – be it in the form of sport-centred mega-events or cultural urban award schemes – has repeatedly proven its power to draw together a disparate bunch of urban policy 'tinkerers' and 'elite dreamers' with various vested interests and competing agendas, resulting in the kind of inchoate 'politics of becoming' (Ponzini and Rossi, 2010) that also shaped Cape Town's designation as the World Design Capital 2014.

## The politics of becoming the first African design capital

In order to illustrate how Cape Town's World Design Capital bid and subsequent designation functioned as a fulcrum for 'extrospective' and introspective politics, the chapter now turns toward the empirical analysis. As alluded to above, the World Design Organization (formerly known as the International Council of Societies of Industrial Design) – the World Design Capital's awarding body – had evidently been inspired by the global rise of the creative city imperative: 'The evolution of the creative industries, their impact and contributions to the global economy, have helped Icsid to position the WDC project as an outlet for design to be recognized as a significant accelerator in city development' (WDO, June 2012). Thus, in 2008 the Montreal-based organization officially launched its biennial World Design Capital project as 'a distinctive opportunity for cities to showcase their accomplishments in attracting and promoting innovative design, as well as highlight successes in urban revitalisation strategies' (WDO, June 2012). Since then, the title has been conferred to Seoul (2010), Helsinki (2012), Cape Town (2014), and Taipei (2016), with Mexico City as the designated WDC for 2018.

In the aftermath of the FIFA World Cup in late 2010, a coalition of high-profile Cape Town designers and academics, politicians, and economic development practitioners, under the stewardship of the influential public-private partnership organization Cape Town Partnership, had come together to devise the city's bid under the tagline 'Live Design.Transform Life'. In October 2011, it was announced that Cape Town had beat 56 other applicant cities – including the other two finalists Dublin and Bilbao – to the title. As I have traced the governmental lineages between the 2010 FIFA World Cup and the WDC designation, as well as the emergence of a new 'design-minded' urban governance regime elsewhere (Wenz, 2014a,b), the following specifically focuses on the inter-scalar governmental dimension of the process. In particular, it exemplifies how Cape Town's World Design Capital process provided a prime opportunity for local policy actors – such as the mayor and members of the bid committee – to engage in what Fraser (2010) has called 'the craft of scalar practices', i.e. the use of scalar discourses and international activities to build local political legitimacy – both *for* and *through* the accolade.

*The WDC as vehicle for 'mayoral brand politics'*

As mentioned above, the initial idea to bid for the World Design Capital title had come from a local growth coalition, which also drove the successful bidding process on behalf of local government since the attention of most officials was bound up by nation-wide local government elections during the same period. Thus, the shortlisting of Cape Town as one of three finalists for the title in June 2011 also coincided with the appointment of Patricia de Lille as the city's new mayor, handing her a prime opportunity for quickly building her profile and asserting her leadership both locally and internationally, i.e. engaging in what Pasotti (2010) has aptly termed 'mayoral brand politics'.

Thus, when Cape Town was announced the winner later that year at the congress of the International Design Association in Taipei, the mayor personally accepted the title with the following words:

> The World Design Capital bid process and title have helped to bring different initiatives together and have made us realize that design in all its forms, when added together, creates human and city development. [...] [this] designation gives cities like Cape Town additional motivation to actively think of transformative design in development plans. We look forward to learning from other cities that are using design as a tool for transformation, including past winners Torino, Seoul and Helsinki and our fellow short-listed cities, Dublin and Bilbao. [...] In 2014 we will truly mature into a broader international course of cities adapting to new realities and in so doing, we will contribute to the quality of life for our city and the larger body of knowledge that aids urban development internationally.
>
> (de Lille, speech transcript, 26 October 2011)

The invocation of the notion of design as a harbinger for urban transformation, and the city's firm self-association with this emerging 'brand identity' is furthermore undergirded by a number of inter-scalar political logics and aspirations that are worth deconstructing: firstly, it is evidently an emotive appeal to her local constituency, serving to 'transmit a message of great leadership courage to innovate, and the ability to steer the city through crisis and change' (Pasotti, 2010, p. 18). Here it is important to recall the strategic role of Cape Town within national politics: While both Cape Town and its encompassing province, the Western Cape, had been initially governed by the ruling African National Congress, both had been won over in 2006 by the biggest national opposition party, the Democratic Alliance (DA). Held by the DA ever since, the Western Cape and particularly Cape Town as its largest metropolis have been perpetually used as poster children to merit the opposition's claim of being more capable than the ruling ANC of running the country. The WDC 2014 title thus provided the mayor with a potent device for presenting her city as the very antithesis to development trajectories

World Design Capital - Interurban Networks

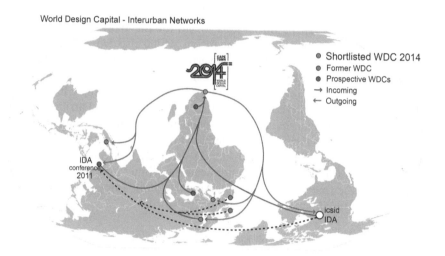

*Figure 11.1* Interurban networks of Cape Town as WDC 2014.
*Source*: Wenz, L. (2014b), p. 396; City of Cape Town External Relations Website, interviews and personal conversations, created by E. Gooris.

in the rest of the country. This strategic motion of 'scale-jumping' is also aptly captured in the following related address by de Lille:

> *Nationally,* there seem to be increasing pressures on us. We are experiencing slower growth than anticipated and nowhere near the rate required to add the amount of jobs our national economy requires. In some places, the compact between business, labour and government appears to be more strained than at any other time in the recent past. We also know that old trading patterns, especially with and within the rest of the African continent, are shifting. Considered together, these factors add to compound challenges facing the country. *And yet, here in Cape Town, we have never been more confident. That is because we know that our performance is not necessarily tied to these national indicators. Cities are forging their own paths in this world.* [...] It is a new story for the world and it is a story in which Cape Town will be a major character. That is one of the reasons why we are honoured to be the World Design Capital for 2014.
>
> (de Lille, 25 February 2013, speech transcript, emphasis added LNW)

The narrative that 'cities are forging their own paths' – ostensibly unperturbed by adverse national trends – also relates to the second political rationale: the expectation that the accolade will function as the nucleus of a burgeoning 'urban learning assemblage' (McFarlane, 2011b, p. 134). In de Lille's acceptance speech the notion of design is presented as a powerful rallying point for inter-urban cooperation and knowledge exchange, especially between previous and incumbent WDCs (see Figure 11.1). This also resonates with the argument of Beal and Pinson (2014), that mayors not only engage in inter-urban competitions for personal prestige and grandstanding but also in order to access additional resources for public policymaking from cities elsewhere. In consequence, becoming part of the WDC 'club' afforded Cape Town's urban leadership with the opportunity to establish a whole set of new inter-urban connections across both the public and private sector. For instance, through its relationship with previous title holder and designated 'mentor city' Helsinki, Cape Town's WDC proponents not only learnt about the Finnish approach to governing the award, they also 'inherited' some international private sector investment in the form of sponsorship by Nokia and Phillips (personal conversation with bid committee member, 21 February 2012), established a new official partnership between the design faculties of Aalto University and the Cape Peninsular University of Technology, and hosted a high-profile mayoral exchange called 'Helsinki meets Cape Town', funded by the Finnish Ministry of Foreign Affairs. Over the course of Cape Town's tenure as World Design Capital, other foreign representatives followed suit in sponsoring similar exchanges, for example Taiwan and the Netherlands. The latter even created a full time position within its local consulate's foreign trade division in order to liaise with local WDC stakeholders and encourage Dutch-South African design collaborations. As such, the accolade provided ample opportunity for local governance elites to build new inter-urban relations through which to mobilize both economic *and* knowledge resources (Lauermann, 2016, p. 14). This strengthening of inter-urban diplomacy and collaboration under the banner of the WDC can also be interpreted as a shift from a 'politics-oriented logic', i.e. building political legitimacy through traditional methods of grassroots campaigning, to a more 'policy-oriented logic' that focuses on the creation (but not necessarily the implementation!) of urban policies (Beal and Pinson, 2014). In fact, with regards to the WDC title, the promotion of such a 'policy-oriented logic' is also already built into the accolade itself: all host cities are required to organize and hold an international Design Policy Conference as one of their six prescribed 'signature events'. Thus, on the occasion of Cape Town's design policy conference, the mayor proclaimed: 'During the City's reign as World Design Capital, it is most fitting that we develop a policy that guides and encourages design in all its facets, not only in South Africa, but on the entire continent of Africa' (de Lille, cited in WDO, 18 October 2014).

In positioning Cape Town's policy advances not only as a gain for the country but for the continent as a whole, the mayor's statement also gives expression to the third political aspiration attached to the award: that within a landscape of inter-urban policy learning still skewed towards North-to-South transmissions, Cape Town would be seen as a 'teacher city' (McCann, 2013, p. 10) – a translational space for continental policy 'solutions', enshrined in its status as the *first* African World Design

Capital. As Beauregard (2003, p. 184) has fittingly pointed out: 'superlatives and "firsts" are rhetorical devices that prepare the reader for a theoretical move – the assertion that the city is a paradigm. Not just unique or exemplary, the city being discussed contains and portends the future'. However, as I will illustrate below, asserting Cape Town's global role as the epitome of a successful design city was by no means a straightforward exercise. Instead, it required considerable discursive work and a particular form of urban policy boosterism to portray Cape Town as *both* a global *and* an African city – two imaginaries that are popularly regarded as incommensurable.

### Relocating Cape Town 'by design'

As the original nucleus of white settler colonialism, it is safe to say that Cape Town historically harboured a rather uneasy relationship with its continental location. After all, Jan van Riebeeck, first administrator of the Cape Colony, not only planted a bitter almond hedge to separate the settlement from its hinterland but at one point even contemplated digging a canal to physically separate the Cape peninsula from the rest of Africa (Hemer, 2012, p. 244). Moreover, having been turned into South Africa's most segregated city during apartheid, the (white) European look and feel of Cape Town's inner city still stands in stark contrast to the vast townships and growing informal settlements, home to the – still predominantly black – urban poor. While Cape Town is demographically 'Africanizing', not least due to steady immigration from other African countries, repeated acts of xenophobic violence since the early 2000s have only exacerbated the city's 'anti-African' image (McDonald, 2008, p. 269f). South African anthropologist Suren Pillay shrewdly captures this schizophrenic relationship between Cape Town and the continent when he asks:

> For us in Cape Town, what does it mean to be aware of location? One response here recognizes our national location as a kind of lament. Our African location is the name of a backwardness marked by how far it doesn't correspond to the ideal. In the Cape this becomes trickier, because we think we are the developed, holding fort and surrounded by the encroaching undeveloped.
>
> (Pillay, 2011)

In sum, for the longest time Cape Town's at best ambiguous at worst hostile relationship with the rest of Africa had reinforced a self-image that was premised on being a city in Africa rather than a truly African city. However, these popular perceptions of Cape Town as a decidedly 'un-African' city posed a critical challenge to local 'policy boosters' (McCann, 2013) who sought to leverage the WDC title as a vehicle to promote their city as a model for the rest of the continent and beyond.

In order to overcome this disjuncture and forge a coherent 'extrospective' politics around the WDC title, I argue that the malleable notion of design functioned as an important 'discursive glue' (Lees, 2004, p. 102). However, in order to be able to claim global legitimacy 'by design', local WDC proponents also needed to make this trope locally intelligible, particularly by showing how this abstract notion could relate positively to the city's divided urban reality. As illustrated earlier by de Lille's acceptance speech, this was done by framing design 'in all its form' as a firm guarantor of 'human and city development'. Furthermore, conscious of the fact that design was commonly associated with luxury objects, conspicuous consumption, and expert knowledge from elsewhere, Cape Town's WDC proponents sought to dispel this exclusive reading of design through a counter-narrative centred on the idea of design as a mundane practice that could already be found in place. As de Lille commented during the Design Policy Conference: 'we see ordinary South Africans making a plan with what they have. Designing and developing innovative and extraordinary things from the most ordinary and often unwanted material' (de Lille, cited in City of Cape Town, 2014). Following this emotive (and evidently cost-effective) narrative, Cape Town's economic, social, and spatial development challenges were also strategically embraced as essential preconditions for – rather than impediments to – innovation. As one bid committee member explains:

> Cape Town has many faces. On the one hand, it is among the most beautiful cities in the world. On the other hand, the legacy of the past created many social problems. While nothing is superficial about Cape Town's beauty, the ugly side is also in-your-face. It is these two extreme conditions that make Cape Town a great incubator for design, motivated by South Africans' resilience and persistence in the face of challenges.
>
> (Tsai, cited in WDC 2014, 26 October 2011)

Thus, in drawing on the evidently simplistic yet emotive truism that 'necessity is the mother of all invention' both statements imply that Cape Town should be seen as a model city – not *in spite* but rather *because of* its violent history, complex urban challenges, and geographic location.

While this might appear counter-intuitive at first, it actually represents a strategic form of policy boosterism, one that is deeply cognizant of the expanding geography of global urban policy markets and the potentials this could hold for a city like Cape Town. As one bid committee member urged even before Cape Town's official designation,

> we should look at the possibility of Cape Town winning not just in terms of us being in a developing country but, rather, as us being part of the 'majority world'. Cities that have won in the past were those that are part of the developed world - and yet those cities form less than 10 percent of the global population; they are part of the 'minority world'. Far more

relevant today is where design is heading for the other 90 percent. [...] In many ways, our bid could be seen as a template for bids in the future. We speak on behalf of the 'majority world' with a powerful voice that resonates across the globe.

(M'Rithaa, cited in Creative Cape Town, 2011, p. 54)

In turn, within this emerging geographic imagination of Cape Town as an aspiring provider of design 'solutions' to a 'majority world context', the city's continental location is no longer reduced to a lament. Instead, being an *African* city (rather than just a city in Africa) is reframed as an indispensable marker of authenticity and legitimacy. Or in the enthusiastic words of another bid committee member:

A win for Cape Town is a win for Africa. An international spotlight on Cape Town's contribution will highlight Africa's contribution to cutting edge transformative design and will also pique the interest of the international design community, allowing us to play host to the world, yet again.

(Cape Town Partnership, 2011)

Moreover, I'd argue that the romanticized celebration of local ingenuity in the face of adversity not only naturalizes Cape Town's – and by extension Africa's – urban crisis, but also exemplifies a 'governmentalization of the local population' (McCann, 2013, p. 22), i.e. generating local pride while simultaneously promoting a self-entrepreneurial spirit. Again, this also resonates with McCann's core argument that extrospective policy boosterism is always co-constituted by more introspective politics of persuasion. Furthermore, it is important to note that the notion of design is evidently flexible enough to hold together divergent entrepreneurial and developmental logics in recombinant ways, making it a powerful 'intellectual technology' (Rose and Miller, 1992, p. 178), whose influence on Cape Town's urban politics continues to present a rich stream for future research. For the purpose of the present argument however, what the above shows is that Cape Town's WDC process was not merely a wholesale downloading of generic world-class city logics but rather constituted an active site for transforming Cape Town's self-image from a 'tourist imagination' (Bickford-Smith, 2009) to one that more genuinely reflected its ambition of becoming an international model city in its own right. That the WDC title furthermore served as important 'discursive glue' for effortlessly portraying Cape Town as an *African* (yet) *global* city is also vividly illustrated by another bid committee member:

The world is very tired of so-called perfect places that aren't actually perfect, in any case scratched underneath the surface. But our advantage is that we are a city on the move and we are using design and innovation also to design solutions to the issues that face all of us. So it's a bold move to claim that space as it kind of challenges your brand at the tip of the African continent, to say that we are a city of the future. We might not be significant in terms of size, numbers, the kind of budget we have to compete with the world, but we absolutely have an incredible story to tell.

(Du Toit-Helmholdt, event transcript, 01 November 2011)

In sum, based on the above empirical vignettes, I'd argue that taking a closer look at Cape Town's designation as the first African World Design Capital provides a pertinent analytical inroad for understanding international urban award schemes as particular *worlding devices*: firstly, because the title has allowed a wide array of local urban governance practitioners to engage in the 'craft of scalar practises' (Fraser, 2010), i.e. the use of scalar discourses and strategic cosmopolitanism for drawing new connections between the local and the global. Secondly, because – under the banner of 'design for development' – the WDC accolade provided a prime opportunity for a variety of urban actors to grapple with the contentious issue of how Cape Town sees itself as an African city (rather than just a city in Africa), and to present their global visions as an emotive counter-narrative to the still overwhelmingly negative representation of the continent. And thirdly, because it has in turn been increasingly used as a lever to promote Cape Town as an emerging 'model city' in its own right, not only for urban Africa but the 'global South' at large. This also falls in line with Ong and Roy's identification of urban modelling as a 'flamboyant feature' of worlding cities that 'indexes the challenges of these cities not only to catch up with one another, but also to create new conceptions of achievable metropolitan standards for the developing world' (Ong, 2011, p. 23).

## Conclusion

With cities looming evermore largely as 'strategic hinges of globalization' (Acuto, 2011, p. 2953), it is timely to scrutinize more closely the wider networks, scalar hierarchies, and broader territories in which they are located, i.e. their specific urban political agency as the introduction suggests. The present chapter has sought to show how emerging 'Southern' cities like post-apartheid Cape Town have become embroiled in global processes of inter-urban circulations, i.e. how they have been 'cast out into the world' (Simone, 2001; Robinson, 2011), not least through their growing exposure to international urban accolades. Hence, framing Cape Town as a *worlding city* and the World Design Capital process as a specific *worlding device* allowed me to uncover the 'politics of connectivity' (Allen and Cochrane, 2014, p. 1641) inherent in the broader political discourse around the aim of being internationally recognized as a model city. After all, while a number of Asian and Latin American cities, such as Singapore and Bogotá, have already emerged as strong global reference points for urban policy innovation, this space still remains to be claimed by an African city.

What then has become clear throughout my analysis is that, on the one hand, the WDC 2014 process offered plenty of opportunities for inter-urban diplomacy and lesson-drawing conducted within different formats of what McFarlane (2011a, p. 134) refers to as 'urban learning assemblages'. These included international summits, study trips, and bilateral intercity exchanges with previous and aspiring World Design Capitals, e.g. Helsinki and Taipei. On the other hand, however, as much as many WDC 2014 proponents displayed an avid inquisitiveness for engaging with 'best practice' examples from elsewhere, there was also an equally strong sense of situated agency, encapsulated in the desire to turn Cape Town into a model city in its own right. Also, as I have shown, the WDC process allowed local urban governance elites to move effortlessly across local, provincial, national, and international scales, for example in claiming to be simultaneously representative of and relevant to a 'majority world context', while at the same time promoting the World Design Capital title to local constituents as 'the best opportunity to design our way out of the old Africa [sic!]' (Creative Cape Town, September 2010)

In sum, Cape Town's convergent and divergent, affirmative and speculative politics of becoming the first African World Design Capital are more than just the result of a globally hegemonic world-class city discourse that has trickled down to the tip of the continent on the back of yet another international award scheme (see also Wenz, 2014b). Rather, as I have sought to demonstrate, Cape Town's burgeoning African design city nexus has been inherently shaped by topological power dynamics and competing introspective and 'extrospective' politics: on the one hand the ardent longing to leverage the accolade for finding idiosyncratic 'solutions' that could address the staggering issue of post-apartheid transformation, while on the other hand using it as a vehicle for presenting the city as an internationally recognized 'truth spot' (Gieryn, 2006) for urban policy knowledge. Precariously holding this composite spectrum of divergent demands and aspirations together has been the use of design as a transversal governmental logic, whose future deployment in and influence on Cape Town's urban politics presents a rich stream for further critical inquiry.

## Note

1 As the term 'global South' has become a popular and important heuristic device in recent urban scholarship, it is important to highlight that this chapter does not use it to refer to a fixed geographic location but rather to highlight its ongoing production through global relations (see Comaroff and Comaroff, 2012).

## References

Acuto, M., 2011. Finding the global city: an analytical journey through the 'invisible college'. *Urban Studies*, 48(14), pp. 2953–2973.

Acuto, M., 2013. *Global cities, governance and diplomacy: the urban link. Routledge new diplomacy studies*. Abingdon and New York: Routledge.

Allen, J. and Cochrane, A., 2014. The urban unbound: London's politics and the 2012 Olympic Games. *International Journal of Urban and Regional Research*, 38(5), pp. 1609–1624.

Beal, V. and Pinson, G., 2014. When mayors go global: international strategies, urban governance and leadership. *International Journal of Urban and Regional Research*, 38(1), pp. 302–317.

Beauregard, R.A., 2003. City of superlatives. *City and Community*, 2(3), pp. 183–199.

Bickford-Smith, V., 2009. Creating a city of the tourist imagination: the case of Cape Town, 'the fairest cape of them all'. *Urban Studies*, 46(9), pp. 1763–1785.

Boyle, P., 2011. Knowledge networks: mega-events and security expertise. In: C.J. Bennett and K.D. Haggerty, eds. *Security games: surveillance and control at mega-events*. London: Routledge. pp. 169–184.

Bunnell, T., 2013. Antecedent cities and inter-referencing effects: learning from and extending beyond critiques of neoliberalisation. *Urban Studies*, August 2015, 52(11), pp. 1983–2000.

Cape Town Partnership, 14 June 2011. *A World Design Capital bid for the continent*. [online] Available at: www.capetownpartnership.co.za/2011/06/a-world-design-capital-bid-for-the-continent/ [Accessed 23 January 2018].

Chattopadhyay, S., 2012. Urbanism, colonialism and subalternity. In: T. Edensor and M. Jayne, eds. *Urban theory beyond the West. A world of cities*. London: Routledge. pp. 75–92.

City of Cape Town, 2014. *Design policy conference report*. Available at: www.wdc-capetown2014.com/profiles/Design-Policy-Conference-Report.pdf [Accessed 20 December 2016].

Comaroff, J. and Comaroff, J.L., 2012. *Theory from the South: or, how Euro-America is evolving toward Africa. [The radical imagination]*. Boulder, CO: Paradigm Publ.

Cook, I.R. and Ward, K., 2011. Trans-urban networks of learning, mega events and policy tourism: the case of Manchester's Commonwealth and Olympic Games Projects. *Urban Studies*, 48(12), pp. 2519–2535.

Creative Cape Town, September 2010. *The new Cape Town design network interim committee speaks out*. [online] Available at: www.creativecapetown.net/the-new-cape-town-design-network-interim-committee-speaks-out-2/ [Accessed 20 December 2016].

Creative Cape Town, 2011. *Creative Cape Town annual 2011*. [online] Available at: http://issuu.com/capetownpartnership/docs/creativecapetownannual2011?e=0 [Accessed 24 August 2014].

Evans, G., 2003. Hard-branding the cultural city - from Prado to Prada. *International Journal of Urban and Regional Research*, 27(2), pp. 417–440.

Franke, S., 2002. *City branding: image building & building images*. Rotterdam: NAI Publ.

Fraser, A., 2010. The craft of scalar practices. *Environment and Planning A*, 42(2), pp. 332–346.

Getz, D., 1991. *Festivals, special events, and tourism*. New York: VanNostrand Reinhold.

Gieryn, T.F., 2006. City as truth-spot: laboratories and field-sites in urban studies. *Social Studies of Science*, 36(1), pp. 5–38.

Gold, J.R. and Gold, M.M., 2008. Olympic cities: regeneration, city rebranding and changing urban agendas. *Geography Compass*, 2(1), pp. 300–318.

Hannam, K., Sheller, M. and Urry, J., 2006. Editorial: mobilities, immobilities and moorings. *Mobilities*, 1(1), pp. 1–22.

Hemer, O., 2012. *Fiction and truth in transition: writing the present past in South Africa and Argentina.* Freiburg studies in social anthropology, Vol. 34. Wien and Münster: LIT.

Hiller, H.H., 1995. Conventions as mega-events: a new model for convention-host city relationships. *Tourism Management*, 16(5), pp. 375–379.

Horne, J., 2007. The four 'knowns' of sports mega-events. *Leisure Studies*, 26(1), pp. 81–96.

Hoyt, L., 2006. Importing ideas: the transnational transfer of urban revitalization policy. *International Journal of Public Administration*, 29(1–3), pp. 221–243.

Jacobs, J.M., 2012. Urban geographies I: still thinking cities relationally. *Progress in Human Geography*, 36(3), pp. 412–422.

Kersting, N., 2012. Local government elections and reforms in South Africa. *Politeia: South African local government election 2011: its significance for democracy and governance*, 31(1), pp. 5–21.

Lauermann, J., 2014. Competition through interurban policy making: bidding to host mega-events as entrepreneurial networking. *Environment and Planning A*, 46(11), pp. 2638–2653.

Lauermann, J., 2016. Municipal statecraft: revisiting the geographies of the entrepreneurial city. *Progress in Human Geography*, 42(2), pp. 205–224

Lees, L., 2004. Urban geography: discourse analysis and urban research. *Progress in Human Geography*, 28(1), pp. 101–107.

Mbembe, A. and Nuttall, S., 2004. Writing the world from an African metropolis. *Public Culture*, 16(3), pp. 347–372.

McCann, E., 2004. 'Best places': interurban competition, quality of life and popular media discourse. *Urban Studies*, 41(10), pp. 1909–1929.

McCann, E., 2013. Policy boosterism, policy mobilities, and the extrospective city. *Urban Geography*, 34(1), pp. 5–29.

McCann, E. and Ward, K. eds., 2011. *Globalization and community: v. 17. Mobile urbanism: cities and policymaking in the global age.* Minneapolis: University of Minnesota Press.

McCann, E. and Ward, K., 2012. Assembling urbanism: following policies and 'studying through' the sites and situations of policy making. *Environment and Planning A*, 44(1), pp. 42–51.

McCann, E., Roy, A. and Ward, K., 2013. Urban pulse - assembling/worlding cities. *Urban Geography*, 34(5), pp. 581–589.

McDonald, D.A., 2008. *World city syndrome: neoliberalism and inequality in Cape Town.* New York: Routledge.

McFarlane, C., 2011a. Assemblage and critical urbanism. *City*, 15(2), pp. 204–224.

McFarlane, C., 2011b. *Learning the city: knowledge and translocal assemblage.* Oxford: Wiley-Blackwell.

Myers, G.A., 2011. *African cities: alternative visions of urban theory and practice.* New York: Zed Books Ltd.

Ong, A., 2011. Introduction worlding cities, or the art of being global. In: A. Roy and A. Ong, eds. *Studies in urban and social change. Worlding cities. Asian experiments and the art of being global.* Malden, MA: Wiley-Blackwell. pp. 1–26.

Pasotti, E., 2010. *Political branding in cities: the decline of machine politics in Bogotá, Naples, and Chicago.* Cambridge studies in comparative politics. Cambridge and New York: Cambridge University Press.

Peck, J., 2012. Recreative city: Amsterdam, vehicular ideas and the adaptive spaces of creativity policy. *International Journal of Urban and Regional Research*, 36(3), pp. 462–485.

Peck, J. and Theodore, N., 2012. Follow the policy: a distended case approach. *Environment and Planning A*, 44(1), pp. 21–30.

Peck, J. and Tickell, A., 2002. Neoliberalizing space. *Antipode*, 34(3), pp. 380–404.

Pillay, S., 2011, June 07. *Thinking Africa from the Cape*. Lecture delivered at the Locations and Locutions Lecture Series at the University of Stellenbosch. http://thoughtleader.co.za/readerblog/2011/06/24/thinking-africa-from-the-cape/

Pillay, U. and Bass, O., 2008. Mega-events as a response to poverty reduction: the 2010 FIFA World Cup and its urban development implications. *Urban Forum*, 19(3), pp. 329–346.

Ponzini, D. and Rossi, U., 2010. Becoming a creative city: the entrepreneurial mayor, network politics and the promise of an urban renaissance. *Urban Studies*, 47(5), pp. 1037–1057.

Quinn, B., 2005. Arts festivals and the city. *Urban Studies*, 42(5), pp. 927–943.

Richards, G., 2000. The European cultural capital event: strategic weapon in the cultural arms race? *International Journal of Cultural Policy*, 6(2), pp. 159–181.

Robinson, J., 2011. Cities in a world of cities: the comparative gesture. *International Journal of Urban and Regional Research*, 35(1), pp. 1–23.

Robinson, J., 2013. 'Arriving at' urban policies/the urban: traces of elsewhere in making city futures. In: O. Söderström, S. Randeria, D. Ruedin, G. D'Amato and F. Panese, eds. *Critical mobilities*. Lausanne: EPFL. pp. 1–28.

Roche, M., 2002. *Mega-events and modernity: Olympics and expos in the growth of global culture*. London: Routledge.

Rose, N. and Miller, P., 1992. Political power beyond the state: problematics of government. *The British Journal of Sociology*, 43(2), pp. 173–205.

Roy, A., 2009. The 21st-century metropolis: new geographies of theory. *Regional Studies*, 43(6), pp. 819–830.

Roy, A., 2011. Slumdog cities: rethinking subaltern urbanism. *International Journal of Urban and Regional Research*, 35(2), pp. 223–238.

Roy, A., 2014. Worlding the south: toward a post-colonial urban theory. In: S. Parnell and S. Oldfield, eds. *The Routledge handbook on cities of the Global South* (1st ed.). London and New York: Routledge. pp. 9–20.

Roy, A. and Ong, A. eds., 2011. *Studies in urban and social change. Worlding cities: Asian experiments and the art of being global*. Malden, MA: Wiley-Blackwell.

Sassen, S., 1991. *The global city: New York, London, Tokyo* (1st ed.). Princeton, NJ: Princeton University Press.

Simone, A., 2001. On the worlding of African cities. *African Studies Review*, 44(2), pp. 15–41.

Simone, A.M., 2010. *City life from Jakarta to Dakar: movements at the crossroads. Global realities*. New York: Routledge.

Söderström, O., 2014. *Cities in relations: trajectories of urban development in Hanoi and Ouagadougou. Studies in urban and social change*. London: Wiley-Blackwell.

Spivak, G.C., 1999. *A critique of postcolonial reason: toward a history of the vanishing present*. Cambridge, MA: Harvard University Press.

Steinbrink, M., Haferburg, C. and Ley, A., 2011. Festivalisation and urban renewal in the Global South: socio-spatial consequences of the 2010 FIFA World Cup. *South African Geographical Journal*, 93(1), pp. 15–28.

Surborg, B., Vanwynsberghe, R. and Wyly, E., 2008. Mapping the Olympic growth machine. *City*, 12(3), pp. 341–355.

Swyngedouw, E., 2011. Interrogating post-democratization: reclaiming egalitarian political spaces. *Political Geography*, 30(7), pp. 370–380.

Swyngedouw, E. and Kaïka, M., 2003. The making of 'glocal' urban modernities. *City*, 7(1), pp. 5–21.

Taylor, P.J., 2006. *World city network: a global urban analysis* (Reprint). London: Routledge.

Temenos, C. and McCann, E., 2013. Geographies of policy mobilities. *Geography Compass*, 7(5), pp. 344–357.

van Donk, M., Swilling, M., Pieterse, E. and Parnell, S. eds., 2008. *Consolidating developmental local government: lessons from the South African experience.* Cape Town: UCT Press.

Vanwynsberghe, R., Surborg, B. and Wyly, E., 2013. When the games come to town: neoliberalism, mega-events and social inclusion in the Vancouver 2010 Winter Olympic Games. *International Journal of Urban and Regional Research*, 37(6), pp. 2074–2093.

Waitt, G., 2008. Urban festivals: geographies of hype, helplessness and hope. *Geography Compass*, 2(2), pp. 513–537.

WDC 2014, 2011. *Cape Town celebrates world design capital win as mayor accepts accolade in Taipei.* [online] Cape Town: WDC 2014. Available at: www.cape-town2014.co.za/home/cape-town-celebrates-world-design-capital-win/ [Accessed 19 April 2013].

WDO (World Design Organization), 2012. *About world design capital.* [online] Available at: http://wdo.org/programmes/wdc/about/ [Accessed 12 October 2017].

WDO (World Design Organization), 18 October 2014. *Cape Town: a design destination.* [online] Available at: http://wdo.org/press-release/cape-town-a-design-destination/ [Accessed 12 October 2017].

Wenz, L., 2014a. The local governance dynamics of international accolades: Cape Town's designation as World Design Capital 2014. In: C. Haferburg and M. Huchzermeyer, eds. *Urbanisation of the earth: Vol. 12. Urban governance in postapartheid cities: modes of engagement in South Africa's metropoles.* Stuttgart and Durban: Schweizerbart; UKZN Press. pp. 251–270

Wenz, L., 2014b. *Worlding Cape Town by design. Creative cityness, policy mobilities and urban governance in postapartheid Cape Town.* Doctoral thesis at the Faculty of Philosophy. University of Münster.

Yeoh, B., 2005. The global cultural city? Spatial imagineering and politics in the (multi)cultural marketplaces of South-east Asia. *Urban Studies*, 42(5/6), pp. 945–958.

# 12 Urban rehabilitation and residential struggles in the post-socialist city of Budapest

*Gergely Olt and Ludovic Lepeltier-Kutasi*

## Introduction

Benjamin Barber suggests in his book *If Mayors Ruled the World...* (2013) that city leaders make pragmatic and un-ideological decisions checked by the democratic control of active urbanites. Barber (2013) highlights the importance of local democratic participation in urban issues and suggests that cities can become global players and at the same time their residents can maintain their democratic control. According to Barber, 'participation endows [people] with a capacity for common vision' (Barber, 2004, p. 232, cited by Costa, 2015, p. 24). However, this assumption is problematic because it 'privileges the benefits of localism without considering that certain interests can use these self-same principles of community and participation to promote inequality and injustice' (Scerri, 2014, p. 410); in other words, it fails to avoid the 'local trap' – the inherent assumption that decisions made on the local scale are necessarily more democratic than on other scales (Purcell, 2006, p. 1921).

Finding consensual solutions for conflicts within cities can be almost impossible because of antagonistic situations (Marcuse, 1998). Strong business interests and state-led rehabilitation efforts often overwrite the interests of lower-status users and inhabitants (for example in cases of gentrification, as shown by Smith, 1996). With the words of Purcell (2006, p. 1921) we can have the 'sense that urban neo-liberalisation threatens urban democracy', and that decentralization of power does not necessarily mean more democratic and inclusive decisions.

Besides the enforcement of the market rule by state power and the attraction of capital investment, in certain contexts other factors are also at work against 'urban democracy'. For example the emergence and effectiveness of urban movements is also determined by the social and political context and specific heritage elements such as widespread clientelism in Southern Europe after WWII (Leontidou, 2010, p. 1191). In the post-socialist context, a low level of citizen involvement and the influence of party politics in the civil society is still typical today (Kębłowski and Van Criekingen, 2014). In Hungary and other post-socialist states – with the terms used by Iván Szelényi

(2016) – neo-patrimonial and later neo-prebendal[1] elements influence – alongside emerging neo-liberalism – political and business relations. In this context neo-patrimonialism means that, as a consequence of the rapid privatization, the legal framework of the economic relations and other institutions of the transformation served the interests of politically connected cliques (see e.g. Stark and Vedres, 2012) and the acquisition of private property depended on political connections. Neo-prebendalism refers to the phenomenon that the legitimacy of property depends on political power relations and the will of the ruling political elite can lead to quick changes in ownership (Szelényi and Csillag, 2015, p. 29). Especially in the illiberal regime that emerged after 2010 – when the right wing party FIDESZ won the elections with a supermajority – more traditional power relations coexist with the legal rational (or liberal) authority (Szelényi, 2016, p. 12), and business plans, rules of the market or bureaucratic processes are often overwritten by political decisions.

The case of Budapest is worth analysing because during the long-term process of post-socialist transformation (see Stenning and Hörschelmann, 2008, p. 329) we could observe different models of urban management and policy that influenced the chances of the articulation of democratic interests. Parallel with the neo-liberal shock therapy of mass privatization of property, market relations were influenced by the interests of political groups and networks and corruption (see Jávor and Jancsics, 2016). Besides market interests, urban projects were determined by the distribution of public spending, and later, by EU funds. A significant part of these sources were channelled directly or indirectly to parties and politicians and this influenced how and for what purposes the money was spent. Especially in the new regime after 2010, civic and grassroots organizations can hardly put pressure on politicians who are personally interested in certain projects while democratic institutions are hollowed out by new legislations and regulations introduced by the power of the two-thirds supermajority of the right-wing ruling party.

We think these consequences of the transformation are contextual features of urban democracy as practiced in Budapest, and illustrate with our case studies that 'urban democracy' cannot be alienated from other scales, especially the national scale of legislation, judicial practice and exercise of political power. We therefore claim that the suggestion to simply turn towards the urban scale for more democratic, progressive and just policies and social practices ignores the embeddedness of urban questions into broader social and power relations.

In the next section of this chapter we outline the theoretical debates concerning participation and urban democracy and discourses of the post-socialist social and political context of Budapest. In the third part of the chapter we describe the context of inner city changes in Budapest and discuss conflicts of urban investments, gentrification and functional change in different parts of the inner city, showing how the interests and rights of the residents were neglected. In the conclusion, we summarize our main findings in light of our research objectives.

# The problems of urban democracy and the post socialist context

Barber (2013) assumes that urban issues can be handled democratically by active urbanites who participate directly in political decisions and planning, or control the power of mayors and other city officials. Later he envisages that the global network of these democratically governed cities can provide an answer to a series of global challenges much better than nation states can. In this chapter we only tackle the first but perhaps more fundamental claim of Barber: that urban democracy is less problematic than on other scales, and cities can overcome the national social and political contexts.

## *The problems of urban democracy*

First, we examine the claim that active urbanites could maintain democratic control in cities through having their opinions heard in local issues. Participation and democratic deliberation is not a panacea (see Sorensen and Sagaris, 2010, p. 298), consensus is not self-evident (Silver, Scott and Kazepov, 2010, p. 453) and residents often cannot exercise their right to the city (Purcell, 2006). Deliberation among unequal power relations can be a tool for the legitimation of the existing hegemonic order: people can make their choices among market- and profit-oriented frameworks presented to them as 'natural', while owners of capital sacrifice some of their gains (Purcell, 2009, p. 146; see also Silver, Scott and Kazepov, 2010, p. 454). In other words: deliberation is not synonymous with empowerment (Silver, Scott and Kazepov, 2010, p. 455). Alternatives could be antagonism, where 'the other [is] an enemy to be destroyed', and 'agonism', where the conflict remains an 'irreducible' part of the society but without actual violence (Purcell, 2009, p. 151). Instead of consensual deliberation these approaches try to undermine the hegemonic order of neo-liberalism (Purcell, 2009, p. 152). Even if people are able to have their say about the issues of the city, it can be necessary to have top-down institutions that force other players to give voice to the most marginalized and disadvantaged (Silver, Scott and Kazepov, 2010, p. 467). In certain contexts even the help of advocates and facilitators can be necessary for their representation (Silver, Scott and Kazepov, 2010, p. 472). Although radical critics condemn deliberation and consensus seeking, examples of successful and long-term urban movements often involve bargaining with the ruling power (see Sorensen and Sagaris, 2010, pp. 305, 307, 310). Empirical data also show that a pragmatic approach and a cooperative attitude towards power (Sorensen and Sagaris, 2010, p. 312) are necessary for the successful and long-term operation of urban movements. However, maintaining the progressiveness of movements without the 'destruction' of the opposing side (also within movements) is indeed a challenge (Teo, 2016, p. 1425).

Second, and as mentioned in the introduction, the urban scale is not independent from other scales and privileging the local as the 'adequate' scale of democracy is misleading. Positioning the decentralization of decision-making as a 'necessary [tool] for democratisation is the essence of the local trap' (Purcell, 2006, p. 1925). Decisions in favour of local residents may have negative consequences for other residents of the whole metropolis (Purcell, 2006, p. 1935) or rural areas of the broader region (Purcell, 2006, p. 1936). However, some critical human geographers maintain that that we can experience problems of neo-liberalism 'most saliently' on the urban level (Jessop, 2002, p. 452), and that state capacities are often transferred downwards to local authorities (Jessop, 2002, p. 454). Therefore Jessop suggests a 'post-national' framework (Jessop, 2002, p. 459), where 'the importance of the national scale of policymaking and implementation is being seriously challenged'. This is a common element in the suggestions of Barber and the analysis of certain critical human geographers. In addition, Sorensen and Sagaris (2010, p. 302) claim that self-organized local community groups can give us examples of how to resist more powerful actors, how the 'right to the city' is 'enacted and made real' and how to practice their collective rights against state or private projects and individual property rights.

Finally, critical scholars argue that the political and social context of these local movements is global capitalism or neo-liberalization, as the 'meta-context' of all other local contexts (Brenner, Peck and Theodore, 2010, p. 202). Therefore market rule and business interests – often directed from a distance – outweigh the interests of locals in different ways all over the world. The issue of the social and political-economic context, however, seems to be even more complicated. A growing number of scholars are raising the awareness in their empirical and theoretical work that neo-liberalism is in fact just one method of governing among many (see Ong, 2007), and economic rationality and market organization can exist independently of each other (Collier, 2005). Moreover, if scholars only acknowledge empirical differences as modifying features of their universal explanation of neo-liberalization, these experiences 'remain varieties of a single genus' (Barnett, 2005, p. 8; see also Collier, 2012, p. 194), instead of being part of theory building (see Robinson, 2016). Robinson (2011) argues that national-level historical and political contexts can also be influential in specific urban cases (Robinson, 2011, p. 1096) and the goals of resistance can be different case by case (Robinson, 2011, p. 1104). Recently Pinson and Morel Journel (2016) also argue that there are other factors at play in urban development projects than neo-liberalism (Pinson and Morel Journel, 2016, p. 176), and criticize Brenner, Peck and Theodore (2010) for their mono-causal explanation (Pinson and Morel Journel, 2016, p. 193). Silver, Scott and Kazepov (2010, p. 457) also argue for the importance of regional and national social and political contexts of participation. In short, urban political decisions are not independent from global economic and power relations and local institutional and social contexts.

### The post-socialist context

In our case studies from the post-socialist Budapest, the framework of neo-liberal market interests against the interests of the locals does not apply perfectly to every case. Besides neo-liberalization, urban changes and conflicts are also determined by more traditional (neo-patrimonial or later neo-prebendal – see above) power and property relations (Szelényi and Csillag, 2015).

Theoretical debates about the validity of the post-socialist category have been going on since the regime changes of 1989/1991. For example, in debates on gentrification, neo-Marxist scholars argue for abandoning the post-socialist perspective and adopting the narrative of global dependency instead (Nagy and Tímár, 2012, p. 122). Others in the same field argue for maintaining the category of post-socialism and taking into account the particular institutional changes in societies under transformation (Bernt, Gentile and Marcinczak, 2015, p. 105; especially Chelcea, Popescu and Cristea, 2015, pp. 128–129; see also Wiest, 2012, p. 838). Arguments against the concept of post-socialism are often expressed as if this concept would entail a totally different, oriental world (see for example Petrovici, 2015) or as if more pronounced neo-liberalization would be inevitable over time (see Sykora and Bouzarovski, 2012). However the 'de-territorialized' concept of post-socialism highlights that post-socialist states and cities are different not in general but rather in certain concrete social institutions and mechanisms (Tuvikene, 2016, pp. 132, 141). Tuvikene (2016, p. 142) suggests the concepts of 'continuities' and 'anti-continuities' in the post-socialist transformation are both the legacies of socialist dictatorships. Either these legacies exist today and fade away only slowly, or they can be sensed by the radical counter reaction. An example of the latter can be the 'patrimonial' (i.e. not market and investor friendly) housing privatization in Budapest that resulted in an ownership structure that differs significantly from the core capitalist cities (see below).

A good example of the relevance of the post-socialist category is a comparative study about two semi-peripheral cities, Cordoba in Spain and Sopot in Poland by Kębłowski and Van Criekingen (2014). The authors analyse participative budgeting (PB) in these cities. In the post-socialist Sopot the civil society proved to be significantly weaker (Kębłowski and Van Criekingen, 2014, p. 14) than in Cordoba, and the political actors only used participation as a tool to inform and educate residents or probe the public sentiment instead of letting them decide on certain questions. Unlike in Cordoba the local power did not consider residents as equal partners (Kębłowski and Van Criekingen, 2014, p. 16), and the mayor had the final word in decisions. The legacy of the centralized power and strict party hierarchy of state socialism (ironically similar to feudal relations) is a 'continuity' of the post-socialist case.

In the next section, we present our long-term ethnographic research of two urban rehabilitation projects in Budapest, and show that besides the market rule, post-socialist political power and property relations are also at odds with the ideas of 'urban democracy'. We try to illustrate our finding that the political and social context of the city strongly determines urban processes and urban democracy as well. Urban movements in this context could hardly affect powerful political players who gain their legitimation through national elections, but control local authorities as well.

## Case studies from post-socialist Budapest

As we already argued in the introduction, Budapest is an interesting case if we try to examine the role of the social and political context in 'urban democracy'. The high sovereignty of local authorities in place of central planning after the transformation – especially the independence of the districts of Budapest from the city; housing privatization (as radical anti-continuities); the EU accession and EU-funded urban rehabilitation initiatives; and since 2010, the re-centralization of power by the right-wing FIDESZ government (that reduced subsidies and took away a large part of the autonomy of local authorities) are all factors that could affect urban democratic processes. The cases of urban rehabilitation initiatives we present here are litmus tests of 'urban democracy', and we try to show how political changes influenced these processes.

After the regime change in 1989, one of the most important institutional transformations in the city was the quick and almost complete privatization of the housing stock (Hegedűs and Tosics, 1998). However, the transformation of property rights was realized differently in every post-socialist country (Sykora, 2005). In Hungary the method was flat-by-flat privatization of the municipality-owned housing stock for the sitting tenants and the transformation of every single building (or even operational parts of a bigger structure) into condominiums.[2] Tenants could gain full property rights for less than 10% of the estimated market value of the flats and they could even get low interest rate credits.

By the end of the 1990s 90% of the whole housing stock and 80% of the inner city housing stock became owner-occupied (Kovács, Wiessner and Zischner, 2013, p. 9). As Smith (1996, p. 176) and later Kovács, Wiessner and Zischner (2013, p. 5) observed as well, the approach towards privatization – carried out in the patrimonial fashion described above – and the condominium structure of dilapidated buildings made investment in the inner city difficult. Upgrading was often almost impossible because of the low status of owners, while a would-be investor in the building had to negotiate with many different owners, all of whom had different motivations. The consequence of this institutional setting was that gentrification was not the most pronounced phenomenon in Budapest during the 1990s and the status of some inner-city neighbourhoods declined even further (Kovács, 1998).

Even after the real estate boom of the 2000s and until the 2008 crisis, gentrification remained limited according to Neil Smith (see his self-critical remarks in Czirfusz et al., 2015, p. 58). However, in areas where municipal-ities proposed rehabilitation or residents were not able or willing to buy their tenements, the housing stock could remain in public hands and larger scale projects were possible. Most of the rehabilitation and reinvestment was initiated by the districts with different methods (Kovács, Wiessner and Zischner, 2015), and it was carried out according to the availability of public or EU funding and not exclusively according to the real estate market ra-tionale. However, it caused significant displacement (Ladányi, 2008) as local authorities used these funds to relocate and buy out residents of municipal social housing (we explain this in more detail in the case studies below). This process was most advanced in the District IX where rehabilitation started before 1989 and continues today (see Jelinek, 2011).

Between 2010 and 2015 symbolic and aesthetical interventions were initi-ated by the national government without the direct involvement of private investors (Kováts, 2014, p. 2) and without any public consultation with the citizens or the city of Budapest and the districts. Of course, these aesthetic and spectacle investments can be used later by entrepreneurs in the hospi-tality and tourism industry. Tourism and place marketing is the rationale behind national-level urban projects like the proposed Museum Quarter in the City Park, the FINA World Championship 2017 and the bidding for the 2024 Olympic Games.

However, besides marketing there are other motivations behind these state-financed mega-projects: overspending and corruption is also notori-ous. According to experts recently, 65% to 75% of all public procurements in Hungary were corrupt (Freedom House, 2011, cited by Jávor and Jancsics, 2016, p. 535) and kickbacks[3] are part of the everyday practice (Freedom House, 2011, cited by Jávor and Jancsics, 2016, p. 541). More up-to-date data also shows that in the illiberal regime corrupt practices became more prev-alent: more than 60% of public procurements were non-transparent (were realized without an open call for competition) compared to the 16% before 2011; and the numbers are even worse in the case of EU funds (CRCB, 2016, pp. 31, 34). The reason for this trend is the change of the legislation of public procurements by the new regime in 2011[4] (CRCB, 2016, p. 31). This means that the highest level of power is involved in 'control deactivation at the inter-organizational level' (Jávor and Jancsics, 2016, pp. 546–547): for ex-ample under the value of 25 million HUF (85,000 EUR) a low transparency procedure can be applied in public procurements. After 2010 higher value public investments were won almost exclusively by the company of a college friend of the PM. After their spectacular break-up in 2014 companies owned by the mayor of the home village of the PM took this role and he became the fifth wealthiest Hungarian by winning billions of Euros in public tenders.[5] The leader of a think-tank with close ties to the PM even expressed that what we see is not corruption but the building of a national capitalist class.[6] Therefore we can look at these mega-projects initiated on the national level on the one hand as neo-liberal projects in favour of entrepreneurs, but on

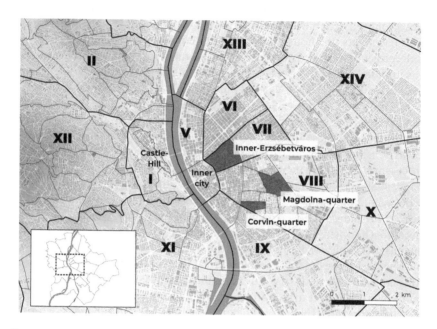

*Figure 12.1* Rehabilitation areas in Budapest.

the other hand, nationally and EU-funded procurements often have no economic rationality at all and market competition is strongly influenced by these corrupt public investments.[7]

## Methodology

Our data stems from two parallel, long-term ethnographic research projects in District VII and District VIII (since 2006 and 2011, respectively) in Budapest where significant social and functional changes have happened since the early 2000s. The research areas are the innermost part of the District VII and the 'social rehabilitation' area of the so-called Magdolna Quarter in District VIII (as indicated on the map in Figure 12.1). We chose these areas of the inner city because there were significant social changes induced by the rehabilitation plans of the local authorities in both areas. In this chapter we concentrate on the struggle of the residents in light of these changing circumstances.

In District VII, we conducted 23 semi-structured interviews with local residents who were involved in the rehabilitation and the functional changes of the area since 2006. Many of these residents lived in local authority owned buildings, and we asked them about their experiences with the rehabilitation, the changing function of the area and the night noise. We also conducted eight interviews with local pub owners about the story of their enterprise in the area and the conflict with the residents and authorities. We also participated in the regular meetings of the civil group of pub owners (called 'Azért') between 2010 and 2014 to help understand their strategy and the ambivalent relations towards the local authority. We also participated in several residents' forums and civil meetings in the area and we recorded them or made field notes. Local politicians were not particularly useful sources of information, but their participation in residential forums or meetings of the entrepreneurs were quite telling about their attitudes. We also recorded over 40 newspaper articles about the changes in the neighbourhood.

In the case of the Magdolna Quarter, we conducted 10 interviews with local active residents who participated in the civic engagement project of the rehabilitation programme and had experience in negotiating with the local authority. Altogether, we interviewed over 40 residents about their overall experiences in the stigmatized neighbourhood and about the effects of reha-bilitation. We also participated in the residents' forums in 2013 and 2014 that were focused on the new phases of the rehabilitation project where attitudes of the local politicians and the divide between groups of residents could be observed (namely those who attended the meetings and those who almost never did). Some local politicians and officials were important sources of information in this research area. We met them repeatedly and asked for insights about the details of negotiations, debates and decisions. All in all, our methods are quite similar to the methods of classical neighbourhood research cases applied for example by Small (2002, p. 47) or Pratt (2009, pp. 1044–1045) where all sorts of data and insider information are collected to understand the stories observed and social mechanisms behind them.

## *Inner-Erzsébetváros – a post-socialist gentrification process*

The Inner-Erzsébetvárs (the most central part of District VII) became very dilapidated during the state socialist era. The proportion of local authority owned buildings that were never turned into condominiums was relatively high in District VII – in the early 2000s more than 10% (Csanádi et al., 2010). This could happen due to the proposed plans of the Madách-promenade (Román, 1997–1998), which meant that dwellings on the planned track of the promenade were not for sale for tenants in the privatization process. Those plans, however, were never realized. In the early 2000s, when real estate prices started to increase rapidly, the local authority decided to vacate the municipality owned buildings. The costs of this action were to be covered by the privatization of entire buildings. Tenants of the municipality dwell-ings did not have much say in this process: they could choose from three exchange flats or take the cash compensation, which was hardly enough to buy another apartment on the outskirts of the city.

This may look like a typical case of state-led gentrification. However, the privatization was realized through a corrupt scheme and the buildings were sold to speculative investors. The neo-patrimonial element in the process was that municipal politicians and bureaucrats were using their political power and connections for personal financial gains in the privatization in the early 2000s. The municipality sold the right to buy the buildings for low prices to firms owned by the straw man of local authority politicians, after which this company – through an offshore company – was sold to a foreign investor. The difference between the two prices was the embezzlement of local politicians (Sipos and Zolnay, 2009). Though the investigation started in 2006, the culprits were only caught after 2009: the mayor was sentenced for two years and the first degree defendant – the former vice mayor of the district – got eight years in jail.

During the privatization, the rules of heritage protection were neglected as well and many buildings were demolished against the regulations, similar to the inner city of Moscow (Badyina and Golubchikov, 2005, pp. 113, 122). This was another aspect of the rampant corruption (see also Jávor and Jancsics, 2016, p. 545): after the demolition of the old buildings, many more new apartments could be squeezed onto the plots. Entrepreneurs with good political connections had no fear of taking these drastic steps. Because of the corrupt privatization, most of the new owners of the buildings were speculative investors, with the intention to resell their property for higher prices, except the 2008 crisis hindered their business plans. Instead of the expected higher status residential area, a party district developed on the ruins of the vacated buildings, which are now used by hospitality entrepreneurs as 'ruin bars' (Lugosi, Bell and Lugosi, 2010; Csanádi, Csizmady and Olt, 2012). Meanwhile, residential apartments have been turned into tourist accommodation, which increased the exchange value of inner city real estate. This case of post-socialist gentrification highlights the importance of the social embeddedness of the gentrification process (Bernt, 2016, pp. 642–643) and the commodification of inner city housing (Csizmady and Olt, manuscript).

### Interest articulation of the municipal tenants

In one particular building, the tenants wanted to privatize their dwellings after 1989 but that was denied because of the plans of the Madách-promenade mentioned above. The tenants discovered in 2004 just by accident that an investor had gained a pre-emptive right to buy the building. As a first reaction, 39 of them decided to file a lawsuit against the local authority that had neglected the pre-emptive privatization rights of the tenants (Somlyódi, 2007). Residents in this building were somewhat higher status than in others and many of the tenants had known each other for decades, two of the tenants were even lawyers. As a tacitly accepted and common reaction of the Hungarian politicians, a council member threatened the tenants that if they did not cooperate, the municipality would not renovate the building for another 15 years. One of my interviewees, who was an important organizer of the resistance in the building, was laid off from her job at a municipal institution without any explanation.

After three years, the investor (a company owned by the straw man of the local politicians) offered rather high prices for residents to leave. They would have taken these generous offers, but because the investigation in the corruption case had already started, the local authority stepped back from the privatization of the building at the last moment. After the political changes in 2010 not much happened in municipal housing except for a 100% rent increase in 2011. Although the prices were still low, this represented a huge problem for poor families. In 2014 the tenants of the building unexpectedly received a letter from the municipality (before the municipal election) that announced the privatization of the apartments for the tenants with the usual discounts for sitting tenants and because of the physical deterioration of the building.

Although the residents had wanted to stay put when they started the lawsuit, many of them bought their apartments in late 2014 and early 2015 at the discounted prices for tenants and left the neighbourhood within weeks after selling their apartments at market prices. Because of the 'party district', most of the residents did not want to stay anymore. The 'Airbnb fever' started in Budapest in 2015 and since then even a bad quality flat in a dilapidated building can be a financial asset instead of the low-use value for owner-occupiers. Because of this commodification process (Csizmady and Olt, manuscript), the market prices were two or three times higher than what the tenants had to pay. In 2016 a co-worker of the municipality turned to the state attorney and the press because he felt this practice of privatization was another case of fraud. However, it turned out to be just privatization according to the rules established in the 1990s. This ironic reaction in 2016 illustrates well the patrimonial nature of housing privatization: municipalities wanted to get rid of their housing stock so badly that it was indeed very similar to a simple case of fraud.

The privatization for tenants was unexpected because real estate investors – as we explained above – are interested in entire buildings. However as suggested by our interviewees, the transcript of the economic committee of the municipal council and newspaper articles,[8] the real motivation of the flat-by-flat privatization was to sell the retail spaces of the building to certain entrepreneurs. This would explain the anomaly: the retail space of a municipality-owned building cannot be procured on its own, but if the apartments are privatized and the building becomes a condominium, the retail space (still owned by the municipality) can be sold. The arbitrariness of the real estate policy is well illustrated by the fact that a similar building nearby was vacated between 2013 and 2016 and is waiting for investors, as the neo-liberal logic would suggest. In the case of our building, however, the neo-patrimonial relations intervened. As in many other cases of privatization, gaining property rights depended on political connections and not on market logic or on whoever has the most capital.

This is why we choose to explain this case in such a detailed manner: besides the neo-liberal real estate policy of the municipality, there are other factors at play in this context, namely the political control of property rights and the political influence of business interests. As a consequence, residents in different situations have different interests, and this divides people in otherwise similarly disadvantaged and disempowered situations.

The weak resistance against the municipality has several other reasons. First, many of these residents are in such a weak financial position that social housing is their only alternative. As they explained to us, discretional bureaucratic decisions are one of the 'continuities' of the socialist era in the social housing sector as well. On paper every tenant has equal rights, but – as was also common in the socialist era – people with connections or means to bribe officials could gain social housing rights much more easily and even choose better quality apartments.[9] In cases of vulnerable social tenants, officials and politicians could exercise grace (like patience with backlogs) for political gains. Many of the tenants had difficulties paying the rent, and it was always a political decision to evict them or not, which is also of course dependent on personal relations and political interests. Uncertainty and lack of transparency because of the corruption was also an obstacle before the successful interest articulation of residents: without getting proper information or without a sound rehabilitation plan it was very difficult to organize resistance in general. Struggles remained on the scale of single buildings, where people knew each other and lawyers and politically active residents were among them, as in our example.

### *The conflict of the party district*

As we mentioned above, the area became a 'party quarter' in the last 15 years (Csanádi, Csizmady and Olt, 2012). The process started with the intermediate use of the buildings that had been vacated by the local authority, but were not yet sold to investors (until the mid-2000s). After the 2008 crisis the real estate development stopped, and the new and mostly speculative private owners had to deal with the buildings, which were in ruins but also gained heritage protection during the mid-2000s (Perczel, 2007). Now these buildings serve as scenery for binge drinking tourism.

The rules for commercial activities changed in 2009[10] in line with EU directives. The law and governmental regulations protect the right for free enterprise more than the rights of the residents according to the ombudsman of civil rights,[11] and also according to the practice of the authorities and courts in these disputes. Real estate investors signed long-term contracts with pub operators. The level of noise at night increased, and residents complained to their local politicians. The only measure in the hands of the local authorities was the restriction of opening hours, which was instantly introduced in the adjacent District VI: after 10pm every commercial activity without special permits had to close. After many modifications the regulation stayed quite strict in District VI, where discretional decisions of a committee are necessary to gain the right to stay open after midnight. In District

VII, right before the municipal elections in 2010, opening hours restrictions were introduced as well. However, the local regulation was actually against national-level laws so it was never really implemented. Because of the rehabilitation (and the above-mentioned corrupt privatization) in District VII, there were more empty buildings and more ruin bars than in District VI, and this was another reason why it was more difficult to handle the situation.

After the political changes in 2010, the proposed solution was to create designated zones within the 0.5 sq km neighbourhood where there are no restrictions in terms of opening hours. The zones included the pubs that were members of an organization of pub owners called 'Azért'. This civil organization included only about 20 of the hundreds of pubs but at least they represented someone to whom the new local authority bureaucrats and politicians could talk. They were typically the entrepreneurs who had started their business before the mid-2000s. Meanwhile, organizations of residents were probed as well. One of them told us: 'They wanted us to say nice things about this regulation in our official communications and in exchange, our buildings would be outside the designated zones ... they wanted to bribe us. I found that so disgusting'. Because pub owners who were not part of the Azért loudly protested against the regulation, it was not implemented.

In late 2012, the national-level regulation changed, which made it possible for the police to restrict opening hours or close down venues if they were 'dangerous' or causing 'too much trouble'. More than 20 pubs in the Inner-Erzsébetváros received a fine or restriction of opening hours within a week. Because the 'ruin bar' scene was already an internationally known tourist phenomenon, a heated press debate followed these verdicts. At that time, the success of the scene also attracted entrepreneurs with much more capital and better connections and they tried to use them against the regulation. The freshly appointed vice-director of the municipal trust went to an 'Azért' meeting and agreed to a solution: for an additional tax, the pubs could be open after midnight. Because a lawyer representing the entrepreneurs participated in the codification process, the tax turned out to be very low (because of a cap that maximized the sum) and difficult to collect.[12] The current local regulation, implemented since 2013, permits opening hours in the Inner-Erzsébetváros until 6am. In the autumn of 2017 a more organized residential protest started with the involvement of 'professional' politicians and managers, and successful demonstrations were held. The local authority did not, however, change the legislation but decided instead to organize a local referendum in the future. This decision postponed the change of regulations, disappointing the residents and leading pub owners to complain about the uncertainty of their situation.

In this conflict, the local authority had to decide between the votes of the local residents and the income from the entrepreneurs' industrial and local taxes. It was not obvious which option they would choose during the entire process (see the adjacent District VI, District VIII or the recent regulation of District IX that prescribe closing times at midnight, 10pm and 11pm, respectively), but the national-level legislation and governmental regulations as well as the larger number of empty buildings pushed the regulation in a market-friendly direction. On the other hand, the rapid changes in the national-level regulation and the codification process showed the importance of political connections, and other inner city districts made steps towards 'maintaining order' instead of supporting tourism and hospitality business interests. According to our informants, officials of the local authority were not even aware of the taxation cap in the regulation. The interest of the residents was neglected after 2013 and though FIDESZ lost their majority in the local council, the mayor supported by the state party still won the local elections in 2014. Under the illiberal regime, a municipality with non-FIDESZ leadership can lose government subsidies and go bankrupt quickly.

### The rehabilitation of the Middle-Józsefváros

District VIII was the lowest status inner city district of Budapest already before the political changes of 1989. Many of the higher status residents left the area after 1989, deepening the 'ghettoization' (Kovács, 1998) and the stigmatization of the area (Czirfusz et al., 2015, p. 64). To improve the reputation of the district, the local authority, together with the city of Budapest and a Hungarian bank, established the Rév8 urban development company in 1997 (Alföldi, 2008, p. 27). First, the company spent 5 million EUR on aesthetic refurbishments, and later as part of a large-scale development plan vacated and demolished 22 hectares of a low status social housing area in the district to make room for a residential development with a 70 million EUR initial private investment (Alföldi, 2008, p. 30) that increased to 850 million EUR by 2010 (Czirfusz et al., 2015, p. 64). The tenants were relocated in social housing within the district, or were compensated in cash to a degree that was hardly enough to buy an apartment on the outskirts of the city. The other major rehabilitation initiative in the district was the social rehabilitation project of the Magdolna Quarter Programme (MQP).[13] Social rehabilitation meant that 85% of the residents had to stay put. Our examples of the residential involvement and struggle are taken from this process.

The right-wing turn in Hungarian politics was preceded by the mid-term mayoral election in the district in 2009 (during the implementation of the second phase of the programme). The new mayor of the district became an important representative of the right-wing FIDESZ party. As a consequence, the 'social' part of the rehabilitation became much less important and the pronounced revanchist politics of the mayor appeared in the rehabilitation project as well (Czirfusz et al., 2015, p. 70). However, application for rehabilitation funds and successful implementation of the project remained a priority since this was a possibility to spend EU money (with the contribution of firms related to the party)[14] and communicate 'results'.

*From community building to political control*

The first phase of the MQP was small-scale and was not even supported by the national-level EU funds distribution agency (Alföldi, 2008, p. 32). It entailed the refurbishment of four buildings, but also 'soft' elements like crime prevention, the facilitation of civil engagement and cooperation, and the establishment of a 'community space' that was to serve social goals. The 2.7 million EUR budget was mostly funded by the city of Budapest and the District VIII itself. Rév8 had high sovereignty in the planning and implementation of the project. The selection of the buildings depended, among other factors, on the application of the tenants and on their will to cooperate in the actual physical work of the refurbishment. This participation was supposed to facilitate communication among tenants within the buildings, and this had unexpected consequences. A very skilled former trade union leader organized the tenants of the four buildings (they called themselves 4 House Association), and criticized the implementation of the programme vehemently, while demanding crime prevention in the buildings and a more transparent social housing policy (drug trafficking in the district happens mostly in illegal sublets). The implementation of the programme was in the authority of Rév8 but crime prevention and housing policy was not. The local authority and the social housing management company never replied properly to these claims.

The second and third phase (MQP2 and MQP3) of the programme were financed by EU funds distributed by the national agency. As our interviewees explained to us, this meant much stricter bureaucratic control and there was much less space for experimentation. After the mid-term elections (mentioned above) the new mayor introduced much stricter control over the Rév8. Especially in MQP3, decisions about the main elements of the programme were made by the local authority and the mayor. Although there were soft programmes in the budgets, the main focus was on physical upgrading, for which contractors were chosen directly by the local authority. This explains why these social rehabilitation programmes were still important: the flow of EU money could be tapped and aesthetic changes became campaign elements. A good example of this attitude can be illustrated by the repeated – and probably unnecessary – renovation of a public square almost every second year. Although the power and independence of the Rév8 was radically reduced, residents were still arguing with them if they were unsatisfied with the results of the renovations in spite of the very limited influence of Rév8 on the local authority officials. This constellation distanced even further the residents from the actual decisions, while the Rév8 had to engage in a two-front battle.

In MQP2 many of the soft programmes were realized by civil organizations, which got a chance 'to do something good' by cooperating with the local political power. Meanwhile, they also tried to criticize the programme, the Rév8 and the local authority. In MQP3 this type of partnership with civic organizations was out of the question and institutions of the local authority (like the Family Support Agency) realized the soft elements. The renovation of a square was realized with a participatory planning approach. A civic organization evolved from this participation (Partners for the Teleki Square), which is eager to maintain 'order' in the square. The association was supported by the local authority and mostly agreed with its revanchist attitudes. Even if they expressed their criticism towards the mayor they believe it is better to criticize within a partnership. Of course this makes them more politically acceptable.

### Illiberal urban democracy

In the case of MQP2, legitimation of the territorially bounded social programme was also provided by the 'professional' engagement of the civil societies and their involvement in the neo-liberal mode of governance (see Silver, Scott and Kazepov, 2010, p. 461). In MQP3, however, the implementation was even more strictly controlled by the 'illiberal' mayor: directly through the employment of the staff. The illiberal attitude is echoed by the supported civil association mentioned above. As our interviewee from this association explained to us he strongly disagrees with the 'extreme liberalism' of other civil organizations or active residents – such as the group of progressive locals called 'KÖZÖD'. This means that active locals are divided: some of them cooperate with the power and try to influence it while others resist directly and try to enforce political changes.

The situation of the tenants is very vulnerable, so their resistance against power is highly unlikely. In an earlier publication (Lepeltier-Kutasi and Olt, 2016), we explained the situation of tenants who asked for reparations and fairer rules of sharing the operational costs of their freshly renovated but half empty building. They were quite afraid to take these steps because some of them already had rent arrears. The local authority and the municipal trust do not have to take into account the sum of the rent arrears, and could start the eviction process without any further debate. Members of KÖZÖD helped these tenants in the legal dispute. In another case a tenant who tried to defend her housing rights in a legal dispute with the local authority was evicted rapidly because of 'anti-social behaviour'. Since it was a made-up accusation to solve the dispute with her with less hassle, she started to collect signatures from other tenants to prove that her family causes no trouble in the building. The local authority now accuses her of forgery and tenants who signed her petition deny that they ever signed anything for the evicted tenant.

Municipality-level decisions about social housing are not controlled by any other authority at all. A housing right group called AVM[15] turned to the human rights court of Strasburg in a case of another accelerated eviction of a family. Meanwhile according to our interviews and a detailed description of the application procedure for municipal housing some local politicians received nice social apartments.[16] Progressive groups like 'KÖZÖD' or

'AVM' can achieve only small and partial success and cannot defend the tenants in general against the uncontrolled power of the municipality trust and the mayor. Tenants who engage in political struggles can lose their home in a blink, and national-level institutional changes would be necessary to exercise civic control over the district-level housing policy.

The increase of political control in MQP fits well with the national-level political changes after 2010. The national-level political context is quite important on the district level as well: the mayor, thanks to his position in the ruling party, can secure resources that are distributed by strictly controlled and politically engaged bureaucrats. The implementation of the National University of Public Service in the district is another symbolic project of the anti-liberal ideology, and it was realized with even less local civil control since it is a 'national project'. The national political power relations are also visible in the local elections, albeit that there are a few council members from opposition parties, the mayor easily won the 2014 elections. In a recent mid-term local authority council election in the constituency of the Magdolna Quarter, the opposition parties could not agree on a common candidate and FIDESZ won the seat again in spite of the fact that their former council member had to resign because of a corruption case. In short, the context of the illiberal Hungarian democracy drastically reduces the chance of political opposition on the local level as well.

## Conclusions

In this chapter we focused on the assumption of Benjamin Barber (2013) that city leaders can democratically represent the interests of urbanites. This democratic legitimation – that supposedly results in a progressive political agenda – is the precondition of fairer and humane solutions to global challenges by the 'parliament of mayors'. However, as the introduction of this volume highlights as well, democratically elected local politicians are just one factor among many in democratic struggles in cities. Global business interests and the social and political context on the national level are just as important.

Our intention was to illustrate how the national political and social context affects the presupposed 'urban democracy'. In post-socialist Hungary, the neo-patrimonial (-prebendal) power and property relations mean that political interests can easily overwrite the legal authority of the state (Szelényi, 2016; Szelényi and Csillag, 2015, p. 26). Obvious cases of corruption are tolerated and end without real political or legal consequences.[17] Even if local politicians are punished by the popular vote every now and then, the political power of FIDESZ gained through national elections can easily be used to control the mayor of the city and the mayors of districts. In this situation the deliberative consensus is either neglected or limited to marginal issues. Agonism or open resistance are also very risky alternatives, since economic actors who depend on EU-funded public spending and people who depend on discretional and non-transparent political decisions can be divided and ruled easily, while their fundamental rights are often shamelessly violated.

Progressive political movements are active in Budapest as well, but their connections with the countryside – where neo-patrimonial (and -prebendal) power relations are even more prevalent and tangible – are rather weak. Struggles of the progressive movements thus seem to be urban issues to which the people of the countryside – the majority of the voters – cannot easily relate. Because the concentrated power of the ruling party is legitimated on the national level and mostly by rural voters, the 'post-national' framework (Jessop, 2002, p. 459) would miss an important factor in the urban political struggles of Budapest.

From a more general perspective we note that market processes are not just enforced by the state but strongly influenced by political relations as well, and capitalist class interests are divided along political networks (see again Stark and Vedres, 2012). This does not fit very well in the mainstream conception of neo-liberalism as a class project (Harvey, 2005 cited by Barnett, 2010, p. 270), and this is why we need to refer to more traditional – feudal – power relations as well. If we try to explain the weakness of the civil society in the post-socialist context (Kębłowski and Van Criekingen, 2014, p. 14) exclusively with the semi-peripheral and subaltern position of these countries, how could we account for the more developed urban movements in similarly semi-peripheral Greece (Leontidou, 2010) or Turkey (Akcali and Korkut, 2015, pp. 86–87)? This is why we argue that the heritage of the state socialist dictatorships and the process of post-socialist transformation are significant contextual features of 'urban democracy' in post-socialist Budapest.

All in all, our case studies suggest that it is necessary to look beyond 'fundamental political-economic rationalities' (Barnett, 2010, p. 269) and understand the interplay of the local contexts and historic trajectories with global forces. This approach could be more helpful to understand how 'more limited forms of rupture, in particular institutional settings' are possible if we accept that the revolutionary overthrow of the whole global system is not plausible (Teo, 2016, p. 1426). There is still a long way to go to achieve 'urban democracy', but local movements can be examples of democratic struggles anywhere. How they could have an effect on other scales, however, is a different question.

## Notes

1 Patrimonialism is defined as 'feudalism' where the vassal is compensated by the 'fief' and the lord appoints the administrative staff. The 'fief' can be inherited but it remains inalienable and cannot be mortgaged (Szelényi, 2016, pp. 14–15). In prebendalism '[t]he member of the staff can remain in office and retain property as long as he or she assures the master of loyalty and offers valuable services to the master' (Szelényi, 2016, p. 14). The neo-patrimonial power and property relations turned into neo-prebendalism, when political bosses like Vladimir

Putin, or 10 years later Viktor Orbán, renationalized and then privatized again certain companies and assets, often with the help of new legislations and the judicial branch of the state accusing or even imprisoning former owners (see Szelényi, 2016; Szelényi and Csillag, 2015).

2 A condominium is a building or complex of buildings containing a number of in-dividually owned apartments or houses. In Hungary if a building is a condomin-ium it also means that the owners of the individual apartments have undivided common property rights over the common spaces of the building such as the basement or the attic, the staircases or the roof. This form of ownership meant that the refurbishment of the undivided common property was the responsibility of the owners of the apartments, who often had no financial means to pay for it.

3 Entrepreneurs who win competitions for public procurements have to give back part of their income (about 25%–30%) to the decision-makers for the 'favour' of winning the competition. We also have to mention here that scandalous public procurements were also prevalent before 2010.

4 Act no. CVIII. of 2011.

5 http://index.hu/gazdasag/2017/04/27/100_leggazdagabb_napi.hu_2017/

6 http://index.hu/belfold/2015/12/21/a_szazadveg_elnoke/

7 Of course there is corruption everywhere in the World, however the extent, mechanisms and acceptance of corruption is highly varied among different con-texts. In Hungary, for example, the global position of the country as a new mem-ber state on the fringe of the European Union and a recipient of massive amount of EU funds combined with the neo-patrimonial relations of the transformation resulted in a situation where a large proportion of EU money landed directly in the pockets of politicians and their 'vassals'.

8 http://nepszava.hu/cikk/1009404-orban-fogorvosa-elintezte

9 For example in District IX relatives and business partners of municipal coun-cil memebers could privatize freshly renovated apartments for extremely low prices: https://tldr.444.hu/2017/05/04/25-eve-vartak-arra-hogy-ne-kelljen-a-folyosora-kimenniuk-vecezni-aztan-jottek-a-fideszesek-es-bekoltoztek [Ac-cessed 1 June 2017].

10 210/2009. (IX. 29.) Government regulation of the 2005. CLXIV. Law.

11 OBH 6327/2008 and AJB 1765/2010.

12 T/11473 proposal accepted in 2013. CXVII. act; see also http://index.hu/belfold/2014/01/09/elszamoltak_a_rogan-fele_romkocsmaadot/

13 The demarcation of quarters was made by the Rév8.

14 www.direkt36.hu/2017/08/22/tortent-buncselekmeny-a-jozsefvarosi-nagyberu-hazasnal-de-a-rendorseg-szerint-nem-lehet-megtalalni-a-tettest/ [Accessed 1 September 2017].

15 The group was established on the model of Picture the Homeless in New York.

16 http://mijozsefvarosunk.blog.hu/2014/03/16/119_lakaspalyazatok_jozsefvarosban see also endnote nr.9

17 Since 2010 the legislative and to a large extent the judicial branch of the state – particularly the state attorney – is controlled by party interests similar to the socialist dictatorship.

## References

Akcali, E. and Korkut, U., 2015. Urban transformation in Istanbul and Budapest: neoliberal governmentality in the EU's semi-periphery and its limits. *Political Ge-ography*, 46, pp. 76–88.

Alföldi, Gy, 2008. Szociális rehabilitáció a Józsefvárosban [Social rehabiliation in Józsefváors]. *Falu Város Régió* [Village, City, Region], 2, pp. 27–34.

Badyina, A. and Golubchikov, O., 2005. Gentrification in central Moscow – a market process or a deliberate policy? Money, power and people in housing regeneration in Ostozhenka. *Geografiska Annaler*, 87B (2), pp. 113–129.

Barber, B., 2004. *Strong democracy*. Berkeley: University of California Press.

Barber, B., 2013. *If mayors ruled the world: dysfunctional nations, rising cities*. New Haven, CT: Yale University Press.

Barnett, C., 2005. The consolations of 'neoliberalism'. *Geoforum*, 36, pp. 7–12.

Barnett, C., 2010. Publics and markets: what's wrong with neoliberalism? In: S.J. Smith, R. Pain, S.A. Marston and J.P. Jones III, eds. *The Sage handbook of social geography*. London: Sage. pp. 269–296.

Bernt, M., 2016. Very particular, or rather universal? Gentrification through the lenses of Ghertner and López-Morales. *City*, 20(4), pp. 637–644.

Bernt, M., Gentile, M. and Marcinczak, S., 2015. Gentrification in post-communist countries: an introduction. *Geografie*, 120(2), pp. 104–112.

Brenner, N., Peck, J. and Theodore, N., 2010. Variegated neoliberalization: geographies, modalities, pathways. *Global Networks*, 10, pp. 182–222.

Chelcea, L., Popescu, R. and Cristea, D., 2015. Who are the gentrifiers and how do they change central city neighbourhoods? Privatization, commodification, and gentrification in Bucharest. *Geografie*, 120(2), pp. 113–133.

Collier, S.J., 2005. The spatial forms and social norms of 'actually existing neoliberalism': toward a substantive analytics. *International Affairs Working Paper*, 2005-04. New York: The New School University.

Collier, S.J., 2012. Neoliberalism as big Leviathan, or ...? A response to Wacquant and Hilgers. *Social Anthropology*, 20, pp. 186–195.

Costa, M.N., 2015. Death of popular sovereignty? Reflections on our (post) democratic condition. *Aufklerung*, 2(1), pp. 11–26.

CRCB, 2016. *Corruption research centre Budapest: strength of competition and risks, the statistical analysis of Hungarian public procurements 2009–2015*. [pdf] 29 February 2016. Budapest: Corruption Research Center Budapest. Available at: www.crcb.eu/wp-content/uploads/2016/03/hpp_2016_crcb_report_2016_hu_160303_.pdf [Accessed 1 June 2017].

Csanádi, G., Csizmady, A., Kocsis, J.B., Kőszeghy, L. and Tomay, K., 2010. *Város tervező társadalom* [Urban planner society]. Budapest: Sík.

Csanádi, G., Csizmady, A. and Olt, G., 2012. *Átváltozóban* [Transformation]. Budapest: Eötvös Kiadó.

Csizmady, A. and Olt, G. (manuscript). Gentrification and commercialization in post-socialist Budapest - the social effects of 'ruin bars'. *Urban Studies*, pp. 1–15.

Czirfusz, M., Horváth, V., Jelinek, Cs., Pósfai, Zs. and Szabó, L., 2015. Gentrification and rescaling urban governance in Budapest-Józsefváros. *Intersections. East European Journal of Society and Politics*, 1(4), pp. 55–77.

Freedom House, 2011. *Nations in transit 2011: Hungary*. [online] Available at: www.freedomhouse.org/report/nations-transit/nations-transit-2011 [Accessed 1 June 2017].

Harvey, D., 2005. *A brief history of neoliberalism*. Oxford: Oxford University Press.

Hegedűs, J. and Tosics, I., 1998. A közép-kelet-európai lakásrendszerek átalakulása [The transformation of the middle-European housing system]. *Szociológia (Sociology)*, (1), pp. 5–33.

Jávor, I. and Jancsics, D., 2016. The role of power in organizational corruption: an empirical study. *Administration and Society*, 48(5), pp. 527–558.

Jelinek, C., 2011. Relocation and displacement in the case of Budapest: the social consequences of gentrification in Ferencváros. *International RC21 conference.* Amsterdam, the Netherlands, 7–9 July 2011.

Jessop, B., 2002. Liberalism, neoliberalism, and urban governance: a state theoretical perspective. *Antipode*, 34, pp. 452–472.

Kębłowski, W. and Van Criekingen, M., 2014. How 'alternative' alternative urban policies really are? *Métropoles*, (15), [online]. Available at: http://metropoles.revues.org/4994 [Accessed 1 June 2017].

Kovács, Z., 1998. Ghettoization or gentrification? Post-socialist scenarios for Budapest. *Netherlands Journal of Housing and the Built Environment*, 13(1), pp. 63–81.

Kovács, Z., Wiessner, R. and Zischner, R., 2013. Urban renewal in the inner city of Budapest: gentrification from a post-socialist perspective. *Urban Studies*, 50(1), pp. 22–38.

Kovács, Z., Wiessner, R. and Zischner, R., 2015. Beyond gentrification: diversified neighbourhood upgrading in the inner city of Budapest. *Geografie*, 120(2), pp. 251–274.

Kováts, B., 2014. Political commodification of the inner city by constructing spectacles: manipulation and gentrification in the contemporary urban development agenda in Budapest. *Paper presented at the MRI conference on 'Social and economic conflicts of transition towards democracy and market economy – Central and Eastern Europe 25 years after, in a comparative perspective'*, November 2014.

Ladányi, J., 2008. *Lakóhelyi szegregáció Budapesten* [Housing segregation in Budapest]. Budapest: Új Mandátum Könyvkiadó.

Leontidou, L., 2010. Urban social movements in 'weak' civil societies: the right to the city and cosmopolitan activism in Southern Europe. *Urban Studies*, 47(6), pp. 1179–1203.

Lepeltier-Kutasi, L. and Olt, G., 2016. Demander réparation(s). Les mobilisations collectives à l'épreuve de leur visibilité. *Culture et conflits*, 101, pp. 81–98.

Lugosi, P., Bell, D. and Lugosi, K., 2010. Hospitality, culture and regeneration: urban decay, entrepreneurship and the 'ruin' bars of Budapest. *Urban Studies*, 47(14), pp. 3079–3101.

Marcuse, P., 1998. Sustainability is not enough. *Environment and Urbanisation*, 10(2), pp. 103–111.

Nagy, E. and Timár, J., 2012. Urban restructuring in the grip of capital and politics: gentrification in East-Central Europe. In: T. Csapó and A. Balogh, eds. *Development of the settlement network in the Central European countries*. Berlin, Heidelberg: Springer-Verlag. pp. 121–136.

Ong, A., 2007. Neoliberalism as a mobile technology. *Transactions of the Institute of British Geographers, New Series*, 32(1), pp. 3–8.

Perczel, A., 2007. Pest régi zsidó negyedének sorsa, jelenlegi helyzet [The fate of the old Jewish Quarter of Pest and the current situation]. *Múlt és Jövő*, 18(2), [online]. Available at: www.multesjovo.hu/hu/2007-2.html [Accessed 1 June 2017].

Petrovici, N., 2015. Framing criticism and knowledge production in semi-peripheries – Post-socialism unpacked. *Intersections. East European Journal of Society and Politics*, 1(2), pp. 80–102.

Pinson, G. and Morel Journel, C., 2016. Beyond neoliberal imposition: state–local cooperation and the blending of social and economic objectives in French urban development corporations. *Territory, Politics, Governance*, 4(2), pp. 173–195.

Pratt, A.C., 2009. Urban regeneration: from the arts 'feel good' factor to the cultural economy: a case study of Hoxton, London. *Urban Studies*, 46(5–6), pp. 1041–1061.

Purcell, M., 2006. Urban democracy and the local trap. *Urban Studies*, 43(11), pp. 1921–1941.

Purcell, M., 2009. Resisting neoliberalization: communicative planning or counter-hegemonic movements? *Planning Theory*, 8(2), pp. 140–165.

Robinson, J., 2011. The travels of urban neoliberalism: taking stock of the internationalization of urban theory -2010 urban geography plenary lecture. *Urban Geography*, 32(8), pp. 1087–1109.

Robinson, J., 2016. Thinking cities through elsewhere: comparative tactics for a more global urban studies. *Progress in Human Geography*, 40(1), pp. 3–29.

Román, A., 1997–1998. Madách Imre, avagy egy sugárút tragédiája [Imre Madách or a tragedy of a promenade]. *Budapesti Negyed* [Budapest Quarter], 18–19, [online]. Available at: http://epa.niif.hu/00000/00003/00015/roman.htm [Accessed 1 June 2017].

Scerri, A., 2014. Should we 'See like a city'? 'If mayors ruled the world,' would it be a better place? *New Political Science*, 36(3), pp. 406–411.

Silver, H., Scott, A. and Kazepov, Y., 2010. Participation in urban contention and deliberation. *International Journal of Urban and Regional Research*, 34(3), pp. 453–477.

Sipos, A. and Zolnay, J., 2009. Ingatlanpanama a régi pesti zsidónegyed elpusztítása [Real-estate corruption the destruction of the old Jewish Quarter]. *Beszélő*, May, 14(5), [online]. Available at: http://beszelo.c3.hu/cikkek/ingatlanpanama [Accessed 1 June 2017].

Small, M.L., 2002. Culture, cohorts, and social organization theory: understanding local participation in a Latino housing project. *American Journal of Sociology*, 108(1), pp. 1–54.

Smith, N., 1996. *The new urban frontier: gentrification and the revanchist city technological mobilities and the urban condition*. London/New York: Routledge.

Somlyódi, D., 2007. Újabb erzsébetvárosi ingatlanügyek: Elegük lett [The latest real estate business in Erzsébetváros: they've had enough]. *Magyar Narancs*, January 11, [online]. Available at: http://magyarnarancs.hu/belpol/ujabb_erzsebetvarosi_ingatlanugylet_eleguk_lett-66574 [Accessed 1 June 2017].

Sorensen, A. and Sagaris, L., 2010. From participation to the right to the city: democratic place management at the neighbourhood scale in comparative perspective. *Planning Practice & Research*, 25(3), pp. 297–317.

Stark, D. and Vedres, B., 2012. Political holes in the economy: the business network of partisan firms in Hungary. *American Sociological Review*, 77(5), pp. 700–722.

Stenning, A. and Hörschelmann, K., 2008. History, geography and difference in the post-socialist world: or, do we still need post-socialism? *Antipode*, 40(2), pp. 312–335.

Sykora, L., 2005. Gentrification in post-communist cities. In: R. Atkinson and G. Bridge, eds. *Gentrification in a global context: the new urban colonialism*. London and New York: Routledge. pp. 91–106.

Sykora, L. and Bouzarovski, S., 2012. Multiple transformations: conceptualising the post-communist urban transition. *Urban Studies*, 49(1), pp. 43–60.

Szelényi, I., 2016. Weber's theory of domination and post-communist capitalisms. *Theory and Society*, pp. 45(1) 1–24.

Szelényi, I and Csillag, T., 2015. Drifting from liberal democracy. Neo-conservative ideology of managed illiberal democratic capitalism in post-communist Europe. *Intersections. East European Journal of Society and Politics*, 1(1), pp. 18–48.

Teo, S.S.K., 2016. Strategizing for autonomy: whither durability and progressiveness? *Antipode*, 48, pp. 1420–1440.

Tuvikene, T., 2016. Strategies for comparative urbanism: post-socialism as a de-territorialized concept. *International Journal of Urban and Regional Research*, 40, pp. 132–146.

Wiest, K., 2012. Comparative debates in post-socialist urban studies. *Urban Geography*, 33(6), pp. 829–849.

# 13 Aspiring global nations? Tracing the actors behind Belgrade's 'nationally important' waterfront

*Jorn Koelemaij*

## Introduction

The 1950s are often referred to as 'the beginning of the end' with regards to the so-called 'era of the nation state', which, seen from a (mainly Western-) European perspective, lasted for more or less 100 years. This statement clearly relates to, on the one hand, the upscaling of power and decision-making autonomy towards international unions or institutions such as the EU or the IMF. On the other hand, the 'end of the nation state' refers to a trend of downscaling governmental regulatory responsibilities towards the urban or regional scale (e.g. Ohmae, 1995; Brenner, 1998). According to Massey (2007), we are witnessing a 'local internationalism' that challenges the dominant geographical imaginary in which the nation deals with the 'important' national and international issues. These changing scales of economic networks and institutional arrangements these days lead to the frequent assumption that urban policymaking practices are the result of interconnected global-local relations, or so-called 'glocal elites' (Swyngedouw, 2004; MacLeod, 2011). Hence, contrary to what used to be common during the Fordist-Keynesian era, the national scale today seems to function less and less as the main geographical basis for political-economic life.

The question remains, however, to what extent these claims apply to areas outside the Euro-American territory. Even within Europe, as this chapter will illustrate, one can witness very different circumstances concerning scalar hierarchies, both in terms of its traditional and its current development. This study focuses on the city of Belgrade, former capital of the Kingdom of Serbia (1882–1918), the Kingdom of Yugoslavia (1918–1945), the Socialist Federative Republic of Yugoslavia (1945–1992), the Federal Republic of Yugoslavia (1992–2003), the State Union of Serbia and Montenegro (2003–2006) and currently the Republic of Serbia. These territorial changes already indicate that the country's institutional past is, to say the least, rather different compared to the Western European founding members of the EU. In order to examine the 'rescaled regulatory capacities-hypothesis' within this peculiar post-socialist context, we have studied the existence of and interaction between global, national and local scales with regards to the announcement of 'Belgrade Waterfront': an urban real estate development project that is being realized through a joint-venture agreement between a globally well-known real estate investor from the United Arab Emirates (UAE) and the Republic of Serbia, which in this context can best be understood as a network of political confidants surrounding president Vučić.

The next section will discuss some key theoretical concepts that relate to contemporary academic debates on variegated neo-liberalism and how this is understood to affect urban management and governance. This theoretical framework will be followed by some reflections on the path-dependent, post-socialist context that is crucial to understand the present-day city of Belgrade. After that, more background information about Belgrade Waterfront is provided, which will be followed by the outcomes of our research, where it is shown how the top-down real estate development project is being executed by an entrepreneurial and authoritarian state, and which tensions arise as a result of it. The conclusion reveals to what extent, as well as the most-probable reasons why, Serbia differs from its Western European counterparts in terms of a supposed 'macro to micro-level' rescaling of political functions, and asserts why general 'universal' assumptions should be questioned.

## Omnipresent 'glocal' neo-liberalism?

Both global- and city-level actors are thus believed to become more and more important as influential decision- and policymakers in the 21st century (Brenner and Theodore, 2002b; Martin, McCann and Purcell, 2003). But how can this observed trend of concurrent up- and downscaling, or *glocalization* for that matter, be explained? What are the underlying causes? Keating (1997), who provides an extensive overview of more historical 're-gionalism tendencies' within the European Union, points out that these can often be attributed to democratization purposes, which indeed still seems to be the case today. In addition to that, it is believed that providing more autonomy to cities can lead to more pragmatic and accurate governance while addressing urgent and universal challenges (see Corijn's chapter in this volume). At the same time, however, contemporary critical social scientists tend to specifically attribute state-rescaling processes to the significant presence of neo-liberalism in today's world (i.e. Peck, 2013). In general, the neo-liberal ideology seems to have been a dominant global force since the 1980s, propagating a belief that 'open, competitive, and unregulated markets, liberated from all forms of state interference, represent the optimal mechanism for economic development' (Brenner and Theodore, 2002a, p. 350).

In the European case in particular, there has been a fair amount of attention for the retrenchment of the welfare state (i.e. MacLeod, 2001a) in this regard, and as a consequence, the increasingly entrepreneurial behaviour of cities, especially those that aspire to be 'global' (i.e. Harvey, 1989;

Golubchikov, 2010). This often translates into public-private partnerships in order to realize eye-catching flagship projects for city marketing purposes, in the interests of increased inter-local competition (i.e. Priemus, Flyvbjerg and van Wee, 2008; Anttiroiko, 2014). Another main neo-liberal feature is that such developments are often facilitated by, and simultaneously voice the interests of financial and/or transnational capital (Jessop, 2002). Smith (2002) confirms all this by arguing that neo-liberal urbanism is an integral part of the wider rescaling of functions, activities and relations, by which he refers to the shifting balances of power in today's globalizing world: 'It comes with a considerable emphasis on the nexus of production and finance capital at the expense of questions of social reproduction' (p. 435).

Brenner and Theodore (2002a) furthermore explain why the urban context is a particularly relevant object of study when looking at neo-liberalization and its consequences:

> In this context, cities – including their suburban peripheries – have become increasingly important geographical targets and institutional laboratories for a variety of neoliberal policy experiments, from place-marketing, enterprise and empowerment zones, local tax abatements, urban development corporations, public-private partnerships, and new forms of local boosterism to workfare policies, property-redevelopment schemes, business-incubator projects, new strategies of social control, policing, and surveillance, and a host of other institutional modifications within the local and regional state apparatus'
>
> (p. 368)

In this light, it is not surprising that many similar-looking policies or projects are currently being conducted in cities across the world. Another product of neo-liberalization and neo-liberal urbanism is the vast increase in the number of mobile and 'fast' policies circulating globally (Peck and Theodore, 2015). This phenomenon is also generally assumed to correspond to the aforementioned *glocalization*-thesis, which thus means that policymaking networks are becoming simultaneously more localized/regionalized and transnational (Swyngedouw, 2004).

Despite the fact that numerous examples of 'progressive mobile policies' undoubtedly exist, the general impression is that these *glocal* neo-liberal urban policy and planning experiments, which are characterized by enhancing competitiveness, are thus dominant practically everywhere. It is also argued that, as a result, the position of the active (democratic) citizen is being threatened by the changing scales of economic networks and institutional arrangements, which leads to 'post-democratic', 'post-political' or depoliticized urban landscapes (Swyngedouw, 2004, 2009; MacLeod, 2011). This tends to be specifically the case with regards to urban megaprojects, which can often be characterized by more diffused, fragmented and flexible modes of governance, while 'less powerful' social groups are generally excluded from the process (Swyngedouw, Moulaert and Rodriguez, 2002; Flyvbjerg, Bruzelius and Rothengatter, 2003).

Whereas the 'actual existence' of neo-liberalism today is hardly being denied, numerous academics still warn against oversimplifying its presence or exaggerating its impact. One should, for instance, be aware that it can be implemented in either aggressive or rather moderate ways, while it can be found in variegated or hybrid forms (e.g. Brenner, Peck and Theodore, 2010; Guarneros-Meza and Geddes, 2010; Collier, 2012). In other words, acknowledging the need for a 'flexible structuralist' approach is, according to Peck (2013), 'more than a poststructuralist tic'. Peck, Theodore and Brenner (2009) therefore stress the importance of taking path dependency into account, as they argue that 'the key point is that these politico-ideological shifts have emerged along a strongly path-dependent evolutionary trajectory' (p. 55), and hence it is important to be aware not only of the inherited institutional legacies, but also of the fact that the implementation of neo-liberalism is an integral part of the dynamics, logics and trajectories of the regulatory transformations that are thereby unleashed. Emphasizing these different path-dependent trajectories, Jessop (2002) states that it is useful to contrast neo-liberalism with other ideal-typical strategies such as neo-statism, which means that the state, instead of adopting a *laissez-faire* attitude, actually plays a vital role in guiding market forces to support a national economic strategy.

While Jessop remains rather ambiguous about whether these alternative '-isms' should exist either alongside or instead of neo-liberalism, it seems that in many occasions archetypical neo-liberal practices on the one hand, and central state interference on the other, do not at all need to be mutually exclusive. This is particularly the case in prestigious urban planning projects, and has often been witnessed in a non-Western context. Golubchikov (2010) namely likewise stresses (also referring to Brenner, 2009) that contemporary neo-liberal urbanism does not necessarily imply a 'hollowing out' of the state, and neither does he signal a widely emerging trend of downscaling: many of today's (post-) authoritarian nation states are actually seeking to establish themselves as dominant actors in the global economy by making their cities visible among the 'truly global cities'. This has also been illustrated by Olds and Yeung (2004), who highlight the significant role of the nation state in the global city-profile building of Kuala Lumpur and Shanghai, constructing these as being 'national projects', while more or less the same applies to for example Nairobi and Luanda, as has been described by Watson (2013) and Cain (2014) respectively. Kennedy et al. (2014), who conducted research on the implementation of various megaprojects in the Global South, likewise state that 'in all of our cases, local governments are not driving the process of economic development' (p. 4).

In addition to this, it should also be addressed that many authors involved in the 'state rescaling' debate have critically questioned the ontological fixity of taken-for-granted scales, since, as Swyngedouw (2004) argues, geographical scale is a deeply heterogeneous and contested process. Also Brenner (2009) acknowledges in much of his later work that the national scale is not losing its significance *per se*, but rather that states are being recomposed while operating increasingly in a more diffuse multi-scalar relational-institutional hierarchy (see also MacLeod, 2001b; Jessop, 2002; Allen and Cochrane, 2010). Furthermore, it could be questioned how a city can possibly be an actor in the first place. As Pinson's chapter in this volume asserts, one has to be aware that alongside urban governments, there is always the existence of a wider network including external supports, relays and gatekeepers allowing the access to other scales. For that same reason, the best strategy to maintain the hypothesis of the city-actor is to observe the struggles and legitimation mechanisms that allow some political leaders, interest groups or technocratic segments to act in the name of a city at certain points. Hence, the sometimes complex configurations of space and the multi-scalar positions in which actors in this case study can find themselves are outlined whenever relevant.

The case study presented in this chapter reflects upon contemporary assumptions regarding state rescaling tendencies. The literature survey above has shown that state rescaling is often regarded as a widespread global phenomenon, while neo-liberalism is nowadays commonly understood as a main driver behind the emergence of *glocal* coalitions. Studies on urban megaprojects in the Global South, however, illustrate that in fact national governments still play a key role in certain contexts. In the meantime, it is also being increasingly stressed that scales are becoming more diffuse in a globalized world, and that actors should, from a social constructivist point of view, be regarded as being 'scaled' in themselves (see also the Introduction to this volume). Besides, many authors argue that neo-liberalism never acts alone (i.e. Peck, 2013), and that it is therefore important to take into account the specific historical-geographic context of places, as well as the existence of alternative 'isms'. On this basis, this research critically assesses how neo-liberalization and state rescaling relate to one another, to what extent their presence can be observed beyond the Western or Euro-American context, and why this could be the case. In order to do so, the decision-making processes and implementation strategies behind a controversial transnational urban development project will be systematically revealed. The next section explains why Belgrade Waterfront has been selected as a case study, after which more details about the research methods that have been used will be provided.

## Case selection and methods

Belgrade Waterfront is an interesting case to study for multiple reasons (Gerring, 2006). Firstly, because it clearly seems to resemble yet another *typical* example of a 'glocal mobile policy'.[1] This becomes particularly obvious if one is aware of other, very similar projects that UAE-based developer Eagle Hills is currently conducting in countries such as Bahrain, Jordan,

Nigeria[2] or Morocco. Related to this, it seems to fit well into the category of neo-liberal urbanism, as will be illustrated and reflected upon later in this chapter. As has been discussed, however, inherited socio-institutional contexts, in this case an eventful and rather specific 'post-socialist' one, and past social struggles and compromises matter. Belgrade Waterfront is, for a variety of reasons, not only an eye-catching but also a contested case. It can be regarded as a 'personal' prestige project of the recently-elected president (and former prime minister) Aleksander Vučić and his close political allies of the SNS-party, who recently came into power within Serbia's autocracy-sensitive political system. Against this backdrop, which will be explained in more detail in section 4, Belgrade Waterfront can be regarded as a *deviant* case of general 'Western' neo-liberal urbanism and state rescaling theories.

Furthermore, the Balkans is a relatively under-researched geographical area in this regard. This is why a critical examination of the neo-liberal character of this particular project, its similarities and differences as compared to international trends, as well as the power relations and interactions between different stakeholders on and across various scales, will be a valuable contribution to the academic debate on these topics. The approach of this research is inspired by Burawoy's (1998) extended case method, which has as a starting point that one can depart from an existing theory, aiming to revise or deepen it rather than seeking confirmations. This also allows us to examine the presence of Jessop's (2002) suggested concept of neo-statism alongside neo-liberalism. Another inspiration for this research has been Peck and Theodore's (2012) *distended* case approach. This approach elaborates on Burawoy's work, but advocates an 'outsider' status for the researcher rather than becoming an active participant, while it also propagates adopting a constructivist approach rather than a disruptive one. Furthermore, the suggested research methods that can be applied are more diverse in this approach. Instead of only using ethnography, a combined approach of conducting in-depth interviews and analysing policy documents alongside observations are needed in order to understand the functioning of global policy networks (Peck and Theodore, 2012).

In order to obtain a holistic and 'as neutral as possible' a view on the project, we have, in addition to our analysis of all available policy documents, interviewed a variety of stakeholders (20 in total), including politicians, civil servants, journalists, academics, activists and businessmen over a long period between August 2015 and 2016. Although only one expressed concern regarding anonymity, the respondents will be referred to by codes (T$x$R$y$, whereby T$x$ indicates the number of the interview and R$y$ the number of the respondent) instead of their names, accompanied by their current occupation. In the selection procedure, we aimed to find a balance regarding their *pro* or *contra* attitude towards the project. While all of our respondents thus had a rather clear-cut opinion about it, it should perhaps be noted as well that, on the contrary, many ordinary citizens of Belgrade seem to have taken a much more ambivalent stance.

During the interviews, it furthermore appeared that some of the respondents clearly hesitated to speak freely about their own involvement in the decision-making process. This was particularly the case in the interviews that were conducted with the Urban Planning Institute of Belgrade and with Eagle Hills, in which respectively four and two employees took part. In the former case, it was striking to observe the non-verbal expressions of the respondents which clearly indicated that they made sure to choose their words very carefully, as well as the mutual behaviour between them as they often changed from English to Serbian and back. Those observations told us that the actors that need to justify the project are somewhat secretive and only limitedly transparent, bearing either a haughty (I9R14, I10R17 & I13R20) or an uncomfortable and insecure (I4R5-9 & I9R15) attitude. The other interviews were conducted with individuals, except the ones with the activist-initiative '*Ne Davimo Beograd*' and with the inner-city municipality of *Savski Venac*, which were both duo-interviews. All in all, given the rather non-transparent and secretive character of the project, it has to be admitted that we are not able to precisely reconstruct which actors made which decisions since 2012, while it is still somewhat speculative to determine what the exact (also obviously speculative) interests of the directly-involved stakeholders are. Nevertheless, we are still able to draw a clear picture of how this project has been implemented, justified and disputed.

In addition, and based on questions that provided us with information about the respondents' individual professional background as well as some of the networks they are currently involved in, it will be addressed on or across which scales the actors are operating. In the remainder of this chapter, this will be explained for several actors. The next section first provides some essential information on the historical political-institutional context of Serbia in general, and the city of Belgrade in particular.

## Serbia's political struggles

Tito's Yugoslavia (1945–1980) already counted as an 'exception to the rule' in many ways, as compared to the former Eastern European member states of the Warsaw Pact. Not being very loyal to Moscow, socialism in this south-eastern corner of Europe was often described as being 'more open' than elsewhere, mainly since there was significantly more self-management among its citizens. During the 1980s, the power vacuum that Tito left behind had to be filled in, and this can be regarded as the beginning of a very chaotic and troubled period for the nation. By the end of this decade, socialist regimes throughout Eastern Europe collapsed, and Yugoslavia stopped existing as a state. This occurrence went along with large-scale privatizations, which went so far – especially in Serbia – that it has been described as a period of wild capitalism (Upchurch and Marinković, 2011). At the same time, one could witness a 're-centralization of government and the weakening of the constitutional role and planning authority of the local communes' (Vujošević and Nedović-Budić, 2006, p. 280). Moreover, the political-economic transition was accompanied by the Yugoslav wars: ethnic conflicts that lasted from 1991 to 2001 and in which the Serbs, led by the 'hybrid author-itarian' regime of Slobodan Milošević (Bieber, 2003), always played a key role. During this period, Serbia was suffering from NATO-bombings, in-ternational sanctions, and a significant public debt (Obradović, 2007; Up-church and Marinković, 2011).

The end of the wars did not quite herald the end of political instability. In the post-Milošević era of the 21st century, plenty of new institutions arose; local authorities received an increased role in development programs and strategies; and civil society had an opportunity to develop itself (Bieber, 2003), although this only occurred to a limited extent (Vujošević and Nedović-Budić, 2006). Meanwhile, the power vacuum that arose, and the continuing presence of extremist and nationalist political forces, resulted in the political climate in Serbia remaining tense. This culminated in the assassination of Prime Minister Zoran Đinđić in 2003. Ever since that mo-ment, multiple parties have alternately won the elections. Many of these were newly established through splits or mergers of parties that came into existence during or shortly after the socialist era. Mutual relationships be-tween the parties are characterized by generally rather hostile attitudes to-wards one another, while several scholars mention the significant amount of nepotism that exists within government institutions (i.e. Obradović, 2007).

This 'us and them-mentality' was proven by one of our respondents who had recently obtained a directorship at the Belgrade Land Development Public Agency: 'We were at a transition period, and now we are at the end of this period, so... I see it as something like planting a flag. Now, we are running!' (I9R14, Director). On the contrary, another respondent, who iden-tifies himself with the opposing Democratic Party, had to quit his position as the director of the Serbian Agency for Spatial Planning, since he 'didn't like the new minister' (I5R9, Emeritus Professor) who was installed after the 2012 elections. After that occurrence, the entire agency was shut down. Belonging or being loyal to the ruling party, in other words, significantly affects one's career opportunities within the state apparatus. Upchurch and Marinković (2011) therefore argue that the capitalist trajectory that Serbia took, although it resembles some 'Western' codes of norms and behaviour, can be characterized by informality, clientelism, corruption, personal polit-ical networking and gangsterism.

Even though a number of political and economic transformations oc-curred during the early 2000s (privatization, attraction of international capital; see Ristić, 2004) that resemble the blueprint of 'Western capitalism', old habits seem to die hard. This is illustrated by Upchurch and Marinković (2011), who observe a 'parallel world of business norms and ethics whereby the conventional practices of western-based multinational corporations become integrated with clientelistic practices of "indigenous" owners of capital' (c/f Peev, 2002, p. 85). In the Serbian case such integration creates 'particularized norms of business behaviour' (Upchurch and Marinković, 2011, p. 14). They furthermore assert that this situation harms putative processes of social dialogue as well as transparency. Notwithstanding the aforementioned initial downscaling of urban planning responsibilities in the early 2000s, Serbia's contemporary political-economic system can thus best be categorized as 'state-centred capitalism' (Lazić and Pešić, 2012).

According to Bieber (2003), the heritage of both the communist party and the Milošević-era can only be reversed by a long process of democratization. At this very moment, president Vučić's Serbian Progressive Party (SNS[3]), which despite its name is generally characterized as populist and conservative, possesses an absolute majority, both on the national level and in the city council of Belgrade. These circumstances are, in addition to what is described in the preceding paragraph, important to be aware of in order to grasp the decision-making processes with regards to Belgrade Waterfront.

Even though Serbia (and Belgrade) can be categorized as having a 'post-socialist' context, one has to be careful in applying this concept, bearing in mind the very different trajectories that cities and nations in Central and Eastern Europe underwent after the collapse of socialism in the late 1980s/early 1990s (Hirt, 2013; Wiest, 2013), or in Serbia's case perhaps, only the 2000s. Furthermore, the fact that Serbia is being politically stuck in between the EU and Russia also has a continuous impact on the sometimes contradictory organization of its political-economy (Upchurch and Marinković, 2011). In the context of this research, it is also relevant to note that, according to Vujošević and Nedović-Budić (2006), 'Belgrade does not have enough indigenous resources to cope alone with its economic, social, physical and environmental revival' (p. 289). This also means that without incoming foreign direct investment, large-scale urban development projects such as Belgrade Waterfront would not be possible. The next section discusses the emergence of Belgrade Waterfront, and analyses how this project is currently being implemented, while determining the various local, national and global actors.

## Belgrade Waterfront and its main actors

The idea of developing a large-scale mixed-use Waterfront area was first announced in 2012, when Serbia's current president Aleksandar Vučić took part in the elections to become the mayor of Belgrade. Despite promises of a €3.5 billion foreign direct investment from the UAE that he was able to arrange to develop the city's derelict amphitheatre site along the river Sava, he did not become mayor that year. Shortly thereafter, however, he did manage to obtain the position of deputy prime minister, which enabled him to keep proposing his project. In more or less the same period, according to several media sources (i.e. Wright, 2015), the UAE and Serbia made other bilateral agreements concerning both the defence and food industries, as well as the national air carrier, Air Serbia. When Vučić became prime minister in 2014, the Waterfront-project was ready to be officially launched. The investor then appeared to be Mohamed Alabbar, a UAE-based businessman with close ties to the country's political elites. Alabbar is mainly known as the founder and the chairman of Emaar Properties, a main player in global real estate development and investment (see also Acuto, 2010; Lowry and Mc-Cann, 2011). The project in Belgrade has thus, as mentioned before, been put under the flag of Eagle Hills, a recently established, Abu Dhabi-based investment company chaired by Alabbar as well. While Alabbar recruited many former Emaar-employees to move to Eagle Hills, it concurrently merged with Al Mabaar, a real estate investment and development company that used to function under the direction of the Abu Dhabi government, thus taking over its ongoing projects such as the aforementioned ones in Jordan and Morocco (Barnard, 2015).

According to the official Belgrade Waterfront brochure that was released shortly after the announcement, Eagle Hills 'shares its expertise' by developing 'flagship city destinations that invigorate aspiring nations, helping countries raise their global profiles to new heights'. The brochure furthermore reveals that the newly developed Waterfront site will contain a more than 200m high tower, 'Kula Belgrade', as well as a large shopping mall and mixed-use spaces for work, living and leisure (namely 5,700 residential units and an office population of 12,700). The site of about 177 hectares, on which Belgrade Waterfront is being developed, is located behind the central railway station and in between the two main bridges that cross the river Sava, and is a very central location in the city (see Figure 13.1). This river became a focal point, especially after the realization of *Novi Beograd,* a large area predominantly consisting of modernist housing estates, developed in the 1960s and 1970s on the river's left bank, opposite the historical city centre. Apart from that, directly adjacent to the construction site is the neighbourhood of Savamala: a formerly dilapidated area that was recently revitalized due to the settlement of numerous creative entrepreneurs. Further away, but still within walking distance, one finds the historical *Kalemegdan* fortress, which overlooks the intersection of the Sava and the Danube, as well as the city's main shopping area and 19th-century boulevards such as *Knez Mihailova* and *Terazije.* During the past, there have been many plans to develop this particular piece of land. For a variety of reasons, however, including flood risk vulnerability, the site has always remained empty, albeit apart from a number of railways and a relatively small number of mainly illegal settlements.

Since the official announcement of Belgrade Waterfront, there have been large-scale marketing campaigns to advertise the project. Apart from the big banners and billboards that were installed at the airport and in Belgrade itself near the construction site (see Figure 13.2), and signs that are put alongside the highways indicating the distance to the future Belgrade Waterfront, the city's mayor Siniša Mali, a close political ally of Vučić, has travelled to international real estate fairs to exhibit the model (Figure 13.3). One example is the 'MIPIM' in Cannes in 2015, while it was also being exhibited in London's luxury shopping mall Harrods over a few weeks in the summer of 2016. This brochure-marketing strategy, along with the symbolic power of the intentionally impressive architectural design of the project, seem to be central elements of the Dubai-model of urbanization. As Acuto (2010) critically notes, 'oppression in Dubai comes with glossy brochures and red-carpeted

*Figure 13.1* Location of Belgrade Waterfront within the city.
*Source*: Map originally created and published by *Places* Journal.

entrances' (p. 281). By this, he means that the political and corporate elites are maintaining the existing gap between socio-economic groups through price barriers, rather than by using police force or physical violence. Another main feature of this model is that real estate projects are often developed as a 'city within the city' and thus become 'gated enclaves' (Acuto, 2010), something which many people fear Belgrade Waterfront is becoming as well.

In April 2015, a contract was signed by the Serbian Minister of Construction, Traffic and Infrastructure Zorana Mihajlović, chair of the Managing Board of Eagle Hills Mohamed Alabbar, simultaneously representing 'Belgrade Waterfront Capital Investment LLC' (the 'Strategic Partner'), 'Al Maabar International Investment LLC' (the 'Guarantor') and Belgrade Waterfront Company director Aleksandar Trifunović. This contract, which was made publicly available a few months later, mainly contains information about how the newly set up public-private partnership 'Belgrade Waterfront

*Figure 13.2* A large billboard next to the historical central train station of Belgrade advertises Belgrade Waterfront.

*Source*: Picture by Jorn Koelemaij.

*Figure 13.3* The Belgrade Waterfront gallery presents the model to anyone who is interested in seeing how the project is supposed to be realized.

*Source*: Picture by Jorn Koelemaij.

Company' is constructed: namely as a joint venture agreement, in which the Republic of Serbia holds 32% of the total share of the company (with a nominal value of €8,100), while the 'Strategic Partner' (a limited liability company incorporated in the United Arab Emirates, thus affiliated to Eagle Hills) holds the rest (currently valued at €22,000). The latter provides €150 million of equity financing to this Company, while at the same time extending its interest-bearing shareholder loans of up to another €150 million. The contract also reveals information about the agreed time schedule and responsibilities: within 20 years' time, 50% of the intended project has to be realized, and the rest of it 10 years later. The Republic of Serbia has to 'lease' the non-public land, develop the public buildings of the project and take care of an environmental clean-up of the area. In addition, they need to prepare all basic utility infrastructure and services, such as water, electricity, gas, roads, sewerage, telecommunication and connections to metro and tram routes, as well as clearing the site by removing old and useless railway tracks, by the year 2019 at the latest. To facilitate both these obligations, the Strategic Partner provides loans of respectively €40 million and €90 million to the Republic of Serbia. Moreover, the Strategic Partner has the right to change the specifications of the project in case they think it is necessary. In the contract, it is furthermore declared that the €150 million investment is on the explicit condition that the Strategic Partner or its affiliates or connected companies will be the sole and exclusive provider of the services to the company, such as project design, development and managing services, sales & marketing, customer relationships, information technology and so on.[4] It is precisely these agreements – the private, 'global' actor getting all sorts of privileges on behalf of the government – that causes frustration among many opposing groups within the city, as the next sections will discuss in more detail.

## A scale-making, re-centralizing project

While the Republic of Serbia thus counts as a main player in this project, it proved to be difficult to make scalar distinctions regarding this project. For the special purpose of Belgrade Waterfront, political allies president Aleksander Vučić and Belgrade mayor Siniša Mali have coupled their resources, and appointed the most important positions through their personal and SNS party networks. This is best exemplified by the newly installed employees of the mayor's office, whose chief of staff is also the vice-chairman of the supervisory board of Belgrade Waterfront Company 'even though it is a project at the Republic-level' (Chief of staff, I13R20). He and his direct colleagues (among whom one fulfils the newly created position of 'chief urbanist') are the determining actors regarding the most important decision-making processes around this project on behalf of the Serbian government. On the opposite side, the activists from the Ne Da(vi)mo Beograd-initiative manage to operate across scales as well. While the initiative sees itself as independent and separate from the existing NGO scene in Belgrade, its expenses are partly being covered by earlier awarded international grants (Morača, 2016). Apart from that, at least some of them are actively involved on an individual basis in wider networks such as the 'International Network for Urban Research and Action' (INURA).

Contrary to these inter-scalar coalitions, city-level civil servants have less operative space. Since the early 2000s, they usually have a fair amount of autonomy when it comes to the creation of urban or strategic plans, but for the development of this particular project, they are only involved to a very limited extent. Even though they are still being consulted with regards to preparing the aforementioned basic infrastructure in cooperation with some 'local' companies, describing themselves as 'subdevelopers' or 'consultants', they also admit that their role and influence on this project is actually rather marginal and that they are in various ways constrained to a purely local and subordinate position. Spatial plans, that are developed by national-level institutions such as the State Ministry, are namely above their '(urban) master plans' in the hierarchy. When they got asked how the plan was developed in the first place, one of the respondents working for the Urban Planning Institute answered:

> It was on the state level. It was on the top level. It was on the level of prime minister, I think, so it was something that had been decided before our plan. We came at the end, when the decision to fulfill this plan was already made. So they said 'we want to fulfill this plan: we want to make this here and this there...'
>
> (I4R6, Civil servant)

After which a colleague further clarified this by asserting:

> The government of Serbia has made a decision to declare this part of the city an area of great significance. Basically this kind of plan goes under the jurisdiction of the Republic, not the city. The city has the master plan, and detailed urban plans: this is the main division of the plans in the city, but this special spatial plan is something that goes under the jurisdiction of the Republic.
>
> (I4R5, Civil servant)

Spatial plans for special purpose areas indeed came into existence within the 'Planning and Construction Act' in 2003, and their hierarchical status on top of everything corresponds to Vujošević and Nedović-Budić' (2006) statements on the earlier observed re-centralization processes in Serbia that occurred both during and after Milošević' regime in the 1990s. The aforementioned contract moreover reveals that the Strategic Partner has provided initial master planning inputs for the project, while the Republic of Serbia ensures that those are incorporated in the Zoning Plan. Additionally, the former is thus also mainly responsible for the aforementioned marketing-strategies. This was revealed and justified by the mayor's office chief of staff, albeit with slight annoyance: 'That is investor-urbanism... ...In this kind of world, you have multinational companies, big companies that have their businesses all over the world. They already have that knowledge, you know, they have that know-how' (I13R20, Chief of staff). Notwithstanding the fact that the project contains a number of crystal-clear neo-liberal features, these observations are in sharp contrast with the earlier discussed state-rescaling trend, since a 'macro-to-micro' shift in terms of decision-making (e.g. Brenner and Theo-dore, 2002b; Smith, 2002; Martin, McCann and Purcell, 2003; Swyngedouw, 2009) cannot be observed here at all. Since Vučić's SNS party gained power, the re-centralization process in Serbia gradually intensified, and in a way, Belgrade Waterfront serves as a mechanism for them to facilitate this process.

The top-down, non-democratic character of the project was also con-firmed by a (non-SNS related) civil servant from the municipality in which Belgrade Waterfront will be developed:

> Nobody asks the municipality anything... Only if we have... when they change some urban plans, all Serbian citizens can give their suggestions; municipalities can also give suggestions, but you know, nobody takes them into consideration.
>
> (Municipal Urbanist & Architect, I2R2)

This lack of participatory democracy of citizens, and the ignoring of 'local' expertise is something that causes disappointment among other respondents as well. Particularly those respondents who had an outspoken critical stance towards the project[5] indicated that it was clear to them that, indeed, ulti-mately not a single one of their suggestions was taken into account by the developers. Many of them moreover mention the fact that it is usually ob-ligatory to organize a public tender for big infrastructural projects like this. Hence, the announcement from national level politicians that they found an investor who instantly came up with a model (see the *maquette*, Figure 13.3), was regarded by many with both surprise and disapproval.

Whereas the Belgrade Waterfront model, accompanied by 3D-visualizations (Figure 13.3) and future-impression images, was presented on billboards and during TV-commercials, concerned citizens were looking in vain for a publically available detailed urban plan justifying the project-design. One of our respondents, a municipal urbanist and former planning consultant, expressed his astonishment about this by stating that:

> usually, like in any state in the world I think, the steps are first to make a plan, to discuss it with the stakeholders, to accept the plan, to make it official, then to make a project, then to get a construction permit, and then to do marketing, yeah? But here, everything was mixed. First there was the model, then the project, then the plan.
>
> (Municipal Urbanist & Consultant, I1R1)

When the Eagle Hills representatives who were interviewed in August 2015 were confronted with the complaints regarding the lack of transparency around the project, they reacted in a twofold manner. On the one hand, they stated that many details were kept secret on purpose, since they had to main-tain the ability to adjust rapidly to a fast-changing market. On the other hand, they admitted that it would perhaps be better for the public opinion on the project if they were to reveal more details about their plan, which eventually happened to a certain extent indeed, through the publishing of the contract.

### Facilitating a project of national importance: need for speed?

Apart from the non-transparency and the altered sequence of the plan, the 'construction permit' that was mentioned in the former quote is also very important to consider. In order to facilitate the project as much as possible, the SNS-party makes use of a so-called *Lex Specialis*: a special law applying to Belgrade Waterfront that overrules existing laws regarding building conditions, while issuing the building permit to start the construction. With this change of legislation, which re-articulates the hierarchical relations between scales by allowing national rules to overrule local rules, Belgrade Waterfront is thus being indicated as a project of special and national importance. Even though this has happened before in Yugoslavia, as well as in other countries across the world, this decision has caused a lot of indignant reactions among people challenging the project, who repeatedly assert the disputed legality of the law that surpasses existing urban plans while exempting the Strategic Partner from usual rules and restrictions:

> They have made three changes: there is no limitation for height of buildings, so right now, instead of twelve floors, they can make thousands of floors – no limits. There is no obligation left for international competition. And the third one is that the government can declare that something is of public interest. It is contrary to the constitution, to all laws in Serbia and also in the world. There is no case in which somebody can declare that something is of public interest, instead of what is proposed by the Constitution. Public interest cannot be commercial buildings in Serbia, considering all other laws. Public interest are hospitals, schools, streets, and what is usually common around the world.
>
> (Municipal Urbanist & Consultant, I1R1)

The flexibility of laws regarding the implementation of megaprojects, often indicated as 'informality' in the literature (e.g. Roy, 2009), is something that has been noted particularly in the Global South: as Follmann (2015) illustrates, for example, the Indian government also bypasses existing environmental laws in order to realize waterfront property development in Delhi.[6] When we, again, confronted Eagle Hills with the people's concerns about the changed legislation around the project, their answer was that they were indeed 'positively surprised by the amount of freedom the Republic of Serbia had provided them with'[7]: 'people are willing to make our job easier' (Head of Business Development and Sales, I10R16), to an extent that they 'usually did not get'. Their justification of this flexibility of the law was that this is inevitable 'according to contemporary market logics'. They stress that in big real estate development projects such as Belgrade Waterfront, one needs to be able to make changes all the time in order to be ready to make adjustments when necessary and to 'speed up the process', because 'otherwise, it will never work': 'The current investor climate in Serbia is not suitable for international business. Changes in the law are necessary for the international property market...' 'You need to be competitive with your tax regime and your visa regime' (All quotes in this paragraph are from the interview with I10R16).

This attitude is inspired by Eagle Hills' experience in the UAE (see earlier) and hence resonates with the 'heightened speed' of contemporary fast mobile policies that is extensively discussed by Peck (2013) and Peck and Theodore (2015). In addition, this extreme form of deregulation, albeit for a particular purpose, clearly corresponds to the neo-liberal fashion. It should be noted here, however, that state-actors representing a variety of scales are obviously still actively involved, which simultaneously confirms the presence of a 'neo-statist' model. Even though such 'exceptionalist practices' are more generally observed in megaproject-implementation, the remarkably all-decisive and authoritative role of the SNS-party elite is important to stress here.

It is furthermore remarkable how some of the key stakeholders that need to justify the project, such as president Vučić himself, deny the fact that the project is being implemented in a very uncommon way, even though it clearly is. During our interview with the Land Development Agency as well as with the Mayor's office chief of staff, the respondents asserted that there was nothing too special about the plan, that 'you have the same thing in Europe' (Acting Director, I9R14), while they even downplayed the importance and size of the project. When we asked our respondents about the aims and strategies behind the project, we got a range of different answers, varying from critical ones: 'there is no such thing as a strategy at all' (Activist, I6R10), 'it is probably all for personal financial benefits' (Journalist, I8R13), and 'they want to create a belief among the people in order to be re-elected' (Activist, I11R18); to attempts at justification: 'we want to attract the wealthy Serbian diaspora to return to their home country, as well as to create an attractive investor climate by providing high-end office space for global high-tech firms' (Marketing Manager, I10R17). Furthermore, the 'Belgrade Investor Guide 2015', a document issued by the City Government, states that 'Belgrade Waterfront should reinstate Belgrade's position on the global real estate map'. These alleged rationales (catching attention, leaving a legacy, making profit) are categorized by Flyvbjerg (2014) as the 'political- and economic sublimes', which appear as important drivers of many megaprojects across the globe in general. The mayor's chief of staff furthermore justifies the project by stating that it will bring employment opportunities to domestic (construction) workers as well as a favourable environment for local economic development, along with the potential increase of tourism. At the same time, he emphasizes the shortcomings of domestic investors and developers as opposed to 'the Arabs', who are able to create 'a new concept of living', while being 'too big to fail' in the event of another economic crisis.

While the 'economic sublime' of making profit and increasing competitiveness corresponds to the neo-liberal capitalist logic, the 'political sublime' that is clearly present, and probably even dominant here, can be understood through Ong's (2011) notion of 'hyperbuilding'. While specifically referring to Asian cities, she understands 'hyperbuilding' as a *worlding* practice whereby spectacular infrastructure facilitated by political exceptions is supposed to attract speculative capital, while proving the political elite's power: '[b]uilding a critical mass of towers in a new downtown zone animates an anticipatory logic of reaping profits not only in markets but also in the political domain' (p. 207). Such urban development projects are thus often meant as a symbolic value, attempting to improve a city's image as well as citizens' hopes and expectations about the nation's future. There are two main reasons why Belgrade Waterfront possesses so many similarities with comparable urban megaprojects situated in the Global South. First of all, the specific historical-political context of Serbia has allowed the SNS-Party to authoritatively relocate decision-making powers to the 'national' scale, thereby reasserting the dominant position of the national scale in scalar hierarchies. These circumstances have made it attractive for the UAE-based investor to launch such a large-scale project in Belgrade, having been guaranteed the right to conduct the project in a way that it is used to in its 'own' context. On the basis of those two combined factors, we can refute the 'universal' assumptions regarding the relation between neo-liberalism and the alleged 'macro to micro' shift of functions and responsibilities. On the contrary, it seems that beyond the Euro-American context, neo-liberal urbanism and neo-statism exist alongside one another, while contemporary 'state-rescaling' should in many geographical areas actually be contrastively understood as the increasing significance of the national state.

## Conclusion

This chapter has investigated the state rescaling processes in Serbia and its relation to an urban real estate development project. Belgrade Waterfront can in some ways be regarded as a *typical case* with regards to the structural theoretical assumptions regarding neo-liberal urbanism in the 21st century. The project can be characterized as an archetypical example of urban boosterism: a large scale flagship project supported by a lot of city-marketing, which eventually has to attract a new business elite that will presumably raise Belgrade's profile as a (semi-)global city. This is furthermore being achieved by an entrepreneurial state that makes use of what can be categorized as a mobile real estate project that is largely funded by transnational capital, while attempting to speed up the process as much as possible by adjusting existing legislation and plans. As a result of this, Eagle Hills, the global private investment company from the UAE, is getting a large amount of freedom with regards to developing the project.

State rescaling is a recurrent phenomenon being caused by neo-liberal forces and practices; it has been mainly understood as a process in which functions are being 'downscaled' from the central government towards lower scalar levels. The example of Belgrade Waterfront, however, teaches us that this understanding of the theory does not apply to contemporary Serbia. Determining the project as one of 'national importance', the rather authoritative government, with a strikingly dominant role for Prime Minister Vučić and his closest political confidants, deals hierarchically with all the necessary issues. Hard measures, such as adjusting existing or introducing new laws, are not avoided. What has mainly been happening is that Vučić and his SNS-party are using an eye-catching urban renewal project that contains many neo-liberal features for state rescaling and recentralization purposes, asserting their recently gained (symbolic) power. For that reason, Belgrade Waterfront can also be considered as a 'scale-making' project. At the same time, the project forces them to further 'scale up' planning responsibilities towards the level of the foreign, 'global' investor, since both the published contract and our interviews illustrate that they implement the project according to 'their' standards and procedures.

This indicates that 'political sublimes' are playing a larger role in this project than economic ones, something which Ong (2011) has previously described as a process of 'hyperbuilding'. Taking into account the institutional context of Serbia, our findings indicate the presence of 'neo-statism' alongside neo-liberalism, which thus indeed never seems to 'act alone' (Jessop, 2002; Peck, 2013). Similar patterns of nationally-induced global city aspirations, thus contrary to the assumed 'state-rescaling' trend, have been noted in Russian, East-Asian and sub-Saharan African cities (see Olds and Yeung, 2004; Golubchikov, 2010 and Watson, 2013 respectively). As previous research has furthermore illustrated, megaprojects' governance practices always appear to be somewhat different from the ordinary. Even though clientelism and non-transparency are also observed in the Western European context to a certain extent (Swyngedouw, Moulaert and Rodriguez, 2002), the 'informal' strategies that are used regarding the implementation of Belgrade Waterfront likewise correspond particularly to a number of projects that are being developed in the Global South (Roy, 2009; Kennedy et al., 2014). There are, at first sight, more than a few striking similarities between the *deviant* post-socialist context of Serbia and other cases that are outside the Euro-American context. The unstable political-institutional past of Serbia and the legacy of the authoritarian Milošević regime of the 1990s, are likely to be a determining factor here. This also explains why the political elite of the Republic of Serbia does not hesitate to conduct deals with a foreign partner without feeling the urge to justify this to its tax-payers in a decent, transparent manner.

Structural theories and assumptions with regards to neo-liberal urbanism and the shifting regulatory responsibilities that supposedly come along with it are very useful, but additional insights into specific contextual differences are needed in order to get a better understanding of the main causes of differences between places. More in-depth and comparative case studies scrutinizing governance and state-rescaling processes and its relation with urban-infrastructural megaprojects in different contexts are needed (see also Brenner, 2009), in order to gain insights into the distinctive features of places. This will be a valuable step to providing essential nuances and additions towards presumed 'universal' assumptions that are derived from a Euro-American perspective. Besides, it would be an interesting follow-up research project to reflect upon the Belgrade Waterfront project from the investor's perspective: it would be valuable to investigate how the mobilization, or 'worlding' of the so-called Dubai-model of urbanization is put into practice. Relevant research approaches could reveal the ties between those investors and their country's political elite, defining their exact interests, uncovering their rationales to invest in certain places and how their projects get implemented elsewhere, analysing their marketing strategies and mapping the network of 'third party actors' – architects, consultants – with whom they decide to work (see also Roy, 2009; McCann and Ward, 2012). Thinking beyond and across concrete scales, in other words, remains important to improve our understanding of the dynamics behind global investment hierarchies and strategies, and the impact they have on 'local' or 'national' political and societal transformations.

## Acknowledgements

This research would not have been possible without the great help of Stefan Janković and Mila Madzarević, who assisted the author during his fieldwork and gave valuable suggestions during the writing process. Furthermore, we would like to thank all our respondents (Table 13.1) for their cooperation, as well as all the people in Belgrade who helped us to get in touch with the right people.

*Table 13.1* List of respondents Belgrade Waterfront research

| Code | Main affiliation | Current institution | Date of interview |
|---|---|---|---|
| I1R1 | Urbanist & Consultant | Municipality of Veliko Gradište | 18/08/2015 |
| I2R2 | Urbanist & Architect | Municipality of Savski Venac | 18/08/2015 |
| I2R3 | Urbanist & Architect | Municipality of Savski Venac | 18/08/2015 |
| I3R4 | Architect | Private Company | 18/08/2015 |
| I4R5 | Civil Servant | Belgrade Urban Planning Institute | 18/08/2015 |
| I4R6 | Civil Servant | Belgrade Urban Planning Institute | 18/08/2015 |
| I4R7 | Civil Servant | Belgrade Urban Planning Institute | 18/08/2015 |
| I4R8 | Civil Servant | Belgrade Urban Planning Institute | 18/08/2015 |
| I5R9 | Emeritus professor | PALGO Center Belgrade | 19/08/2015 |
| I6R10 | Activist | Ne da(vi)mo Beograd | 19/08/2015 |
| I6R11 | Activist | Ne da(vi)mo Beograd | 19/08/2015 |

*(Continued)*

| Code | Main affiliation | Current institution | Date of interview |
|------|------------------|---------------------|-------------------|
| I7R12 | DS Politician | Belgrade City Assembly | 20/08/2015 |
| I8R13 | Journalist | BIRN | 21/08/2015 |
| I9R14 | Acting Director | Belgrade Land Development Public Agency | 21/08/2015 |
| I9R15 | Civil Servant | Belgrade Land Development Public Agency | 21/08/2015 |
| I10R16 | Head of Business Development & Sales | Eagle Hills/ Belgrade Waterfront Company | 21/08/2015 |
| I10R17 | Marketing Manager | Eagle Hills/ Belgrade Waterfront Company | 21/08/2015 |
| I11R18 | Activist | Ne da(vi)mo Beograd | 29/07/2016 |
| I12R19 | Journalist | Istinomer | 03/08/2016 |
| I13R20 | Chief of Staff | Mayor's Office | 03/08/2016 |

## Notes

1 Or, 'mobile real estate megaproject' for that matter, depending on howw one were to define a 'policy'.
2 This project, which looked strikingly similar to Belgrade Waterfront in terms of design and marketing, was abruptly withdrawn in early 2017.
3 *Srpska Napredna Stranka* in Serbian.
4 For the design of this model or master plan, spokespersons of the project report having cooperated with a number of international consultancy firms, including SOM (design Kula Belgrade), RTKL (master plan), Woods Bagot (general urban design), SWA (public spaces), COWI (traffic) and Arcadis (flood protection). Re-markably, however, apart from SWA, none of these firms mention the project on their respective websites.
5 Generally opposition party politicians, academics, activists, journalists and architects.
6 In which, by the way, Emaar is also involved.
7 Paraphrased.

## References

Acuto, M., 2010. High-rise Dubai urban entrepreneurialism and the technology of symbolic power. *Cities*, 27(4), pp. 272–284.
Allen, J. and Cochrane, A., 2010. Assemblages of state power: topological shifts in the organization of government and politics. *Antipode*, 42(5), pp. 1071–1089.
Anttiroiko, A.-V., 2014. *The political economy of city branding.* New York: Routledge.
Barnard, L., 2015. Emaar boss uses Eagle Hills property firm to revive push into Africa. *The National*, [online], 13 January. Available at: www.thenational.ae/business/property/emaar-boss-uses-eagle-hills-property-firm-to-revive-push-into-africa-1.114107 [Accessed 25 October 2017].

Bieber, F., 2003. The Serbian opposition and civil society: roots of the delayed transition in Serbia. *International Journal of Politics, Culture and Society*, 17(1), pp. 73–90.

Brenner, N., 1998. Global cities, glocal states: global city formation and state territorial restructuring in contemporary Europe. *Review of International Political Economy*, 5(1), pp. 1–37.

Brenner, N., 2009. Open questions on state rescaling. *Cambridge Journal of Regions, Economy and Society*, 2(1), pp. 123–139.

Brenner, N. and Theodore, N., 2002a. Cities and the geographies of 'actually existing neoliberalism'. *Antipode*, 34(3), pp. 349–379.

Brenner, N. and Theodore, N. eds., 2002b. *Spaces of neoliberalism: urban restructuring in North America and Western Europe*. Oxford: Blackwell.

Brenner, N., Peck, J. and Theodore, N., 2010. Variegated neoliberalization: geographies, modalities; pathways. *Global Networks*, 10, pp. 182–222.

Burawoy, M., 1998. The extended case method. *Sociological Theory*, 16(1), pp. 4–33.

Cain, A., 2014. African urban fantasies: past lessons and emerging realities. *Environment and Urbanization*, 26(2), pp. 561–567.

Collier, S.J., 2012. Neoliberalism as big Leviathan, or...? A response to Wacquant and Hilgers. *Social Anthropology*, 20(2), pp. 186–195.

Flyvbjerg, B., 2014. What you should know about megaprojects and why: an overview. *Project Management Journal*, 45(2), pp. 6–19.

Flyvbjerg, B., Bruzelius, N. and Rothengatter, W., 2003. *Megaprojects and risks: an anatomy of ambition*. Cambridge: Cambridge University Press.

Follmann, A. 2015. Urban mega-projects for a 'world-class' riverfront – The interplay of informality, flexibility and exceptionality along the Yamuna in Delhi, India. *Habitat International*, 45, pp. 213–222.

Gerring, J., 2006. *Case study research: principles and practices*. New York: Cambridge University Press.

Golubchikov, O., 2010. World-city-entrepreneurialism: globalist imaginaries, neoliberal geographies, and the production of new St Petersburg. *Environment and Planning A*, 42(3), pp. 626–643.

Guarneros-Meza, V. and Geddes, M., 2010. Local governance and participation under neoliberalism: comparative perspectives. *International Journal of Urban and Regional Research*, 34(1), pp. 115–129.

Harvey, D., 1989. From managerialism to entrepreneurialism: the transformation in urban governance in late capitalism. *Geografiska Annaler. Series B, Human Geography*, 71(1), pp. 3–17.

Hirt, S., 2013. Whatever happened to the (post)socialist city? *Cities*, 32, pp. 29–38.

Jessop, B., 2002. Liberalism, neoliberalism, and urban governance: a state–theoretical perspective. *Antipode*, 34(3), pp. 452–472.

Keating, M., 1997. The invention of regions: political restructuring and territorial government in Western Europe. *Environment and Planning C: Government and Policy*, 15(4), pp. 383–398.

Kennedy, L., Robbins, G., Bon, B., Takano, G., Varrel, A. and Andrade, J., 2014. *Megaprojects and urban development in cities of the south*. Bonn: Chance2Sustain.

Lazić, M. and Pešić, J., 2012. *Making and unmaking state centered capitalism in Serbia*. Belgrade: Čigoja štampa.

Lowry, G. and McCann, E., 2011. Asia in the mix. In: A. Roy and A. Ong, eds. *Worlding cities: Asian experiments and the art of being global*. Chichester, West Sussex: Wiley-Blackwell. pp. 182–204.

MacLeod, G., 2001a. Beyond soft institutionalism: accumulation, regulation, their geographical fixes. *Environment and Planning A*, 33, pp. 1145–1167.

MacLeod, G., 2001b. New regionalism reconsidered: globalization and the remaking of political economic space. *International Journal of Urban and Regional Research*, 25(4), pp. 804–829.

MacLeod, G., 2011. Urban politics reconsidered: growth machine to post-democratic city? *Urban Studies*, 48(12), pp. 2629–2660.

Martin, D., McCann, E. and Purcell, M., 2003. Space, scale, governance, and representation: contemporary geographical perspectives on urban politics and policy. *Journal of Urban Affairs*, 25(2), pp. 113–121.

Massey, D., 2007. *World city*. London: Polity.

McCann, E. and Ward, K., 2012. Policy assemblages, mobilities and mutations: toward a multidisciplinary conversation. *Political Studies Review*, 10(3), pp. 325–332.

Morača, T., 2016. Between defiance and compliance: a new civil society in the post-Yugoslav space? *Osservatorio Balcani e Caucaso Transeuropa*, Occassional Paper, 1–24.

Obradović, M., 2007. The socio-historical consequences of privatisation in Serbia. *SEER: Journal for Labour and Social Affairs in Eastern Europe*, 10(1), pp. 39–60.

Ohmae, K., 1995. *The end of the nation state. The rise of regional economies.* New York: The Free Press.

Olds, K. and Yeung, H., 2004. Pathways to global city formation: a view from the developmental city-state of Singapore. *Review of International Political Economy*, 11(3), pp. 489–521.

Ong, A., 2011. Hyperbuilding: spectacle, speculation and the hyperspace of sovereignty. In: A. Roy and A. Ong, eds. *Worlding cities: Asian experiments and the art of being global.* Chichester, West Sussex: Wiley-Blackwell.

Peck, J., 2013. Explaining (with) neoliberalism. *Territory, Politics, Governance*, 1(2), pp. 132–157.

Peck, J. and Theodore, N., 2012. Follow the policy: a distended case approach. *Environment and Planning A*, 44(1), pp. 21–30.

Peck, J. and Theodore, N., 2015. *Fast policy: experimental statecraft at the thresholds of neoliberalism.* Minneapolis: University of Minnesota Press.

Peck, J., Theodore, N. and Brenner, N., 2009. Neoliberal urbanism: models, moments, mutations. *SAIS Review*, 29(1), pp. 49–66.

Peev, E., 2002. Ownership and control structures in transition to 'crony' capitalism: the case of Bulgaria. *Eastern European Economics*, 40(5), pp. 73–91.

Priemus, H., Flyvbjerg, B. and van Wee, B., 2008. *Decision-making on mega-projects.* Cheltenham: Edward Elgar Publishing Limited.

Ristić, Z., 2004. Privatisation and foreign direct investment in Serbia. *SEER: Journal for Labour and Social Affairs in Eastern Europe*, 7(2), pp. 121–136.

Roy, A., 2009. The 21st century metropolis: new geographies of theory. *Regional Studies*, 43(6), pp. 819–830.

Smith, N., 2002. New globalism, new urbanism: gentrification as global urban strategy. *Antipode*, 34(3), pp. 427–450.

Swyngedouw, E., 2004. Globalisation or 'glocalisation'? Networks, territories and rescaling. *Cambridge Review of International Affairs*, 17(1), pp. 25–48.

Swyngedouw, E., 2009. The antinomies of the postpolitical city: in search of a democratic politics of environmental production. *International Journal of Urban and Regional Research*, 33(3), pp. 601–620.

Swyngedouw, E., Moulaert, F. and Rodriguez, A., 2002. Neoliberal urbanization in Europe: large–scale urban development projects and the new urban policy. *Antipode*, 34(3), pp. 542–577.

Upchurch, M. and Marinković, D., 2011. Wild capitalism, privatisation and employment relations in Serbia. *Employee Relations*, 33(4), pp. 316–333.

Vujošević, M. and Nedović-Budić, Z., 2006. Planning and societal context – The case of Belgrade, Serbia. In: S. Tsenkova and Z. Nedović-Budić, eds. *The Urban Mosaic of Post-Socialist Europe*. Heidelberg: Physica Verlag.

Watson, V., 2013. African urban fantasies: dreams or nightmares? *Environment and Urbanization*, 26(1), pp. 215–231.

Wiest, K., 2013. Comparative debates in post-socialist urban studies. *Urban Geography*, 33(6), pp. 829–849.

Wright, H., 2015. Belgrade Waterfront: an unlikely place for Gulf petrodollars to settle. *The Guardian*, [online], 10 December. Available at: www.theguardian.com/cities/2015/dec/10/belgrade-waterfront-gulf-petrodollars-exclusive-waterside-development [Accessed 10 October 2016].

# Index